Lecture Notes in Computer Science

Lecture Notes in Artificial Intelligence   **15052**

Founding Editor

Jörg Siekmann

Series Editors

Randy Goebel, *University of Alberta, Edmonton, Canada*
Wolfgang Wahlster, *DFKI, Berlin, Germany*
Zhi-Hua Zhou, *Nanjing University, Nanjing, China*

The series Lecture Notes in Artificial Intelligence (LNAI) was established in 1988 as a topical subseries of LNCS devoted to artificial intelligence.

The series publishes state-of-the-art research results at a high level. As with the LNCS mother series, the mission of the series is to serve the international R & D community by providing an invaluable service, mainly focused on the publication of conference and workshop proceedings and postproceedings.

M. Nazmul Huda · Mingfeng Wang ·
Tatiana Kalganova
Editors

Towards Autonomous Robotic Systems

25th Annual Conference, TAROS 2024
London, UK, August 21–23, 2024
Proceedings, Part II

 Springer

Editors
M. Nazmul Huda 🆔
Department of Electronic and Electrical
Engineering
Brunel University London
London, UK

Mingfeng Wang 🆔
Department of Mechanical and Aerospace
Engineering
Brunel University London
London, UK

Tatiana Kalganova 🆔
Department of Electronic and Electrical
Engineering
Brunel University London
London, UK

ISSN 0302-9743 ISSN 1611-3349 (electronic)
Lecture Notes in Artificial Intelligence
ISBN 978-3-031-72061-1 ISBN 978-3-031-72062-8 (eBook)
https://doi.org/10.1007/978-3-031-72062-8

LNCS Sublibrary: SL7 – Artificial Intelligence

This Springer imprint is published by the registered company Springer Nature Switzerland AG
The registered company address is: Gewerbestrasse 11, 6330 Cham, Switzerland

If disposing of this product, please recycle the paper.

Preface

This volume contains the papers presented at TAROS 2024, the 25th Towards Autonomous Robotic Systems (TAROS) Conference, held at the College of Engineering, Design and Physical Sciences, Brunel University London, UK, during August 21–23, 2024 (https://taros-conference.org/).

TAROS is the longest running UK-hosted international conference on robotics and autonomous systems (RAS), and is aimed at the presentation and discussion of the latest results and methods in autonomous robotics research and applications. The conference offers a friendly environment for robotics researchers and industry to take stock of current developments and plan for future progress. It welcomes senior researchers and research students alike, and specifically provides opportunities for research students and young research scientists to present their work to the scientific community.

TAROS 2024 was held at Brunel University London, a world-class university with an outstanding reputation for industry engagement in education, technology, research, and innovation. The papers in this volume were selected from 69 submissions under a single-blind peer review process. Out of these, 54 full papers and 11 short papers were selected for the conference. The conference programme included an academic conference, industry exhibitions, robot demonstrations, and other social events. TAROS 2024 covered topics of robotic systems, human–robot interaction, soft robotics, robot navigation and planning, robot control, and industrial robots, and highlights included – Keynote lectures by world-leading experts in robotics, including Robert Deaves from Dyson Ltd., UK; Ana Cavalcanti from University of York, UK; Asoke Nandi from Brunel University London, UK; Josie Hughes from École Polytechnique Fédérale de Lausanne, Switzerland; Kaspar Althoefer from Queen Mary University London, UK; Marc Hanheide from University of Lincoln, UK; Radhika Gudipati from Advanced Research and Invention Agency, UK; Cigdem Sengul from Brunel University London, UK; and Zakareya Hussein from Touchlab, National Robotarium, UK – Oral and Poster presentations of papers, covering various topics of robotics and autonomous systems. This year the conference also included contributions from Brunel's Robotics and AI for Healthcare group and the EPSRC Centre for Doctoral Training: AgriFoRwArdS, ALBERT, and FARSCOPE.

The TAROS 2024 Organizing Committee would like to thank all the authors, the reviewers, and the conference sponsors, including IET, MathWorks, RoboSavvy, Dobot, Unitree, Queen Mary University of London, UK-RAS Network, Springer, and MDPI for their support to the conference.

July 2024

M. Nazmul Huda
Mingfeng Wang
Tatiana Kalganova

Organization

General Chair

M. Nazmul Huda   Brunel University London, UK

Program Committee Chairs

Mingfeng Wang   Brunel University London, UK
Tatiana Kalganova   Brunel University London, UK

Programme Committee

Barry Lennox   University of Manchester, UK
Andrew Weightman   University of Manchester, UK
John Oyekan   University of York, UK
Marisé Galvez Trigo   Cardiff University, UK
Yohan Noh   Brunel University London, UK
Pedro Machado   Nottingham Trent University, UK

Web Chair

Farzana Sharmin Mou   Brunel University London, UK

Administrator

Surjeet Khurana   Brunel University London, UK

TAROS Steering Committee

Fumiya Iida   University of Cambridge, UK
Ana Cavalcanti   University of York, UK
Robert Richardson   University of Leeds, UK
M. Nazmul Huda   Brunel University London, UK

Mingfeng Wang	Brunel University London, UK
Chenguang Yang	University of Liverpool, UK
Antonia Tzemanaki	University of Bristol, UK
Farshad Arvin	Durham University, UK
Guodong Zhao	University of Manchester, UK
Barry Lennox	University of Manchester, UK
Marisé Galvez Trigo	Cardiff University, UK
Yang Gao	King's College London, UK

Invited Speakers

Robert Deaves	Dyson Ltd., UK
Ana Cavalcanti	University of York, UK
Asoke Nandi	Brunel University London, UK
Josie Hughes	École Polytechnique Fédérale de Lausanne, Switzerland
Kaspar Althoefer	Queen Mary University London, UK
Marc Hanheide	University of Lincoln, UK
Radhika Gudipati	Advanced Research and Invention Agency, UK
Cigdem Sengul	Brunel University London, UK
Zakareya Hussein	Touchlab, National Robotarium, UK

Additional Reviewers

Oneeba Ahmed	Brunel University London, UK
Sarfraz Ahmed	Coventry University, UK
Ibrahim Alispahic	University of Sarajevo, Bosnia-Herzegovina
Xiangyu An	Brunel University London, UK
Gerardo Aragon Camarasa	University of Glasgow, UK
Farshad Arvin	Durham University, UK
Kawsar Arzomand	Brunel University London, UK
Arjun Badyal	University of York, UK
Lingfan Bao	University of Leeds, UK
Joshua Bettles	University of Manchester, UK
Jordan Bird	Nottingham Trent University, UK
Pablo Borja	University of Plymouth
Matt Butler	Harper Adams University, UK
Daniele Cafolla	Swansea University, UK
Abu Bakar Dawood	Queen Mary University of London, UK
Sairaj R. Dillikar	Cranfield University, UK

Feliciano Domingos	Nottingham Trent University, UK
Ubada El Joulani	Brunel University London, UK
Laurenz Elstner	Western Norway University of Applied Sciences, Norway
Juan Pablo Espejel Flores	University of Lincoln, UK
Omar Faris	University of Lincoln, UK
Charles Fox	University of Lincoln, UK
Mohamed Gaballa	Brunel University London, UK
Ebony Harker-Bannister	Brunel University London, UK
Yuanzhi He	Cardiff University, UK
Henry Hickson	University of Bristol, UK
Ningzhe Hou	University of Oxford, UK
Junyan Hu	Durham University, UK
Marianne Huchard	University of Montpellier, France
Joseph Humphreys	University of Leeds, UK
Hoda Ibrahim	Brunel University London, UK
Kennedy Ihianle	Nottingham Trent University, UK
Katherine James	University of Lincoln, UK
Fahad Khan	Cranfield University, UK
Mohit Kumar	Software Competence Center Hagenberg, Germany
Seyonne Leslie-Dalley	University of Manchester, UK
Cunjia Liu	Loughborough University, UK
Haowen Liu	University of York, UK
Pengcheng Liu	University of York, UK
Qi Lu	University of Texas Rio Grande Valley, USA
Nan Ma	Northwestern Polytechnical University, China
Pedro Machado	Nottingham Trent University, UK
Soumya Kanti Manna	Canterbury Christ Church University, UK
Fraser McGhan	Cardiff University, UK
Nahal Memar Kocheh Bagh	University of York, UK
Hongying Meng	Brunel University London, UK
Dennis Monari	Nottingham Trent University, UK
Juan Sebastian Mosquera Maturana	Universidad Autónoma de Occidente, Colombia
Fama Ngom	University of Montpellier, France
Yohan Noh	Brunel University London, UK
Matthew Parris	University of Buckingham, UK
Tianhu Peng	University College London, UK
Vishnu Rajendran Sugathakumary	University of Lincoln, UK
Abhra Roy Chowdhury	Indian Institute of Science, India
Matteo Russo	University of Rome Tor Vergata, Italy

Contents – Part II

Mechanism Design

Soft Robotics

Swarms and Multi-Agent Systems

Contents – Part I

Robotic Modeling, Sensing and Control

Machine Vision

Human-Robot
Interaction/Collaboration

Online Robust Robot Planning for Human-Robot Collaboration

Yang You[1]([✉]) [iD], Vincent Thomas[2] [iD], Francis Colas[2] [iD], Robert Skilton[1] [iD], and Olivier Buffet[2] [iD]

[1] Remote Applications in Challenging Environments (RACE), United Kingdom Atomic Energy Authority, Culham, UK
yang.you@ukaea.uk

[2] Université de Lorraine, INRIA, CNRS, LORIA, Nancy, France

Abstract. Human-robot collaboration often necessitates the robot to adapt to the uncertainty of human objectives and their induced behaviors. This may require the robot to have a human model to anticipate human partners' objectives and predict their actions, which is typically learned by the robot through available human data. However, in complex collaboration tasks, a chicken-and-egg problem arises because human data cannot be collected without a collaborative robot policy in the first place. In this article, we describe the human-robot collaboration task with Markov decision models and solve the chicken-and-egg problem by proposing an online robot planning algorithm. This online framework can automatically derive a human model without real human data and plan robust robot actions to support human partners with respect to their uncertain objectives and behaviors. Through experiments with a human-robot co-working scenario, we demonstrate that our online method outperforms the previous offline approach in terms of scalability and the ability to plan robot actions within a bounded time.

Keywords: Human-Robot Collaboration · Autonomous Robots · Decision Making & Planning

1 Introduction

Human-robot collaboration (HRC) is a hot topic with a wide range of applications in manufacturing [4,12,14,15]. In HRC, a collaborative team of humans and robots works together to achieve a shared goal. HRC is also part of the broader field of human-robot interaction (HRI), where the various partners do not necessarily benefit from the interaction [1]. However, unlike in a multi-agent system (MAS) where optimal policies are computed for all agents, in HRC problems, only the robot is controlled to provide support to its human partner. Moreover,

---

This work has been supported by the EPSRC Energy Programme under UKAEA/EPSRC Fusion Grant 2022/2027 No. EP/W006839/1 and by French National Research Agency (ANR) through the "Flying Coworker" Project under Grant 18-CE33-0001.

M. N. Huda et al. (Eds.): TAROS 2024, LNAI 15052, pp. 3–13, 2025.
https://doi.org/10.1007/978-3-031-72062-8_1

there may be multiple solutions for collaboration, especially if the human has several possible objectives and considers some sub-optimal behaviors. Therefore, from the robot's point of view, the challenge lies in how to be robust against uncertain human objectives and related uncertain behaviors.

Many previous works [2,3,8,11,16] rely on Markov decision models, especially utilizing the partially-observable Markov decision process (POMDP) framework to represent robot decision-making for HRC. Additionally, a mathematical model of the human partner [2,7,8] is commonly integrated into the environment dynamics. This integration enables the robot to reason about its actions considering potential human behaviors and their consequences. This model is also referred to as a mental model of the human, which may behave differently from the real human partner. However, developing such human mental models typically requires expert knowledge or collected data [2,8]. In specific applications [5,6], the human partner is assumed to consider only themselves when taking actions. This simplification allows for the design of a human mental model for the robot by reasoning on a single-agent problem, with the robot's role akin to a helper or assistant who facilitates ongoing tasks. However, in more complex HRC problems, coordination actions may be necessary for simultaneous execution by the entire human-robot team. In such cases, designing a human model for the robot necessitates consideration of the robot's behaviors, which may not be initially available. To address this challenge in complex HRC problems, an offline robot planning framework was proposed for HRC in [13]. In this framework, the robot automatically generates a human mental model from a centralized MAS solution and then computes the robot's best-response policy. Although experiments suggest that this framework can achieve encouraging results with real humans regarding uncertain human objectives and behaviors, its offline structure suffers from heavy computations and poor scalability.

In this article, we propose a new algorithm called *Online Robust Robot Planning* (ORRP) for human-robot collaboration problems. ORRP improves the offline planning framework presented in [13] by transitioning it to an online structure, thereby mitigating heavy offline computations and enabling real-time interaction between the robot and human within a bounded time. Moreover, as a sampling-based online algorithm, our contribution has the potential to scale up to large collaboration problems. The structure of this article is organized as follows: Technical backgrounds regarding Markov decision models are discussed in Sect. 2. Our main contribution is presented in Sect. 3 and evaluated in Sect. 4. Finally, potential improvements and conclusions are discussed in Sect. 5.

2 Background

2.1 Markov Decision Models

In this article, we employ decentralized POMDPs (Dec-POMDPs) as a modeling tool for representing the human-robot collaboration problem, consistent with the methodology presented in [13]. However, we do not directly solve this Dec-POMDP to obtain a joint policy for the human-robot team. This choice is guided by the concern that imposing a pre-computed policy on the human partner may not be suitable for practical implementation.

Definition 1. *A Dec-POMDP with $|\mathcal{I}|$ agents is represented as a tuple $M \equiv \langle \mathcal{I}, \mathcal{S}, \mathcal{A}, \Omega, T, O, R, b_0, H, \gamma \rangle$, where: $\mathcal{I} = \{1, \ldots, |\mathcal{I}|\}$ is a finite set of agents; $\mathcal{S}$ is a finite set of states; $\mathcal{A} = \times_i \mathcal{A}^i$ is the finite set of joint actions, with $\mathcal{A}^i$ the set of agent i's actions; $\Omega = \times_i \Omega^i$ is the finite set of joint observations, with Ω^i the set of agent i's observations; $T : \mathcal{S} \times \mathcal{A} \times \mathcal{S} \rightarrow \mathbb{R}$ is the transition function, with $T(s, a, s')$ the probability of transiting from s to s' if a joint action a is performed; $O : \mathcal{A} \times \mathcal{S} \times \Omega \rightarrow \mathbb{R}$ is the observation function, with $O(a, s', o)$ the probability of observing a joint observation o if a joint action a is performed and the next state is s'; $R : \mathcal{S} \times \mathcal{A} \rightarrow \mathbb{R}$ is the reward function, with $R(s, a)$ the immediate reward for executing a joint action a in s; b_0 is the initial probability distribution over states; $H \in \mathbb{N} \cup \{\infty\}$ is the time horizon; $\gamma \in (0, 1)$ is the discount factor applied to future rewards.*

An agent's i action *policy* π^i maps its possible action-observation histories to actions. The objective is then to find a joint policy $\pi = \langle \pi^1, \ldots, \pi^{|\mathcal{I}|} \rangle$ that maximizes the expected discounted return from b_0:

$$V_H^\pi(b_0) \stackrel{\text{def}}{=} \mathbb{E}\left[\sum_{t=0}^{H-1} \gamma^t r(S_t, A_t) \mid S_0 \sim b_0, \pi \right].$$

POMDP models can be viewed as special cases of Dec-POMDPs when the system has only one agent ($\mathcal{I} = \{1\}$), and the MDP is a special case of the POMDP where states can be directly observed without any noise. In the multi-agent setting, if there exists a centralized controller that governs all agents' actions and receives all agents' observations at each time step, then we can employ a multi-agent POMDP (MPOMDP) [9] to formulate such a problem. An MPOMDP is essentially a POMDP, as its action space comprises the joint action space of all agents, and its observation space is the joint observation space of all agents. All properties and algorithms developed for POMDPs can be directly applied to MPOMDPs.

2.2 Monte-Carlo Planning

In this article, our main contribution relies on solving Markov decision models through Monte-Carlo planning, particularly based on partially observable Monte-Carlo planning (POMCP) [10]. POMCP is an online planning algorithm for solving large POMDPs. Instead of deriving a complete optimal policy, which may be computationally expensive, POMCP searches for the best action given the current state distribution (also known as a belief). More specifically, POMCP combines Monte-Carlo sampling during belief state updates with Monte-Carlo tree search during planning. This approach helps address the curse of dimensionality and allows real-time planning in large POMDPs.

3 Online Robust Robot Planning

In this article, we propose an online algorithm for deriving robust robot policies, to address the heavy computation costs issue in offline settings [13]. The key ideas are as follows:

- We temporarily relax the HRC model (presented as a Dec-POMDP) to a centralized decision model (an MPOMDP). We solve this MPOMDP and compute a distribution of human actions online for each possible situation. This human action distribution will be considered by the robot to predict the real human partner's behaviors (which are hidden from the robot) at each time step;
- For the robot's own planning process, we develop a POMDP generative model G_e with an extended state space. Then, we apply online POMDP planning to G_e to make robust robot decisions within a bounded time.

The process for generating the robot's extended generative model G_e is presented in Sect. 3.1, and the main algorithm is discussed in Sect. 3.2.

3.1 Robot Extended Generative Model

A generative model (black box simulator) is typically necessary for online planning. For a Dec-POMDP, its generative model G is defined as follows:

$$s', o, r \leftarrow G(s, a),$$

where G takes the current state s and joint action a as input, and returns the next state s', joint observation o, and instantaneous reward r. However, in HRC tasks, the robot cannot directly use G to make its plans since it should reason about its actions using its own observations. Therefore, to construct an online robot algorithm, the question arises: how do we create a *robot generative model*?

In [13], an offline framework was proposed for constructing an explicit extended POMDP for the robot, which accounts for the uncertainty of human objectives and behaviors by integrating a mathematical model of the human. In this article, we extend this framework to the online setting; the robot should then infer the human's objective θ_H and internal belief b_H^t online. Thus, the robot's extended state is now defined as:

$$s_e^t = \langle s^t, \theta_H, b_H^t, o_R^t \rangle,$$

where s^t is the real current state, θ_H is the human's objective, b_H^t is the human's current belief, and o_R^t is the robot's observation. With this definition of the extended state, we can derive the dynamics of the robot's extended POMDP as follows:

$$T_e(s_e^t, a_R^t, s_e^{t+1}) = Pr(s_e^{t+1} | s_e^t, a_R^t)$$
$$= \frac{Pr(s^{t+1}, b_H^{t+1}, \theta_H, o_R^{t+1}, s^t, b_H^t, o_R^t, a_R^t)}{Pr(s^t, b_H^t, \theta_H, o_R^t, a_R^t)}$$

$$= \frac{\sum_{a_H^t} \sum_{o_H^{t+1}} Pr(s^{t+1}, b_H^{t+1}, \theta_H, o_R^{t+1}, o_H^{t+1}, s^t, b_H^t, o_R^t, a_H^t, a_R^t)}{Pr(s^t, b_H^t, \theta_H, o_R^t, a_R^t)}$$

$$= \sum_{a_H^t} \sum_{o_H^{t+1}} Pr(b_H^{t+1}|o_H^{t+1}, a_H^t, b_H^t) \cdot Pr(s^{t+1}|s^t, \langle a_H^t, a_R^t \rangle) \cdot$$

$$Pr(\langle o_R^{t+1}, o_H^{t+1} \rangle | s^{t+1}, \langle a_H^t, a_R^t \rangle) \cdot Pr(a_H^t|\theta_H, b_H^t)$$

$$= \sum_{a_H^t} \sum_{o_H^{t+1}} \eta(b_H^t \langle a_H^t, o_H^{t+1} \rangle, b_H^{t+1}) \cdot T(s^t, \langle a_H^t, a_R^t \rangle, s^{t+1}) \cdot$$

$$O(s^{t+1}, \langle a_R^t, a_H^t \rangle, \langle o_R^{t+1}, o_H^{t+1} \rangle) \cdot \psi_H(b_H^t, \theta_H, a_H^t),$$

$$O_e(s_e^{t+1}, a_R^t, o_R^{t+1}) = Pr(o_R^{t+1}|s_e^{t+1}, a_R^t)$$

$$= Pr(o_R^{t+1}|\langle s^{t+1}, \tilde{o}_R^{t+1}, b_H^{t+1}, \theta_H \rangle, a_R^t)$$

$$= \mathbf{1}_{o_R^{t+1} = \tilde{o}_R^{t+1}},$$

$$r_e(s_e^t, a_R^t) = \sum_{a_H^t} r(s^t, \langle a_H^t, a_R^t \rangle) Pr(a_H^t|\theta_H, b_H^t)$$

$$= \sum_{a_H^t} r(s^t, \langle a_H^t, a_R^t \rangle) \cdot \psi_H(b_H^t, \theta_H, a_H^t),$$

where we define ψ_H is a human action selection function such that $\psi_H(b_H^t, \theta_H, a_H^t) = Pr(a_H^t|\theta_H, b_H^t)$ (further details provided in Algorithm 1). Following this concept, we build an extended generative model G_e as presented in Fig. 1 with the following specifications:

– Inputs: current extended state s_e^t and robot action a_R^t,
– Outputs: next extended state s_e^{t+1}, robot observation o_R^{t+1}, and reward r.

Algorithm 1 details the internal processes of G_e. First, G_e takes as input the current extended state s_e^t and robot action a_R^t. Then, the algorithm decomposes s_e^t (line 3) and computes the Q values for b_H^t using POMCP, considering the Dec-POMDP is an MPOMDP for the human. A softmax distribution P_{A_H} of human actions is computed based on obtained Q values (line 4), enabling the robot to consider different human actions. Next, a human action a_H^t is sampled from P_{A_H} (line 5), and the algorithm passes the state s^t and joint action $\langle a_H^t, a_R^t \rangle$ to G (line 6). This step samples the next state s^{t+1}, all agents' observations $\langle o_H^{t+1}, o_R^{t+1} \rangle$, and an instant reward r. Then, to build the next extended state s_e^{t+1}, a particle filter is used to compute an updated human belief b_H^{t+1} using his observation o_H^{t+1} (corresponding to the function η_H in Fig. 1). It samples a state $\tilde{s}$ from b_H^t at line 10 and passes it to G with the same joint action $\langle a_H^t, a_R^t \rangle$, if the sample observation $\tilde{o}_H$ is identical to o_H^{t+1}, the generated next state $\tilde{s}'$ is kept and added as a particle to b_H^{t+1} until the maximum particle size (lines 12–14) is reached. At this stage, all the elements for building the next extended state s_e^{t+1} are obtained (line 15). The algorithm then returns the step results.

At time $t = 0$, the extended state $s_e^0 = \{s^0, b_H^0, \theta_H, \varnothing\}$ where $s^0 \in b^0$, $\theta_H \in P_\theta$, and $b_H^0 = b^0$. The robot's initial belief b_R^0 is the probability distribution over s_e^0, which is defined as follows:

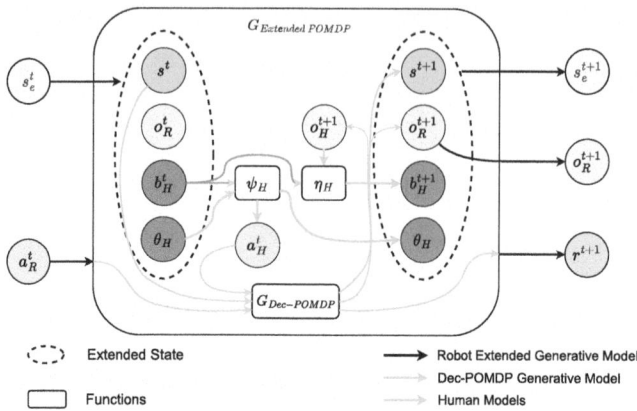

Fig. 1. Structure of the extended generative model G_e for the robot: The black arrows represent the inputs and outputs of the robot's extended generative model; The blue arrows show the inputs and outputs of a Dec-POMDP simulator G; The green arrows give the evolution of the human model. (Color figure online)

Algorithm 1: Robot's Extended Generative Model G_e in HRC

---

1 [Input:] s_e^t: extended state | a_R^t: robot's action
2 [Parameter:] G: Dec-POMDP simulator | T: softmax parameter
3 $\langle s^t, b_H^t, \theta_H, o_R^t \rangle \leftarrow s_e^t$
4 $P_{A_H} \leftarrow \psi_H(\cdot | b_H, \theta_H) T$ // POMCP solves MPOMDP with b_H and get Q values
5 $a_H^t \leftarrow Sample(P_{A_H})$
6 $s^{t+1}, \langle o_H^{t+1}, o_R^{t+1} \rangle, r \leftarrow G(s^t, \langle a_H^t, a_R^t \rangle, \theta)$
7 $k \leftarrow 0$
8 $b_H^{t+1} \leftarrow \varnothing$
9 **while** $k < k_{particles}$ **do**
10     $\tilde{s} \leftarrow Sample(b_H^t)$
11     $\tilde{s}', \langle \tilde{o}_H, \tilde{o}_R \rangle, \tilde{r} \leftarrow G(\tilde{s}, \langle a_H^t, a_R^t \rangle)$
12   **if** $\tilde{o}_H = o_H^{t+1}$ **then**
13        $b_H^{t+1} \leftarrow b_H^{t+1} \cup \{\tilde{s}'\}$
14        $k \leftarrow k + 1$

15 $s_e^{t+1} \leftarrow \langle s^{t+1}, b_H^{t+1}, \theta_H, o_R^{t+1} \rangle$
16 **return** s_e^{t+1}, o_R^{t+1}, r // return step results

$$
\begin{aligned}
b_R^0(s_e^0) &= Pr(s, b_H^0, \theta_H, \varnothing) \\
&= Pr(s, b_0, \theta_H, \varnothing) \\
&= Pr(s|b_0) Pr(\theta_H) \\
&= b_0(s) P_\theta(\theta_H),
\end{aligned}
$$

where P_θ is the probability distribution over human objectives. We have thus built a valid POMDP generative model G_e for the robot, allowing online planners to plan robot actions that account for possible human objectives and behaviors. Moreover, one can adjust the parameter T in the softmax function (line 4) to allow the robot to consider different human types (more or less rational).

3.2 Nested Online Planning

The method described in Algorithm 1 constructs an extended generative model G_e for the robot. As mentioned in the previous section, one can then use an online POMDP solver to plan robot actions using G_e. This involves two levels of online planning:

- First, the robot plans its own actions using G_e at the top level;
- Second, inside G_e, another online planning is used to compute the probability distribution over human actions (see line 4 in Algorithm 1).

Using POMCP as an online POMDP solver, we propose an online solver for the robot in our HRC tasks, outlined in Algorithm 2. Specifically, each robot's extended state s_e contains a human belief b_H, which is used within G_e by the bottom-level POMCP to compute a policy tree. When reaching a leaf node, instead of conducting rollouts using a random policy to estimate the node's value in the robot POMDP, we employ the value of b_H (for the MPOMDP guiding the human) as a heuristic estimate, highlighted in red in line 17. To avoid recomputations, this is achieved by reusing the policy tree constructed in the last call to the bottom-level POMCP, which stores estimated values for all visited human beliefs.

4 Experiments

4.1 Experiment Setting

In this article, we utilize the same experimental setup as in [13] and evaluate our contribution using the same dataset of human policies. All our experiments were conducted on a laptop equipped with a 2.3 GHz i9 CPU.

The experiment HRC scenario is illustrated in Fig. 2 (a). In this scenario, One device requires the robot's maintenance (presented in grey), situated at its designated cell. Repairing either of the two broken devices (presented in red) necessitates simultaneous actions from both the human and the robot at the device location, with the human selecting a component beforehand. Limited environmental observation is granted to both agents. This HRC task is modeled as a Dec-POMDP. The state space (S) comprises the locations of both agents, device statuses, and the human's possession of a component. Human observations (Ω_H) include his own location, robot presence, and device status, while robot observations (Ω_R) encompass robot location, human location, and device status. Both agents have movements (*Up, Down, Left, Right*) and repair actions, but only the human has a component picking action, and only the robot has a maintenance action. Rewards (R) are assigned for successful repairs of all

---

Algorithm 2: Online Robot Search For HRC

---

1 $N()$: counts visits to a history or history-action pair | $\mathcal{T}$: policy tree | b_0: initial belief of the problem | $B(h)$: estimated belief given an action-observation history h

2 **Fct** *Search(h)*

3   **repeat**

4   **if** $h = empty$ **then**

5           $s_e \sim b_0$

6   **else**

7           $s_e \sim B(h)$

8   Simulate$(s_e, h, 0)$

9   **until** *Timeout()*

10 **Fct** *Simulate(s_e, h, depth)*

11   **if** $\gamma^{depth} < \epsilon$ **then**

12   **return** 0

13   **if** $h \notin \mathcal{T}$ **then**

14   **for** $a \in \mathcal{A}$ **do**

15           $\mathcal{T}(h,a) \leftarrow (N_{init}(h,a), Q_{init}(h,a), \varnothing)$

16        $b_h \leftarrow s_e.b_h$

17        $V(b_h) \leftarrow POMCP_M$

18   **return** $V(b_h)$

19     $a \leftarrow \arg\max_a [Q(h,a) + c\sqrt{\frac{\log N(h)}{N(h,a)}}]$

20     $(s_e', o, r) \sim G_e(s_e, a)$

21     $R \leftarrow r + \gamma Simulate(s_e', hao, depth + 1)$

22     $N(h) \leftarrow N(h) + 1$

23     $N(h,a) \leftarrow N(h,a) + 1$

24     $Q(h,a) \leftarrow Q(h,a) + \frac{R - Q(h,a)}{N(h,a)}$

25   **return** $\arg\max_a Q(h,a)$

---

devices ($+100$), with penalties for all actions (-2) except waiting (-1), encouraging efficient coordination to help the human finish the task early. If the robot performs a invalid action (*e.g.*, the robot hits a wall), an additional penalty (-20) is given. This formulation involves 2304 states, 49 joint actions, and 5400 joint observations.

The uncertainty about the human's objectives is reflected in the variations of the reward function. Figure 2 (b) illustrates two variants: one where the human prefers to repair the left device first and another where the preference is for the right device. These variants generate different human policies, as the human's objectives influence their decision-making process. Initially, the robot only possesses a prior belief over the human's reward functions, with each variant having a probability of 0.5. This uncertainty about the human's objectives adds complexity to the collaborative task, requiring the robot to adapt its actions to accommodate different potential human preferences.

In our experiments, the robot has to deal with a total of 3×50 different human trajectories. This includes 50 human policies $\pi_{H,l}^{Sample}$ that consistently favor the left repair device first, another 50 policies $\pi_{H,r}^{Sample}$ that always favor the right device first, and an additional 50 policies $\pi_{H,l+r}^{Sample}$ that are a uniform mixture of both.

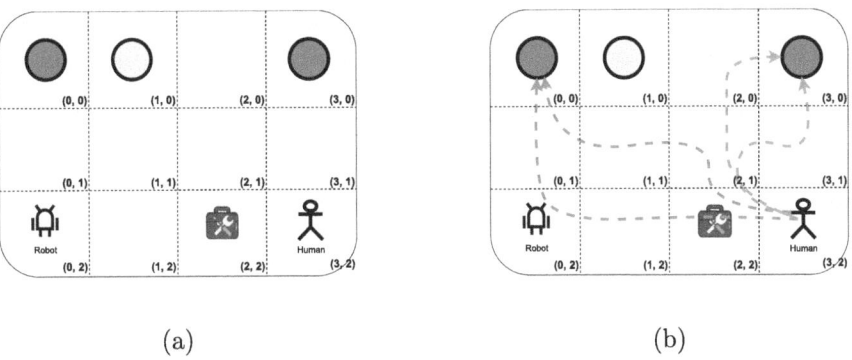

(a)   (b)

Fig. 2. Figure (a) presents a collaboration task: A robot and a human evolve in a 4×3 grid world. The top-left cell $(0,0)$ and the top-right cell $(3,0)$ both contain a broken device. A device to be maintained is also located in cell $(1,0)$. A toolbox is located in cell $(2,2)$ where the human can pick components. Figure (b) illustrates of the uncertainties about the human partner in this HRC task: 1. Uncertainty arises regarding the human objective, particularly regarding the preference for the first repairing device (presented with blue and orange colors); 2. Uncertainty on human policies (presented with different trajectories). (Color figure online)

4.2   Results

We use Algorithm 2 to plan the robot's actions online with a softmax temperature $T = 0.5$, and we test three different timeouts ($10\,s$, $30\,s$ and $60\,s$) in the experiments. We evaluate our online approach using the same human policy dataset as in the offline method [13], and the results are presented in Table 1. Firstly, we observe that:

- Our online method provides relatively good solutions even with a short timeout. For example, with a timeout of $10\,s$, our online approach achieves a 82% success rate on both objectives (evaluated with $\pi_{H,l+r}^{Sample}$);
- The performance of our online method improves with the timeout. From the timeout of $10\,s$ to $60\,s$, the online method enhances its success rate from 82% to 90% on sets $\pi_{H,l+r}^{Sample}$, and the average value increases from 6.5 to 9.7.

Secondly, we observe that in the *prefer-right* scenario, our online method can achieve the same success rate as the offline method (90%), but with a lower average value (9.1 compared with 10.0). However, in the *prefer-left* scenario, the online approach can reach a success rate of 90% with a value of 10.2, which outperforms the offline method. An explanation for this phenomenon is that, in the *prefer-left* case, the offline method cannot cover all possible human policies, and thus may perform poorly in some situations, while the online method keeps predicting human behaviors and replanning robot actions that makes good decisions in more situations. In the *prefer-right* scenario, although both methods can achieve the same success rate, the offline approach is equipped with an explicit POMDP model that properly covers the human behavior and, therefore, can compute a better policy thanks to offline POMDP solvers with a higher value compared with the online algorithm.

Table 1. Average value and success rates (in parentheses) of various robot policies vs. various synthetic human behaviors. For comparison, we select the offline robot policy $\pi_{R,Offline}$ which has the highest average value from [13] and report its total computation time. For our online method, we present the different timeouts used in each $\pi_{R,Online}$.

	3 × 50 Human Policies					
	$\pi_{H,l}^{Sample}$		$\pi_{H,r}^{Sample}$		$\pi_{H,l+r}^{Sample}$	
$\pi_{R,Offline}(time = 6465\,\mathrm{s})$	9.5 ± 2.2	(84%)	10.0 ± 2.0	(90%)	9.8 ± 2.1	(87%)
$\pi_{R,Online}(timeout = 10\,\mathrm{s})$	5.6 ± 2.4	(80%)	7.3 ± 2.3	(84%)	6.5 ± 2.3	(82%)
$\pi_{R,Online}(timeout = 30\,\mathrm{s})$	7.0 ± 2.4	(84%)	6.7 ± 2.2	(86%)	6.9 ± 2.3	(85%)
$\pi_{R,Online}(timeout = 60\,\mathrm{s})$	10.2 ± 2.0	(90%)	9.1 ± 2.1	(90%)	9.7 ± 2.1	(90%)

5   Conclusion

In this article, we propose an online robust robot planning algorithm (ORRP) designed for HRC tasks. Through experiments, we demonstrate that ORRP is robust to uncertain human behaviors with different objectives and achieves competitive performance against the previous offline method in the given HRC task. Moreover, compared with the previous offline method, our online approach only requires a generative model and, therefore, can scale up to large problems. We demonstrate the process of constructing a valid POMDP generative model for the robot with an extended state space that includes the human's internal state. We also provide an online robot planning algorithm based on POMCP. Using this POMDP generative model, the robot plans its actions at each time step by estimating the human's actual objectives and actions online.

For future work, we plan to conduct further experiments on more HRC problems interacting with real humans. In particular, we would like to validate our approach in real-world scenarios, and further improve the robot policy by combining learning techniques with the real human data collected during interactions.

References

1. Bauer, A., Wollherr, D., Buss, M.: Human-robot collaboration: a survey. Int. J. Humanoid Rob. **5**, 47–66 (2008)
2. Chen, M., Nikolaidis, S., Soh, H., Hsu, D., Srinivasa, S.: Planning with trust for human-robot collaboration. In: 2018 13th ACM/IEEE International Conference on Human-Robot Interaction (HRI), pp. 307–315 (2018)
3. Görür, O.C., Rosman, B., Sivrikaya, F., Albayrak, S.: Social cobots: anticipatory decision-making for collaborative robots incorporating unexpected human behaviors. In: Proceedings of the 2018 ACM/IEEE International Conference on Human-Robot Interaction, pp. 398–406 (2018)
4. Guerin, K.R., Lea, C., Paxton, C., Hager, G.D.: A framework for end-user instruction of a robot assistant for manufacturing. In: 2015 IEEE International Conference on Robotics and Automation (ICRA), pp. 6167–6174 (2015)
5. Karami, A.B.: Modèles Décisionnels d'Interaction Homme-Robot. Theses, Université de Caen (2011)
6. Karami, A.B., Jeanpierre, L., Mouaddib, A.I.: Partially observable Markov decision process for managing robot collaboration with human. In: 2009 21st IEEE International Conference on Tools with Artificial Intelligence, pp. 518–521 (2009)
7. Nikolaidis, S., Kuznetsov, A., Hsu, D., Srinivasa, S.: Formalizing human-robot mutual adaptation: a bounded memory model. In: 2016 11th ACM/IEEE International Conference on Human-Robot Interaction (HRI), pp. 75–82. IEEE (2016)
8. Nikolaidis, S., Zhu, Y.X., Hsu, D., Srinivasa, S.: Human-robot mutual adaptation in shared autonomy. In: Proceedings of the 2017 ACM/IEEE International Conference on Human-Robot Interaction, HRI 2017, pp. 294–302. Association for Computing Machinery (2017)
9. Pynadath, D.V., Tambe, M.: The communicative multiagent team decision problem: analyzing teamwork theories and models. J. Artif. Intell. Res. **16**, 389–423 (2002)
10. Silver, D., Veness, J.: Monte-Carlo planning in large POMDPs. In: Advances in Neural Information Processing Systems, vol. 23 (2010)
11. Unhelkar, V.V., Li, S., Shah, J.A.: Decision-making for bidirectional communication in sequential human-robot collaborative tasks. In: Proceedings of the 2020 ACM/IEEE International Conference on Human-Robot Interaction, HRI 2020, pp. 329–341. Association for Computing Machinery (2020)
12. Wang, X.V., Kemény, Z., Váncza, J., Wang, L.: Human-robot collaborative assembly in cyber-physical production: classification framework and implementation. CIRP Ann. **66**(1), 5–8 (2017)
13. You, Y., Thomas, V., Colas, F., Alami, R., Buffet, O.: Robust robot planning for human-robot collaboration. In: 2023 IEEE International Conference on Robotics and Automation (ICRA), pp. 9793–9799. IEEE (2023)
14. Zanchettin, A.M., Ceriani, N.M., Rocco, P., Ding, H., Matthias, B.: Safety in human-robot collaborative manufacturing environments: metrics and control. IEEE Trans. Autom. Sci. Eng. **13**(2), 882–893 (2016)
15. Zanchettin, A.M., Rocco, P.: Path-consistent safety in mixed human-robot collaborative manufacturing environments. In: Proceedings of the IEEE/RSJ International Conference on Intelligent Robots and Systems, pp. 1131–1136 (2013)
16. Zheng, W., Wu, B., Lin, H.: POMDP model learning for human robot collaboration. In: 2018 IEEE Conference on Decision and Control (CDC), pp. 1156–1161. IEEE (2018)

The Dilemma of Decision-Making in the Real World: When Robots Struggle to Make Choices Due to Situational Constraints

Khairidine Benali[(✉)] and Praminda Caleb-Solly

School of Computer Science, University of Nottingham, Nottingham, UK
{Khairidine.Benali,Praminda.Caleb-Solly}@nottingham.ac.uk

Abstract. In order to demonstrate the limitations of assistive robotic capabilities in noisy real-world environments, we propose a Decision-Making Scenario analysis approach that examines the challenges due to user and environmental uncertainty, and incorporates these into user studies. The scenarios highlight how personalization can be achieved through more human-robot collaboration, particularly in relation to individuals with visual, physical, cognitive, auditory impairments, clinical needs, environmental factors (noise, light levels, clutter), and daily living activities. Our goal is for this contribution to prompt reflection and aid in the design of improved robots (embodiment, sensors, actuation, cognition) and their behavior, and we aim to introduces a groundbreaking strategy to enhance human-robot collaboration, addressing the complexities of decision-making under uncertainty through a Scenario analysis approach. By emphasizing user-centered design principles and offering actionable solutions to real-world challenges, this work aims to identify key decision-making challenges and propose potential solutions.

Keywords: Decision-Making · Human-Robot Collaboration · Assistive Robots

1 Introduction

In recent years, considerable progress has been made in developing robotics and autonomous systems for use within assistive scenarios [1]. Despite these advancements, a major challenge remains: building robots that can collaborate seamlessly with humans in an error-free, intuitive, and comprehensible manner. Achieving this requires utilizing high-level multi-modal recognition systems to efficiently manage communication and understand human requirements and goals in noisy real-world environments, which often result in ambiguity and uncertainty.

The growth of the Internet of Things (IoT) has led to a widespread presence of devices and sensors in daily life [2], making them integral components

© The Author(s), under exclusive license to Springer Nature Switzerland AG 2025
M. N. Huda et al. (Eds.): TAROS 2024, LNAI 15052, pp. 14–26, 2025.
https://doi.org/10.1007/978-3-031-72062-8_2

of human-robot interaction (HRI). Traditional one-way interfaces focused on a single sensory modality-such as visual, tactile, auditory, or gestures-are no longer the exclusive options.

In assistive robotics, visual and auditory modes of communication are intuitive and effective [3], thanks to advances in modeling that enable more natural language-based interactions [4]. Nevertheless, there is still a need to improve the reliability of these interactions, which requires a high level of situational awareness. The primary objective of multimodal HRI [5,6] in assistive scenarios is to enable individuals with accessibility needs to interact effectively with robots, thereby increasing the demand for contextual understanding to improve performance in ambiguous or uncertain environments.

Advancements in human-system communication have the potential to revolutionize human-robot interactions, leading to increased collaboration, user satisfaction, and efficiency. Enhanced communication can help robots better understand human instructions, reduce errors, and improve overall task performance and user experience. Additionally, improved communication enables robots to respond more effectively to unforeseen situations, adapt to new tasks with greater ease, and better handle unexpected changes in their environment.

In real-world HRI, robots often face challenges [7] where certain factors are not covered in their decision-making algorithms or training [8]. Communication and interaction with humans can potentially improve a robot's performance in these challenging situations [9], enhancing the repairability of communication failures. However, several constraints due to disabilities and impairments (visual, physical, cognitive, and auditory) are not always considered in the design of HRI hardware and software [10].

Our objective is to tackle these challenges and explore potential strategies for achieving safe management of failures and learning in the context of human-robot collaboration in complex and dynamic real-world environments. This work can serve as a valuable basis for analyzing and reflecting on the highlighted challenges, planning comprehensive trials, and evaluating different scenarios in user studies. A better understanding of how interactions can be personalized for people with various disabilities and clinical requirements, as well as considering environmental factors like noise, lighting, and clutter, can inform the design of robots. This includes considerations of their physical form, sensing capabilities, mechanisms for movement, and overall behavior.

2   Human-Robot Interaction

Improving human-robot collaboration to resolve potential repairable failures [11] and learn how to manage safely in a "messy" real-world environment is a challenging endeavor. The space around, and the distance between, the robot and the person can also impact the interaction, particularly when considering accessibility needs, person-environment interaction guidelines should also be considered [12]. One potential avenue for progress can be achieved by designing more adaptable and customizable robots. Furthermore, providing developers and designers with a variety of benchmarks [13], for their methods and solutions, and to help refine and optimize their approaches.

2.1 Optimal Solutions in New Scenarios

In healthcare assistive technologies, human-robot interaction is evolving into a critical sector where the nuances of human factors require proactive consideration and prioritization to ensure effective collaboration, avoid errors, and optimize patient care [14]. However, it is evident that the majority of failure situations in the healthcare sector are caused unexpected situations for the system, in which small discrepancies or variations in the system's inputs or conditions can impact on the system's performance or behavior. In addition, the absence of personalized and adaptable human interaction, underscores the significant impact of human factors on resolving the consequences of failure. Consequently, it is crucial to better understand situations can result in failure, and prioritize safety and identify mitigations to minimise risk ensure safer and resilient human-robot interaction. Furthermore, safe and efficient human-robot interaction is crucial in a Human-Robot Collaborative systems [15], particularly when designing the robot as a general assistant that offers flexibility and freedom to users. Given the critical role of feedback and trust within such systems, novel approaches are necessary to facilitate seamless communication between humans and robots. To realize this objective, the development of dedicated Human-Robot Interaction Modules is imperative. These modules should be designed with consideration for potential unforeseen circumstances, necessitating advanced and multimodal interfaces that cater to individuals with impairments. Furthermore, it is crucial to comprehend the influence of failures on trust and perception, and to address their impact through strategies such as failure recovery and explainability.

In various scenarios, robots are often designed to provide assistance with daily activities. However, in many cases, these robots are trained to make decisions based on specific scenarios and feedback from sensors. The robot may face several different types of limitations that prevent it from carrying out the next action, such as an obstacle blocking its path, noise in the environment affecting the quality of voice-based interaction or an unidentifiable target. Developers often rely on identifying a range of various use cases or scenarios to try and expose possible limitations or gaps in functionality in their applications. Often hazard assessment approaches, such as SHARD-UML (Software Hazard Analysis and Resolution in Design, a hazard analysis technique which is a variant of HAZOP -Hazard and Operability Analysis) and STPA (Systems-Theoretic Process Analysis) are used to analyze safety and can help to highlight what errors the robot can encounter [16]. STPA is a holistic, systems-based approach to investigating accident causation. It departs from traditional reliability-based models, focusing on design flaws and unsafe interactions among non-failing components. This comprehensive framework considers a broader range of variables, making it adaptable to complex technologies. The SHARD-UML framework identifies hazards in human-robot interaction systems by applying guide words (Omission,

Commission, Early, Late, Value) to UML diagrams, evaluating them based on severity levels from no hazard to high risk of user annoyance or injury.

In literature, several methodologies have been used to identify errors in tasks that require human-robot interaction. One such technique is Root Cause Analysis (RCA) [17], which identifies the underlying causes of system failures by systematically examining the sequence of events leading to the failure. Similarly, Fault Tree Analysis (FTA) [18,19] focuses on visualizing possible failure scenarios using a logic diagram, helping to identify potential failures and prioritize system improvements. Other studies have applied general techniques for identifying human errors in robotics. Human Error Analysis (HEA) investigates and categorizes errors made by humans interacting with robots to better understand failure mechanisms and develop countermeasures. Event Tree Analysis (ETA) [20] examines and maps out possible event chains triggered by an initial failure, enabling the identification of potential failure scenarios and their resulting consequences. Formal methods, such as model checking [21] and temporal logic [22], provide rigorous error detection through mathematical verification of system behaviors, but can be resource-intensive and challenging to apply to complex systems. Behavior-based methods [23], which analyze real-time human and robot behaviors using sensors and cognitive modeling, can detect deviations and predict errors based on human cognitive states. While effective in dynamic scenarios, these methods require extensive sensor infrastructure and accurate behavioral models. However, real-world environments can change dynamically, and not all possible scenarios can be covered in terms of decision-making algorithms. In this situation, the robot may need to interact with humans to ask for help, when the decision-making algorithm fails to find the optimal solution, especially in cases where physical obstructions are encountered, or instructions are ambiguous. By training robots through adjusting their decision-making based on human guidance, we can also improve their performance.

2.2 Coping with Changing Preferences

Humans can be unpredictable and may change their preferences, leading to inconsistent behavior in similar situations [24]. In the same scenario, an individual may alter their preferences, such as a shift in their goals for a similar situation or changes in the environment itself. In an assistive scenario, the human user might have a illness or long-term condition, causing fatigue, memory or cognitive changes, difficulty in speaking etc., due to which they might want to change how they want a task completed, or change the way that they want to interact with the robot. In such scenarios, ability for explicit or implicit communication that helps the robot adjust quickly and easily to the change can provide a user with greater control over the situation. So rather than perceiving robots as having capabilities limited to repetitive scenarios, considering means to adapt with agility to the changes is needed.

2.3 Socio-economic Situations, Cultures and Attitudes Affecting Preferences

Robots can be trained using various datasets to make decisions in specific tasks and scenarios. However, the dataset used for training could be generated based on specific socio-economic situations, attitudes and cultures, which may result in decisions that depend on the attitudes present in the training dataset.

One potential area where culture plays a role is cooking or meal preparation by robots. For example, preparing a cup of tea can depend on the cultural background that the designers and developers drew on during the learning phase, but it may still not be sufficient as individuals, even from the same culture, have their own preferences. Similar issues can arise from differences in responding within different social situations, where there might be concerns about how formal or informal the behaviour of robot should be, privacy concerns when providing reminders, when it is polite to interrupt and when it is not, and behaviours that might undermine the dignity of the person being supported.

3 Video Exploring Some Real-World Challenges

In order to expose some of the breakdowns in communication, we created videos (Watch the video) that help to highlight several scenarios (Table 1) where a robot can face dilemmas in decision-making when it encounters difficulties in completing a task.

In Scenario 1, a person asks the robot to deliver an item using voice and hand gestures. The robot moves towards the target, but struggles to access the item, which is located on the table. When encountering physical obstacles like furniture, the robot activates its LIDAR and camera sensors to detect the obstructions and recalculates its route using obstacle avoidance algorithms. If the obstruction cannot be overcome, the robot re-attempts the process to find

Table 1. Challenging situations for robot decision-making and the incorporation of human feedback to enhance assistance

Example	Scenario	What puzzled the robot	Human Feedback
1	The robot struggles in successfully delivering an item to a human as per the desired target due to a physical obstruction	Should I follow the same approach as in a previous similar scenario, or should I seek assistance, or should I alter my course of action, or tell the user?	The user suggests moving the rollator out of the way to make it easier for the robot to access the object
2	The necessity for multiple communication modalities, as some individuals may have speech impairments	Should I keep asking or should I provide more options, or should I ask for help?	The user is engaging with the robot through an alternative means of communication rather than verbal communication during the task
3	The user in facing away from the robot when it arrives and doesn't respond to the robot when it speaks to it	Why is the person not responding, how many times should I repeat the question or should I go away and try again?	There is no response from the user, so the robot just goes away after a while
4	The robot was instructed to bring a pack of medicine, but instead, it found three similar-sized boxes	Should I follow the same approach as in the previous similar scenario (bring the green box), or should I seek assistance, or Should I modify my plan to take all the boxes?	The robot takes into account the user's preferences and attempts to retrieve the correct item again

an optimized path. However, the robot tries out various options to resolve the issue, but it is unable to determine the most effective course of action. It is unsure whether to follow the same approach as used previously, seek assistance, or alter its course of action. The person response could generate further ambiguity, as there are two rollators present blocking its path. Furthermore the robot lacks functionality to resolve the problem, this is often the case with assistive robots designed to handle small payloads, a trade-off of being able to fit in small spaces.

In scenario 2 the robot employs a range of communication models, including voice and vision, to engage with users. It intuitively selects the most suitable approach, adjusting its response strategy based on user feedback to facilitate effective and clear interaction. However, the scenario demonstrates the need for multimodal communication options, as the robot is unable to understand the feedback from the person. It can often be the case, if the person has a stroke, or problem with their teeth, or be drowsy, that their speech patterns change. There is also a danger of mis-hearing the person, which can result in an erroneous action on the part of the robot. Integration with other smart home or internet of things devices can be considered here. This can also improved reliability of the input in some cases. We could also consider cases where the background noise or light levels affect interaction.

In Scenario 3, the need for advanced situational awareness is illustrated. Also, this is where person's preference, or priority of the task needs to be considered. Upon arrival, if the robot finds the user facing away and unresponsive, it initially employs verbal cues to capture their attention. Then, using camera data, it identifies the user's orientation and adjusts its approach accordingly. For example the person might not want to be disturbed when sleeping and the robot should wait until they are awake. For this, it needs additional data to find out whether the person is sleeping, and might actually accidentally wake them up by checking. On the other hand, it might be important to wake the person up, if the task is urgent, such as regular doses of medication. It could also be that the user has lost consciousness, and this brings in the need for more consideration of being able to sense breathing patterns and heart-rate using wearable sensors or advanced imaging.

In scenario 4 (Fig. 1), the person asks the robot to bring their medicine, the robot utilizes its vision system to detect and assess distinct characteristics such as labels, shapes, or colors, enabling it to identify objects and selectively retrieve the desired one. However, the robot finds more than one box in the same place and gets confused. The robot is unsure whether to follow the same approach it might have used in a previous similar scenario, seek assistance, or modify its plan to take all the boxes. This scenario also highlights the need for including other information regarding knowledge of specific items and their use, to ensure safety of the person, based on their cognition or vision.

In all cases, the robot seeks assistance and follows the person's suggestions. The scenarios demonstrate how the optimal solution could be achieved through collaboration, which could also result in the user feeling more engaged and in control.

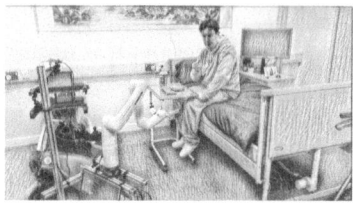

(a) Person asks the robot to bring their medicine from a designated location

(b) Robot arrives at the location but finds multiple boxes of medicine

(c) Robot pauses, realizing the importance of specific knowledge about the medicines for safety

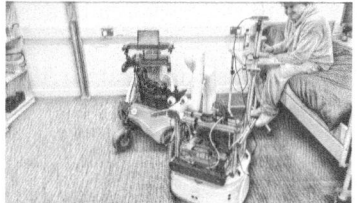

(d) Robot decides to seek assistance and modify its plan to ensure the correct medicine is delivered

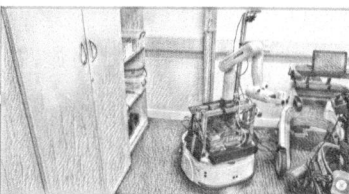

(e) Robot selects the correct medication and is prepared for delivery

(f) Robot delivers and hands over the correct medicine, completing the task

Fig. 1. Medicine Retrieval Dilemma: Navigating Uncertainty for Safety

4   Safety Analysis State for Robot-Assisted Patient Recovery Situation

In developing a robotic system to provide assistance for patients in the recovery period, a crucial initial step for ensuring safety involves reviewing existing methods for identifying safety concerns and potential failures during a task. For assessing the operational safety of the human-robot interaction, we utilized STPA to examine safety concerns arising from human actions, integrating insights from the Human-Robot Failure Taxonomy (Fig. 2) to comprehensively address potential failure modes. Our selection of this technique was driven by its relevance to our specific application and its potential applicability to other similar physically assistive human-robot interaction tasks.

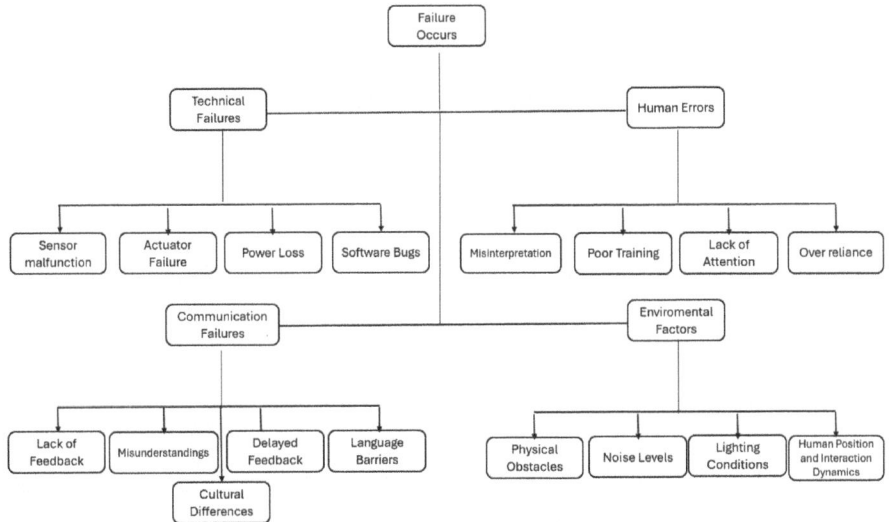

Fig. 2. Human-Robot Failure Taxonomy

4.1   STPA: User Error Analysis

While most existing methodologies in the literature focus on identifying and anticipating component failure, STPA (Systems-Theoretic Process Analysis) takes a novel approach by adopting a holistic, systems-based framework to investigate accident causation. This departure from traditional reliability-based models allows STPA to surface a growing class of accidents caused by design flaws or unsafe interactions among non-failing components, such as those highlighted in [25].

STPA is considered a comprehensive framework that encompasses the hazards and their causes identified by traditional methods. This approach was developed to address the growing complexity of technology. However, it is essential to recognize that STPA requires customization and completion for each specific task, which necessitates considering a broader range of variables compared to traditional methods. Within a given context, the range of possible human actions is finite but diverse. To identify the essential actions and potentially unsafe behaviors, we have selected task scenario 4 and apply the STPA method to provide an initial abstraction.

Fig. 3. (A)User error analysis in a medicine handover scenario 4 (B) User errors common to any task scenario element.

For our specific use case, we conducted a detailed analysis of user errors for Scenario 4, which is illustrated in the top row of Fig. 3A. Our analysis focused on situations where users take actions that could lead to potential hazards. Every task component is assigned a unique identifier for User Error Analysis purposes, facilitating hazard identification. This aspect of hazard analysis takes on particular importance, given the potential for older adults with cognitive or physical impairments, such as memory loss or mobility issues, to interact with the system in unpredictable ways, including through unintended movements, gestures, or verbal expressions.

5   Discussion

This paper explores scenarios related to healthcare and assistive robots, with the goal of paving the way for additional applications and potential failures in various contexts that consider both hardware and software from the robot's perspective, as well as human error in the methodology of interaction.

From analysing the context from a robot's perspective, various methodologies from the literature have shown promising results. One such approach is the application of Unified Modeling Language (UML) for system software analysis. UML is a widely used tool for software applications, which enables a thorough breakdown of possible outcomes by creating a diagram that defines the logical connections between blocks and the reasoning flow. This methodology can help identify user-related errors that can occur at different stages of a robotic task. Furthermore, the SHARD (Robot Hazard and Error Analysis) methodology can be applied to identify potential errors that the robot may encounter. This allows for a focused examination of potential failure points based on the logical sequence of actions the robot may take. By applying the SHARD technique at each node or point of the design, as defined by the UML (Unified Modeling Language), researchers can systematically analyze potential errors and develop strategies to mitigate them.

The SHARD analysis framework, integrated with Unified Modeling Language (UML), can be used to conduct a thorough examination of potential hazards in human-robot interaction systems. By applying a set of guide words, including Omission, Commission, Early, Late, and Value, to each node of the UML diagram, we can generate a comprehensive list of potential hazards. Each hazard is then evaluated based on its severity level, ranging from no hazard to high hazard, according to the risk of user annoyance or injury. This evaluation is aligned with established severity standards, such as the Abbreviated Injury Scale (AIS) [26], to ensure consistency in assessing potential risks. The identified hazards encompass a range of issues, including communication delays, unclear sensor interpretation, user distraction, and interventions by secondary actors. While the SHARD+UML approach is effective for analyzing software flow, it is acknowledged that certain human-related aspects are not adequately addressed. Therefore, the System-Theoretic Process Analysis (STPA) methodology is explored to provide additional insights into human factors and potential hazards in human-robot interactions.

Investigating robot accidents is crucial, especially with the growing presence of social robots in various settings. As highlighted in [27], robot accidents are inevitable, and a framework for thorough accident investigations is essential, similar to those used in aviation and rail industries. The proposed framework aims to collect comprehensive data, identify key facts, and conduct a structured analysis of social robot accidents. This study provides valuable insights into the challenges and processes involved in investigating robot accidents, emphasizing the importance of a proactive approach to ensure the safe deployment of social robots in society.

However, real-world settings are subject to dynamic changes, and decision-making algorithms may not encompass every conceivable scenario. Consequently, the robot might require human interaction to seek assistance when the algorithm falls short of identifying the optimal solution, particularly in instances involving physical obstacles. To advance the robotic system's ability to handle similar situations when it becomes stuck with decision-making, it is essential to consider how best to adapt the algorithms and design to suit specific applications and collect more data on multimodal detection of errors and failures in Human-Robot interactions, which will help to confirm error analysis, identify patterns, and train models to mitigate failures.

Translating these insights into actionable guidelines for developers and practitioners involves several key steps. Future research directions could involve the development of sophisticated error detection systems that effectively integrate data from multiple sensors to accurately analyze intricate interactions, thereby enabling the detection of nuanced indications of errors of or abnormal behaviors, including visual, auditory, and tactile inputs. Enhancing multi-modal communication interfaces enables robots to engage users in a highly personalized and adaptable manner, effortlessly transitioning between spoken language, textual interactions, and gestures. Furthermore, thoughtful design incorporates visually and aurally captivating alerts to ensure user attention and understanding. By prioritizing user-centric design principles, interaction flows can be tailored to individual preferences, user feedback can be seamlessly integrated, and accessibility features can be incorporated to accommodate diverse user needs. By leveraging machine learning techniques to extract and combine features from various sources, thereby enabling the seamless integration of diverse information streams and the detection of potential indicative of errors or unusual behaviors. In addition, pattern recognition techniques can allow researchers to uncover common error patterns, which can be used to refine interaction models and train error-prevention strategies, ultimately enhancing the reliability and efficiency of human-robot interactions. In addition, user feedback and iteration are essential to perfecting the system, accomplished through routine user testing, adoption of an iterative development methodology, and provision of post-deployment support to resolve issues, implement new features, and ensure sustained system improvement.

6   Conclusion

Our video, and an example hazard analysis of a specific scenario, seek to prompt designers and developers of assistive robots to think more deeply about how improved human-robot collaboration can help to address the challenges in real-world scenarios, particularly where individuals are likely to have a range of disabilities (including visual, physical, cognitive, and auditory impairments) and clinical requirements. There are also likely to be a multitude of varying environmental factors like noise, light levels, and clutter, as well as a range of daily living activities where the user will require assistance. We hope these videos will encourage other researchers to develop other such failure scenarios and draw focus to the need for more accessible and collaborative human-robot interaction, as well as guiding user studies which incorporate situations focused on different types of failures. This will help to guide the design of robots and the underlying machine learning and AI algorithms, with more creative thinking regarding their interaction, physical form, sensors, actuators, and behavioral strategies.

Acknowledgments. This work is supported by the Metrics Project, an EU Horizon 2020 research and innovation program under grant agreement No 871252, and the EPSRC Health Technologies Network+ Emergence, EP/W000741/1. Special thanks to all contributors and collaborators for their valuable insights and support throughout the Metrics project. Thanks also to Gabriel Leach for his help in making the video.

References

1. Martinez-Hernandez, U., Metcalfe, B., Assaf, T., Jabban, L., Male, J., Zhang, D.: Wearable assistive robotics: a perspective on current challenges and future trends. Sensors **21**, 6751 (2021). https://www.mdpi.com/1424-8220/21/20/6751

2. Kumar, S., Tiwari, P., Zymbler, M.: Internet of Things is a revolutionary approach for future technology enhancement: a review. J. Big Data. **6**, 111 (2019). https://doi.org/10.1186/s40537-019-0268-2

3. Arulkumaran, K., et al.: A comparison of visual and auditory EEG interfaces for robot multi-stage task control. Front. Robot. AI. **11**, 1329270 (2024). https://www.frontiersin.org/articles/10.3389/frobt.2024.1329270

4. Li, Z., Mu, Y., Sun, Z., Song, S., Su, J., Zhang, J.: Intention understanding in human-robot interaction based on visual-NLP semantics. Front. Neurorobotics **14**, 610139 (2021). https://www.frontiersin.org/articles/10.3389/fnbot.2020.610139

5. Andronas, D., Apostolopoulos, G., Fourtakas, N., Makris, S.: Multi-modal interfaces for natural Human-Robot Interaction. Procedia Manuf. **54**, 197–202 (2021). https://www.sciencedirect.com/science/article/pii/S2351978921001669, 10th CIRP Sponsored Conference on Digital Enterprise Technologies (DET 2020) - Digital Technologies as Enablers of Industrial Competitiveness and Sustainability

6. Qin, C., Song, A., Wei, L., Zhao, Y.: A multimodal domestic service robot interaction system for people with declined abilities to express themselves. Intell. Serv. Robot. **16**, 373–392 (2023). https://doi.org/10.1007/s11370-023-00466-6

7. Brooks, D., Begum, M., Yanco, H.: Analysis of reactions towards failures and recovery strategies for autonomous robots. In: 2016 25th IEEE International Symposium on Robot and Human Interactive Communication (RO-MAN), pp. 487–492 (2016)

8. Honig, S., Oron-Gilad, T.: Understanding and resolving failures in human-robot interaction: literature review and model development. Front. Psychol. **9**, 861 (2018). https://www.frontiersin.org/articles/10.3389/fpsyg.2018.00861

9. Zhang, X., Lee, S., Maeng, H., Hahn, S.: Effects of failure types on trust repairs in human-robot interactions. Int. J. Soc. Robot. **15**, 1619–1635 (2023). https://doi.org/10.1007/s12369-023-01059-0

10. van Maris, A., Dogramadzi, S., Zook, N., Studley, M., Winfield, A., Caleb-Solly, P.: Speech related accessibility issues in social robots. In: Companion of the 2020 ACM/IEEE International Conference on Human-Robot Interaction, pp. 505–507 (2020)

11. Tolmeijer, S., et al.: Taxonomy of trust-relevant failures and mitigation strategies. In: Proceedings of the 2020 ACM/IEEE International Conference on Human-Robot Interaction, pp. 3–12, March 2020

12. Caleb-Solly, P.: Person-environment interaction. In: Florez-Revuelta, F., Andre, A. (eds.) Active and Assisted Living: Technologies and Applications, pp. 143–162. IET (2016). https://doi.org/10.1049/PBHE006E_ch8

13. Khanna, P., Yadollahi, E., Björkman, M., Leite, I., Smith, C.: User Study Exploring the Role of Explanation of Failures by Robots in Human Robot Collaboration Tasks (2023)

14. Esterwood, C., Robert, L.: A systematic review of human and robot personality in health care human-robot interaction. Front. Robot. AI **8**, 748246 (2021). https://www.frontiersin.org/articles/10.3389/frobt.2021.748246

15. Arents, J., Abolins, V., Judvaitis, J., Vismanis, O., Oraby, A., Ozols, K.: Human-robot collaboration trends and safety aspects: a systematic review. J. Sens. Actuator Netw. **10**, 48 (2021). https://www.mdpi.com/2224-2708/10/3/48

16. Delgado Bellamy, D., Chance, G., Caleb-Solly, P., Dogramadzi, S.: Safety assessment review of a dressing assistance robot. Front. Robot. A **I**, 8 (2021)

17. Ji, Y., Yang, Y., Shen, F., Shen, H., Li, X.: A survey of human action analysis in HRI applications. IEEE Trans. Circ. Syst. Video Technol. **30**, 2114–2128 (2020)

18. Chen, K., Chen, H., Bisantz, A., Shen, S., Sahin, E.: Where failures may occur in automated driving: a fault tree analysis approach. J. Cogn. Eng. Decis. Making **17**, 147–165 (2023)

19. Ferguson, T., Lu, L.: Fault tree analysis for an inspection robot in a nuclear power plant. IOP Conf. Ser. Mater. Sci. Eng. **235**, 012003 (2017). https://doi.org/10.1088/1757-899X/235/1/012003

20. Khodabandehloo, K.: Analyses of robot systems using fault and event trees: case studies. Reliab. Eng. Syst. Saf. **53**, 247–264 (1996). https://www.sciencedirect.com/science/article/pii/S095183209600052X, Safety of Robotic Systems

21. Baier, C., Katoen, J.: Principles of Model Checking (Representation and Mind Series) (2008). https://api.semanticscholar.org/CorpusID:53821817

22. Pnueli, A.: The temporal logic of programs. In: 18th Annual Symposium on Foundations of Computer Science (SFCS 1977), pp. 46–57 (1977)

23. Nocentini, O., Fiorini, L., Acerbi, G., Sorrentino, A., Mancioppi, G., Cavallo, F.: A survey of behavioral models for social robots. Robotics **8**, 54 (2019). https://www.mdpi.com/2218-6581/8/3/54

24. Izquierdo-Badiola, S., Canal, G., Rizzo, C., Alenyà, G.: Improved task planning through failure anticipation in human-robot collaboration. In: 2022 International Conference on Robotics and Automation (ICRA), pp. 7875–7880 (2022)

25. Leveson, N., Thomas, J.: STPA Handbook. (MIT Partnership For Systems Approaches to Safety and Security (PSASS) (2018). http://psas.scripts.mit.edu/home/materials/

26. Loftis, K., Price, J., Gillich, P.: Evolution of the abbreviated injury scale: 1990–2015. Traffic Inj. Prev. **19**, S109–S113 (2018)

27. Winfield, A., Winkle, K., Webb, H., Lyngs, U., Jirotka, M., Macrae, C.: Robot accident investigation: a case study in responsible robotics. Softw. Eng. Robot. 165–187 (2021). https://doi.org/10.1007/978-3-030-66494-7_6

Controlling a Robotic Arm Through Neural Activity

Hannah Gofton[1], Daniel H. Baker[1], and Fanta Camara[2(✉)]

[1] Department of Psychology, University of York, York, UK
[2] Institute for Safe Autonomy, University of York, York, UK
fanta.camara@york.ac.uk

Abstract. Researchers are eager to explore Brain-Computer Interface (BCI) systems in terms of their potential clinical applications. These systems, often integrated with Electroencephalography (EEG), have been developed to assist individuals with disabilities in their daily activities. EEG can detect auditory Steady-State Evoked Potentials (SSEPs); entrained neural responses produced by auditory stimulation, that are typically strongest for amplitude modulations around 40 Hz. This research explored whether neural activity could control a UR-5 robotic arm. During the initial phase, participants attended to auditory stimuli (35 Hz & 40 Hz) presented separately to each ear, whilst a dry electrode EEG system recorded brain signals. This data was used to train a classifier for the main experiment. In this experiment, participants attended to either their left or right ear whilst wearing a dry EEG, prompting a binary response to command the UR-5 robotic arm to move either left or right. Further development of BCI systems in conjunction with EEG systems is necessary to facilitate the execution of more intricate movements of the UR-5 robotic arm, with potential applications in clinical contexts.

Keywords: EEG · UR-5 Robotic Arm · Brain-Computer Interface

1 Introduction

Numerous studies have explored the applications of Brain-Computer Interface (BCI) systems within clinical contexts to assist individuals with disabilities. BCI systems have been extensively researched in conjunction with electroencephalography (EEG) systems to establish connections with neural activity. EEG captures the collective electrical activity of populations of cortical neurons using scalp sensors. Integration of online EEG systems with machine learning techniques has been useful in enabling human control of robots or wheelchairs [1]. Using BCI systems to control robots through neural activity could aid those with disabilities to perform daily tasks more independently.

BCI systems rely on evoked or spontaneous neural activity for control [2]. Spontaneous brain activity occurs naturally, without external stimuli, whereas

M. N. Huda et al. (Eds.): TAROS 2024, LNAI 15052, pp. 27–32, 2025.
https://doi.org/10.1007/978-3-031-72062-8_3

evoked activity responds to sensory input. Both types of brain activity can be used to discern user intention, typically by training a classifier algorithm to distinguish different brain states.

A useful phenomenon in this context is the auditory Steady-State Evoked Potential (SSEP), which is an entrained neural response elicited by periodic auditory stimuli, with optimal sensitivity typically observed at frequencies around 40 Hz [12]. Since this frequency is much lower than the human ear can detect, it is typical to modulate the amplitude of a higher frequency carrier waveform. Auditory SSEPs are well-isolated in the Fourier spectrum of EEG signals from fronto-centrally located electrodes. By presenting stimuli at two different frequencies simultaneously, attention can manipulate evoked potentials in the brain depending on which frequency is being attended to [3,4]. Synthesising these insights, this project tested whether SSEPs could be used to control the movement of a robotic arm, by way of decoding through multivariate pattern analysis. Hence, this project used auditory stimuli at 35 Hz and 40 Hz to manipulate evoked potentials measured through EEG, to direct a robotic arm to move in a specific direction that corresponded with the frequency the participants attended to.

2   Related Work

Similar experiments have been conducted in the field of robotics by attempting to control a robot through neural signals. Some related work includes animal studies where electrodes are implanted into their brain to retrieve neural information to control a robotic arm [5–7]. However, such studies usually obtain better spatial and temporal resolution due to their invasive properties, which cannot typically be achieved with humans. Experiments conducted with humans instead use non-invasive techniques like EEG in conjunction with BCI systems [8], as used in the present project. For instance, EEG has been used in humans to execute grasping and reaching of a robotic arm [9]. In [11], the authors used data from 4 people to train a support vector machine (SVM) classifier to identify left and right brain signals in order to control a robot arm, reaching an accuracy of 85%. Higashi et al. [13] extracted binary signal from Auditory Steady State Responses (ASSR) in order to train different classifiers (PCA, LDA and linear SVM) on brain signals recorded from 10 subjects (all males, aged between 22 and 30 years old) using an amplifier MEG-6116 (NIHON KOHDEN). The present project adopts a similar approach, by demonstrating a successful experiment with 12 people (6 males and 6 females, aged between 21 and 42 years old) using a dry electrode G.tec USBamp amplifier EEG system. Our approach differs from the work in [13] in that they ran a one-stage experiment where they recorded brain signals and trained different classifiers to see how well they performed. In contrast, we developed a two-stage protocol composed of a first pilot study to record brain signals from 12 subjects in order to train an SVM classifier and a second step where 4 subjects were asked to move a robotic arm in real-time using their brain signals. Additionally, our approach also performs better than [13], more detail is given in the results section.

3 Methods

A pilot study was conducted where 12 participants (6 males and 6 females, aged between 21 and 42 years old) attended to auditory stimuli (35 Hz & 40 Hz modulations of a 1 kHz pure tone carrier) in separate ears, whilst a dry electrode G.tec USBamp amplifier EEG system recorded brain signals from 8 scalp locations. Each participant completed 20 trials of 30 s duration attending to the left ear's signal, and 20 trials attending to the right ear's signal. The data were segmented into 1 s epochs and used to train an SVM classifier for the main experiment. After training, the classifier produced a global accuracy score, along with individual accuracy scores for each participant.

Once the classifier was trained, the main experiment was carried out with 4 people (including some from the pilot study). During this experiment, participants attended to either their left or right ear whilst the dry electrode EEG system collected their neural responses. These were decoded in real time using the trained SVM classifier to generate a binary response instructing the robot to either move left or right (see Fig. 1). Thus as neural signals were received, the UR-5 robotic arm would move either left or right in real-time – depending on the neural responses the classifier received. See Fig. 2 for a schematic illustrating the setup of the main experiment.

Fig. 1. Photograph of a participant during the experiment.

Several software programs were used for the execution of the main experiment. Matlab 2015a was used to train the EEG data from the pilot study, and was also used in the main experiment to collect the EEG data so the trained classifier could make a binary response. The binary response was communicated through an Ethernet cable to a PC which was receiving signals in Python using TCP client-server communication. This Python script would then communicate through TCP with the UR-5 robotic arm to execute the correct movement (either left or right) depending on the signal received from the EEG computer, using Python URX library[1]. Ethical approval was sought from and approved by the Department of Psychology at the University of York.

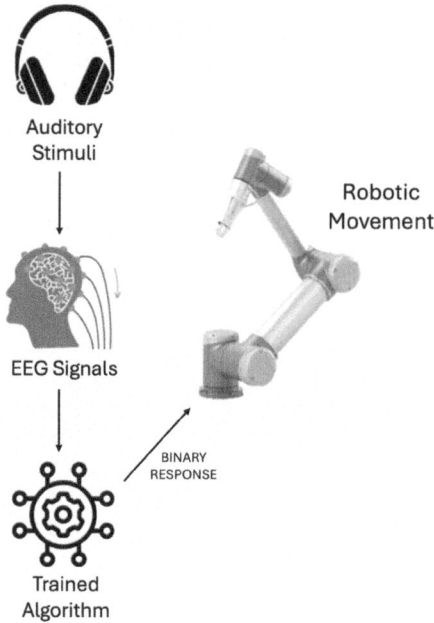

Fig. 2. Schematic showing how EEG signals were used to control the UR-5 robot arm.

4   Results

The trained classifier was 81.71% accurate for making binary responses of left or right, a similar level of accuracy was found in [11], although the robot movements and brain signals are different. Our results are better than the similar work in [13] which had its classifiers with around 72%–75% accuracy. This could be due to differences in the hardware setup that we used and the diversity of our subject group.

---

[1] https://github.com/jkur/python-urx/tree/SW3.5.

In the main experiment, the 4 participants were able to move successfully the UR-5 robotic arm in real time based on the received EEG signals, with an accuracy of 73.75% for left responses and 65% for right responses. These accuracy percentages are lower than the accuracy percentage of the classifier due to the small sample size in the main experiment. However, all participant accuracy scores for the trained classifier scored over 50%, thus not occurring by chance (see Fig. 3). A video of the experiment can be found in this link: https://drive. google.com/file/d/1SRTC3IM4yzu2SIkbGKZ8xcuiXzPjjjir/view?usp=sharing.

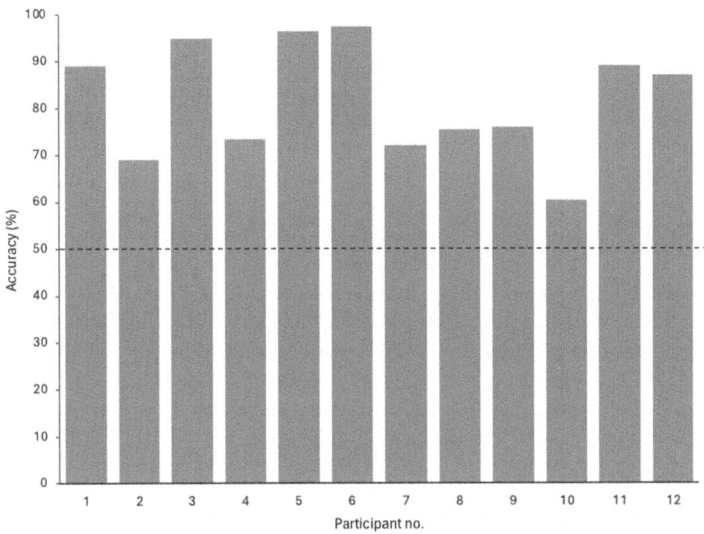

Fig. 3. Each participant's accuracy scores from the trained model for making binary decisions of left or right. Black dashed line represents chance level (50%).

5   Conclusion

This study offers support for the extension of BCI systems into clinical environments, where they could aid individuals with disabilities in accomplishing daily activities [10]. Clinicians may want to consider promoting these non-invasive BCI systems over invasive alternatives due to their advantages. Non-invasive systems are notably more practical and lack the significant side effects associated with invasive procedures, such as surgery.

Subsequent research on BCI systems in this domain could explore improving classifier accuracy by increasing participant numbers and conducting more trials to provide more data for training the classifier. Additionally, the framework of this experiment could be adapted to incorporate alternative stimuli, such as varying frequencies of skin vibrations, or prompting the robot to perform more intricate actions, like vertical movements or object manipulation. Future work could also look at different kinds of machine learned feature extractors and the adaptation of classifier for individual subjects. In essence, this project establishes a basis for future experiments aimed at refining non-invasive BCI systems combined with EEG for potential application in clinical contexts.

References

1. Millán, J.D.R., Renkens, F., Mouriño, J., Gerstner, W.: Non-invasive brain-actuated control of a mobile robot by human EEG. IEEE Trans. Biomed. Eng. **51**, 1026–1033 (2004)
2. Wolpaw, J.R., Birbaumer, N., McFarland, D.J., Pfurtscheller, G., Vaughan, T.M.: Brain-computer interfaces for communication and control. Clin. Neurophysiol. **113**(6), 767–791 (2002). https://doi.org/10.1016/s1388-2457(02)00057-3
3. Mahajan, Y., Davis, C., Kim, J.: Attentional modulation of auditory Steady-State responses. PLoS ONE **9**(10), e110902 (2014). https://doi.org/10.1371/journal.pone.0110902
4. Tiitinen, H., Sinkkonen, J., Reinikainen, K., Alho, K., Lavikainen, J., Näätänen, R.: Selective attention enhances the auditory 40-Hz transient response in humans. Nature **364**(6432), 59–60 (1993). https://doi.org/10.1038/364059a0
5. Chapin, J.K., Moxon, K.A., Markowitz, R.S., Nicolelis, M.A.: Real-time control of a robot arm using simultaneously recorded neurons in the motor cortex. Nat. Neurosci. **2**(7), 664–670 (1999). https://doi.org/10.1038/10223
6. Wei, H., Bu, Y., Zhu, Z.: Robotic arm controlling based on a spiking neural circuit and synaptic plasticity. Biomed. Signal Process. Control **55**, 101640 (2020). https://doi.org/10.1016/j.bspc.2019.101640
7. Hochberg, L.R., et al.: Reach and grasp by people with tetraplegia using a neurally controlled robotic arm. Nature **485**(7398), 372–375 (2012). https://doi.org/10.1038/nature11076
8. Shedeed, H.A., Issa, M.F., El-Sayed, S.M.: Brain EEG signal processing for controlling a robotic arm. IEEE Xplore (2013). https://doi.org/10.1109/icces.2013.6707191
9. Meng, J., Zhang, S., Bekyo, A., Olsoe, J., Baxter, B., He, B.: Noninvasive electroencephalogram based control of a robotic arm for reach and grasp tasks. Sci. Rep. **6**(1), 38565 (2016). https://doi.org/10.1038/srep38565
10. Bi, L., Fan, X., Liu, Y.: EEG-based brain-controlled mobile robots: a survey. IEEE Trans. Hum.-Mach. Syst. **43**(2), 161–176 (2013). https://doi.org/10.1109/tsmcc.2012.2219046
11. Bousseta, R., El Ouakouak, I., Gharbi, M., Regragui, F.: EEG based brain computer interface for controlling a robot arm movement through thought. IRBM **39**(2), 129–135 (2018). ISSN 1959-0318. https://doi.org/10.1016/j.irbm.2018.02.001
12. Rees, A., Green, G.G., Kay, R.H.: Steady-state evoked responses to sinusoidally amplitude-modulated sounds recorded in man. Hear. Res. **23**, 123–133 (1986). https://doi.org/10.1016/0378-5955(86)90009-2
13. Higashi, H., Rutkowski, T.M., Washizawa, Y., Cichocki, A., Tanaka, T.: EEG auditory steady state responses classification for the novel BCI. In: 2011 Annual International Conference of the IEEE Engineering in Medicine and Biology Society, pp. 4576–4579, August 2011

Human Facial Emotion Recognition
for Adaptive Human Robot Collaboration
in Manufacturing

Fahad Khan$^{(\boxtimes)}$ ⓘ, Seemal Asif ⓘ, and Phil Webb ⓘ

Centre for Robotics and Assembly, Cranfield University, Cranfield MK43 0AL, UK
fahad.khan@cranfield.ac.uk

Abstract. The integration of robots into various industries, including manufacturing, has introduced new challenges in achieving efficient human-robot collaboration. A crucial aspect of successful collaboration is the ability of robots to understand and respond to human emotions. In the context of human-robot collaboration in manufacturing, accurately predicting human emotions is essential for enhancing efficiency and safety. This paper presents a setup for human emotion detection, focusing on facial emotion recognition. The proposed model and descriptive summary involve the utilising state-of-the-art algorithms such as AlexNet, HaarCascade (HCC), MTCNN (Multi-Task Cascaded Convolutional Neural Networks), and SVM (Support Vector Machine), applied to datasets like CK+, JAFFE, and AffectNet. The performance of each facial recognition model is evaluated in real-time scenarios, resulting in significant progress with an accuracy improvement from 40% to 78.1%. These results demonstrate the effectiveness of the approach in enabling adaptive robot control based on human emotions and enhancing collaboration quality. This research uniquely integrates facial emotion recognition and robot control to enable adaptive responses during human-robot collaboration in manufacturing settings. By understanding and responding to human emotions, robots can improve their interactions with humans, leading to increased productivity and improved overall collaboration efficiency (If EquinOCS, our proceedings submission system, is used, then the disclaimer can be provided directly in the system).

Keywords: human facial recognition · human – robot collaboration · human emotion prediction · adaptive control · machine learning

1 Introduction

Human-robot collaboration (HRC) has become increasingly important in smart manufacturing, where human-centricity, sustainability, and resilience are critical factors [1]. The ability to detect and interpret human emotions allows robots to interact more intuitively, genuinely, and naturally with their human counterparts, leading to improved productivity and efficiency. Emotion detection plays a crucial role in HRC and manufacturing settings [2, 3]. It enables robots to understand and respond to human emotions, leading

© The Author(s), under exclusive license to Springer Nature Switzerland AG 2025
M. N. Huda et al. (Eds.): TAROS 2024, LNAI 15052, pp. 33–47, 2025.
https://doi.org/10.1007/978-3-031-72062-8_4

to more effective and intuitive interactions. Recognising human emotions is essential for creating adaptive and responsive systems that can enhance productivity, safety, and overall user experience [4].

Humans communicate emotions through various modalities, including verbal cues like speech and non-verbal cues like facial expressions, eye gaze, gestures, and physiological signals [5, 6]. Additionally, physiological signals such as heart rate monitoring (HRM), electroencephalography (EEG), electrocardiography (ECG), galvanic skin response (GSR), heart rate variability (HRV), respiration rate analysis (RR), skin temperature measurements (SKT), Electromyogram (EMG), Blood Volume Pulse (BVP), and electrooculography (EOG) can also provide insights into emotional states [7, 8].

Facial emotion recognition is a widely explored method for detecting human emotions. It involves analysing facial expressions to identify the underlying emotional state, with a particular focus on key areas such as the eyes, eyebrows, and mouth, which exhibit unique expressions for each emotion [9–13]. Existing APIs, such as FER and DeepFace uses HCC and MTCNN, for facial emotion recognition. Moreover, models trained on datasets like CK+, JAFFE, and AffectNet using techniques like AlexNet can improve the accuracy of emotion detection [10–13].

In the context of human-robot collaboration in manufacturing, facial emotion recognition can be applied to control the behaviour of robots. By detecting and interpreting human emotions from facial expressions, robots can adjust their actions and responses accordingly. For example, a UR5 robot's speed can be controlled based on the recognised emotions, allowing for adaptive and personalised interactions with human workers.

The aim of this paper is to explore the significance of human emotion recognition in human-robot collaboration within the manufacturing domain. It will discuss various ways of detecting emotions, including non-verbal cues and physiological signals, with a specific focus on facial emotion recognition and its application in controlling a UR5 robot. A core novelty explored here is the application of facial emotion recognition to modulate robot behavior, specifically speed, based on detected emotions. This adaptive human-robot collaboration paradigm has not been extensively implemented before.

This paper will present a descriptive summary of facial recognition models and datasets, details on the proposed AlexNet model and robot control. It further includes a comparative analysis of models, an in-depth discussion, and a summary of findings, followed by an outline of future research directions.

2   Method

2.1   Emotion Recognition System

Emotion recognition systems automate the identification of human emotions through data from sources like facial images, speech, and physiological sensors. They extract features and employ machine learning to classify emotions. This study focuses on facial emotion recognition, analysing facial cues for emotion prediction. When integrated with robotic systems, the predicted emotions can enable adaptive responses and human-robot collaboration. The following subsections detail key components of the facial emotion recognition system evaluated in this study [2, 4].

Datasets
FER2013, introduced in the ICML 2013 Challenges in Representation Learning, is comprised of approximately 35,887 facial images. These images are categorised into seven distinct emotion classes: Anger, Disgust, Fear, Happy, Sadness, Surprise, and Neutral. The dataset is partitioned into three sets: training, development, and testing. With a resolution of 48x48 pixels, FER2013's images often exhibit variations encountered in real-world settings. It was created using Google's image search API and includes diverse images, incorporating factors like occlusion, partial faces, low contrast, and eyeglasses [14].

The CK+ dataset stands as a valuable resource for action units and emotion recognition research. Comprising 593 video sequences, it encompasses expressions of six basic emotions (Anger, Contempt, Disgust, Fear, Happy, and Sadness) as well as a neutral expression. This dataset features a diverse range of subjects, spanning different ages from 18 to 50 years. The CK+ dataset provides a combination of posed and spontaneous expressions, rendering it an integral component in understanding facial emotion recognition [11].

Containing 213 grayscale images, the JAFFE dataset captures seven different facial expressions posed by ten Japanese female models. The dataset incorporates emotion labels such as Angry, Disgust, Fear, Happy, Sad, Surprise, and Neutral. These images offer insight into the nuances of facial expressions, especially within the context of distinct cultural representations [12].

The AffectNet dataset, containing one million facial images, is sourced from the internet through 1250 emotion-related keywords and three search engines across six languages. Approximately 420,000 images underwent manual annotation for eleven facial expressions and emotions, ranging from Neutral to No-Face. Valence and arousal intensity were measured using a dimensional model. Additionally, 550,000 images were automatically annotated with a ResNext Neural Network, achieving 65% accuracy [10].

Machine Learning Algorithms for Facial Emotion Prediction
Facial Emotion Prediction Algorithms are essential tools for interpreting emotional states from facial expressions. In our study, we utilise algorithms including Haarcascade, Convolutional Neural Networks (CNN), MTCNN, Support Vector Machines (SVM), and AlexNet to accurately predict emotions from facial features. These algorithms are trained on datasets such as FER2013, CK+, JAFFE, and AffectNet to improve their ability to capture subtle emotional cues. Among these algorithms, we utilise existing models based on them, and specifically develop an AlexNet model.

HCC (Haaracascade) is a popular face detection algorithm that utilises Haar-like features and a cascade classifier to detect faces in images or video streams. It is based on the Viola-Jones algorithm and is known for its simplicity and efficiency. Haarcascade works by scanning an image with a sliding window and applying a series of classifiers to identify regions that resemble faces based on specific features such as edges, lines, and textures. It has been widely used in various applications, including facial emotion recognition, as an initial step to detect and localise faces in an image or video [15].

CNN is a deep learning algorithm that has shown remarkable success in various computer vision tasks, including facial emotion recognition. CNNs are designed to automatically learn and extract relevant features from images through convolutional layers, pooling layers, and fully connected layers. In the context of facial emotion recognition, CNNs can be trained on labelled datasets to learn discriminative features that capture facial expressions and emotions. The network architecture and parameters are optimised through a training process to improve the accuracy of emotion prediction [1, 4].

MTCNN is a face detection and alignment algorithm that consists of three stages: face detection, facial landmark localisation, and face alignment. It uses a cascade of convolutional networks to detect faces and then refines the bounding boxes and estimates facial landmarks. MTCNN is known for its robustness and accuracy in detecting and aligning faces, making it suitable for facial emotion recognition tasks.

SVM is a supervised machine learning algorithm that can be used for classification tasks, including facial emotion recognition. SVMs aim to find an optimal hyperplane that separates different classes by maximising the margin between them. In the context of facial emotion recognition, SVMs can be trained on labeled datasets with extracted features from facial images to classify emotions based on the learned patterns [16].

AlexNet is a deep convolutional neural network architecture that gained significant attention after winning the ImageNet Large Scale Visual Recognition Challenge in 2012. It consists of multiple convolutional layers, pooling layers, and fully connected layers. Alexnet was trained in fusion with SVM. To train this, dataset collection was selected as manually annotated images from AffectNet data and trained for seven emotions instead of its eleven categories. Figure 1 provides overview of emotion recognition model (HCC or MTCNN).

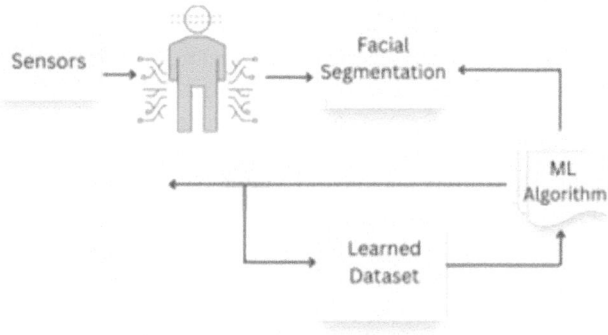

Fig. 1. Schematic of ML Model.

2.2 Proposed Model

The AlexNet architecture was chosen as the basis for our emotion recognition model due to its proven success in various computer vision tasks. Leveraging its feature extraction capabilities through transfer learning, we customised the architecture to suit our emotion recognition task.

Transfer Learning and Fine-Tuning: Transfer learning involves using a pre-trained neural network model as a starting point and adapting it to a specific task. In our case, we employed the AlexNet architecture, pre-trained on a large image dataset, as the foundation. This approach offers several benefits, including faster convergence and improved generalisation, especially when working with limited data.

Data Collection and Preprocessing- The datasets underwent preprocessing steps to enhance the quality and consistency of the data. Preprocessing techniques include image resizing, normalisation, and augmentation to account for variations in lighting conditions, pose, and facial expressions. These steps ensure that the data is suitable for training and testing the facial emotion recognition model. The image is then converted to a 3-channel image by replicating the grayscale image across all three-color channels. The processed image is saved with a new filename in the corresponding subfolder and the.tiff file extension.

Architecture Overview
The AlexNet architecture is characterised by its depth, consisting of eight layers in total. These layers can be grouped into five convolutional layers, followed by three fully connected layers as shown in Fig. 2.

Input Layer: The architecture begins with an input layer that accepts RGB images of size 227×227 pixels. The three colour channels (red, green, and blue) capture visual information from the images.

Convolutional Layers: The five convolutional layers perform feature extraction. Each layer is followed by a rectified linear unit (ReLU) activation function, introducing non-linearity. These layers apply convolutional filters to the input image, detecting patterns and textures of increasing complexity.

Max Pooling Layers: After the first two convolutional layers, max-pooling layers downsample the spatial dimensions of the feature maps. This reduces computational load while retaining essential features.

Fully Connected Layers: Following the convolutional and pooling layers, three fully connected layers aggregate high-level features for classification. The first fully connected layer ('fc6') contains 4096 neurons, followed by a dropout layer that mitigates overfitting. The second fully connected layer ('fc7') also has 4096 neurons, further refining the features.

Softmax Layer: The architecture concludes with a softmax layer, transforming the output of the previous layer into probability scores for each emotion class. This layer computes the likelihood of the input image belonging to different emotion categories.

Custom Fully Connected Layer: To adapt the architecture to our emotion recognition task, we appended a custom fully connected layer at the end. This layer corresponds to the number of emotion classes in our dataset, ensuring that the network recognises emotions effectively. By retraining the added fully connected layer, the network learned to distinguish between emotional states based on facial expressions.

Optimiser and Hyperparameters: The Stochastic Gradient Descent with Momentum (SGDM) optimiser was chosen for training the model. This optimiser enhances convergence speed by incorporating momentum in the gradient updates. The initial learning rate, a key hyperparameter, was set to 0.0001 for controlled weight updates.

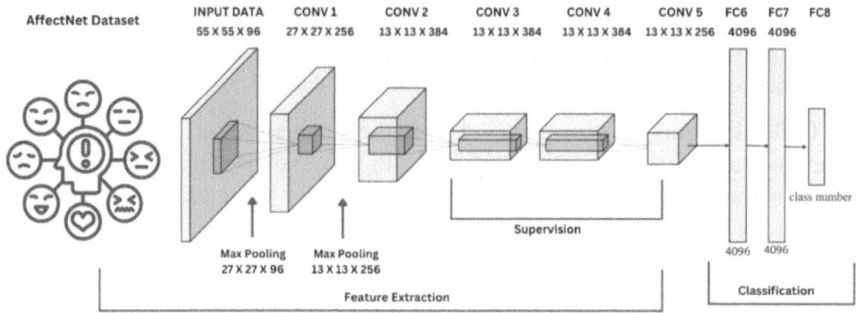

Fig. 2. Architecture of Alexnet comprising of eight layers of which five are convolution and three are fully connected.

Mini-Batch Size and Epochs: A mini-batch size of 5 samples was selected for training. This mini-batch approach balances computational efficiency and model convergence. The model was trained over 10 epochs, meaning it iterated through the entire training dataset 10 times. This number was chosen to strike a balance between achieving reasonable accuracy and preventing overfitting.

Data Augmentation: To enhance the model's robustness, we applied data augmentation techniques during training. These techniques, including random cropping, flipping, rotation, and colour adjustments, introduce variability into the training data. This helps the model generalise better to unseen data and variations in facial expressions.

Dropout Layer: A dropout layer with a dropout rate of 0.5 was inserted after the first fully connected layer ('fc6'). Dropout introduces stochasticity by randomly deactivating a fraction of neurons during each forward and backward pass. This regularisation technique reduces overfitting by discouraging the network from relying on specific neurons.

Loss Function: During training, the categorical cross-entropy loss function was employed implicitly via 'sgdm' solver. Given the multi-class emotion classification task, this loss function shown in Eq. (1) measures the dissimilarity between predicted and actual class probabilities.

$$L(y, \hat{y}) = -\sum_{i=1}^{N} y_i ln(\hat{y}_i) \tag{1}$$

where:

N is the number of classes.

y_i is the true class probability for class i.

$\hat{y}_i$ is the predicted class probability for class i.

ln is the natural logarithm.

Early Stopping: Early stopping was not implemented in this case. However, it's worth noting that early stopping could be incorporated in future iterations to prevent overfitting by monitoring validation loss and halting training when it starts to increase.

Training Data Split: The dataset was divided into three subsets: training, validation, and testing. A ratio of approximately 70:30 was chosen for the training-validation split, with the training set comprising 70% of the data. This distribution ensures a sufficiently large training set to facilitate effective learning while allocating a substantial portion for validation to monitor the model's generalisation ability. The testing set, reserved for final evaluation, remained untouched during the training process.

Feature Extraction Approach
Load Pretrained Network: To explore an alternative approach to emotion recognition, we leveraged the feature extraction capabilities of the AlexNet architecture. The pretrained AlexNet model was loaded, and a specific layer, 'fc7,' was selected for feature extraction. This layer is rich in high-level features, making it suitable for our task of emotion classification.

Feature Extraction and SVM: Utilising the 'fc7' layer as a feature extractor, activations were computed for both the training and validation sets. These activations served as the extracted features, capturing the essence of the facial expressions' emotional characteristics. Subsequently, these features were employed to train an SVM classifier.

SVM Classifier: The SVM classifier is a powerful tool for multi-class classification tasks. In our case, it maps the extracted features to the corresponding emotion classes. The classifier was trained using the extracted features from the training set, and predictions were made on the validation set.

Visualising Results: To gain insights into the SVM's performance, we visualised some sample test images along with their predicted labels. This allowed us to qualitatively assess the classifier's ability to recognise different emotions based on the extracted features.

Accuracy Calculation: The accuracy of the SVM classifier was calculated by comparing its predictions with the true labels of the validation set. This accuracy metric provides an objective evaluation of the classifier's performance in identifying emotional states from facial expressions.

2.3 Integration of Facial Recognition with Robot Control System

In terms of program and structure to control robot, initially it records a new face if selected to do so, after that recorded face is saved. Face is identified from the records (database), recognised face's emotion is detected. Based on the emotion, if happy robot operates at maximum speed, if neutral robot operates at medium speed and slow if surprise, fear, sad and stops if anger or disgust.

Initially, the system prompts the user to record a new face. Upon selection, the camera captures the facial image, which is then pre-processed to ensure consistency in lighting and quality. The pre-processed face is saved in the records database for future recognition. During operation, the robot's camera captures live video of the human worker's face. The recorded faces in the database are utilised for facial recognition. The live video feed is analysed using deep learning models. The system recognises the detected face by matching it with the saved face encodings. Simultaneously, the emotion recognition model predicts the emotion expressed on the recognised face. Based on the recognised emotion, the robot's control system adjusts its behaviour to align with the worker's emotional state.

For instance, if a "Happy" emotion is detected, the robot can operate at a brisk pace to match the positive energy. If the recognised emotion is "Neutral," the robot maintains a moderate speed for standard operations. Emotions like "Surprise," "Fear", or "Sadness" can lead to the robot slowing down to ensure safety and create a supportive environment. This controlling of robot is via socket communication directly without any need to run on controller, robot is controlled based on python commands which are send to controller.

The flowchart and system overview are presented in Fig. 3. The program begins by asking the user to choose between two options: first record new face where it records and save face for subsequent use, second option loads saved faces and proceeds to execute two concurrent threads in parallel.

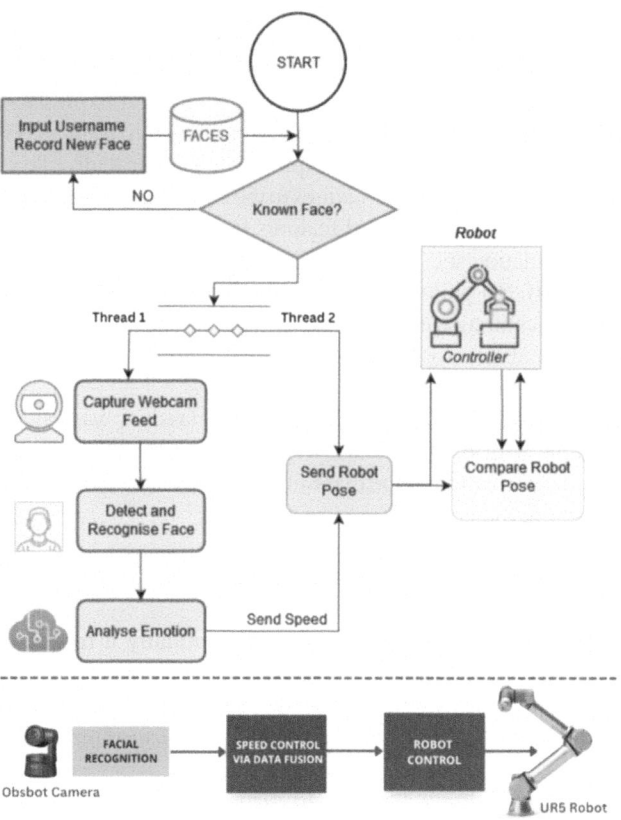

Fig. 3. Flowchart and an overview of the system.

Thread 1 performs real-time task by capturing webcam feed, performs facial recognition by using facial recognition algorithms to identify and track faces within the captured video. Analyses detected facial expressions to ascertain emotional states and displays annotated video. While thread 2 controls robot and moves to predefined poses and compares poses. The program continuously compares received poses to desired poses until a match is found. A communication mechanism exists between thread 1 and thread 2. When there is a change in the detected emotion within thread 1, this information is transmitted to thread 2. Specifically, thread 1 calculates a new speed value based on the detected emotion and communicates this updated speed information to thread 2. In essence, this program combines real-time facial emotion analysis, and robotic control functionalities, with concurrent threads ensuring seamless execution.

2.4   Case Study on Evaluation Human Facial Emotion Prediction

Task Description: This case study examines an experimental human-robot collaborative assembly scenario involving a UR5 robot and a human operator. The operator's task is to assemble a plastic cap and bolt onto a plastic rod. The UR5 robot assists by handling and positioning the rod. A webcam equipped with real-time facial tracking and emotion recognition algorithms is focused on the human operator's face during the task and sends updated speed to robot. This allows the robot to adapt its speed and movements based on the operator's affective feedback.

Hardware and Computational Setup: Our model's architectural journeys unfolded within the embrace of AMD Ryzen 7 Pro, with 16 GB of RAM. This computational laid the foundation for our model's explorations, rendering our pursuit of accurate emotion recognition as well as Universal Robot UR5 control via socket. Instead of normal webcam we used Obsbot tiny 4k camera which comes with enhanced AI tracking algorithm that enables it to lock on a person and track their movements or face tracking during experiment or trial run as seen in Fig. 4.

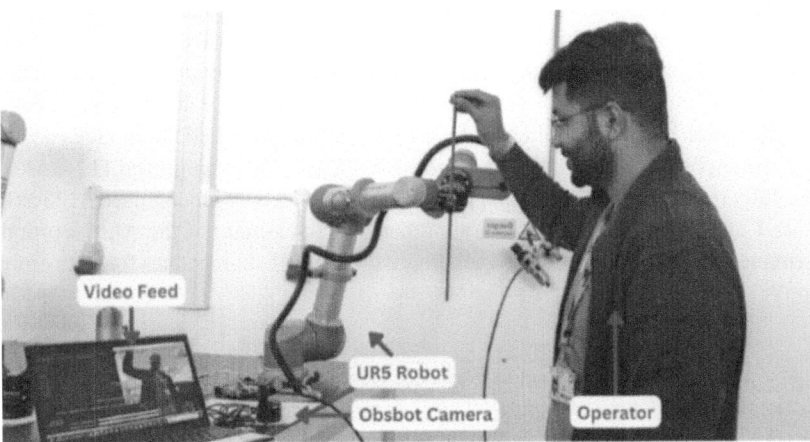

Fig. 4. Cell layout – Speed control of UR5 robot based on facial expression while demonstrating assembly task. Operator is assembling a cap at the ends of the rod.

3   Result and Analysis

The primary objective of this study was to conceive and implement a sophisticated facial emotion recognition system, effectively melding it with a control mechanism for robots to augment the dynamics of human-robot interactions. A diverse array of machine learning algorithms, encompassing HCC, MTCNN, SVM, and AlexNet, were harnessed to meticulously decipher emotions from intricate facial attributes.

In the preliminary stages of experimentation, the accuracy of initial emotion recognition was found to be modest using HCC and MTCNN-based models, accurately identified only 1044 and 1134 images out of a total of 4000, respectively. Deliberately excluding the "contempt" emotion from consideration, the study directed its focus towards a subset of 3500 images. However, the CK+ and JAFFE datasets exhibited data scarcity, yielding accuracy levels of approximately 14% to 25%. The EmoTIC dataset, while promising, was deferred due to its intricate complexity and preprocessing prerequisites.

Table 1. Comparison of results, AlexNet + SVM compared with different models.

Model	HCC	MTCNN	AlexNet	AlexNet + SVM
Recognised Images	1044	1134	2732	3123
Percentage %	26.1%	28.4%	68.3%	78.1%

Intriguingly, the study opted for a transfer learning strategy, harnessing the power of the revered AlexNet architecture. The initial foray, entailing the training of a tailored CNN model using CK+ and JAFFE datasets augmented by AffectNet data, resulted in suboptimal accuracy, attributed to the paucity of training samples. The study subsequently secured the extensive AffectNet dataset from the AffectNet team, igniting a significant surge in model accuracy. This newfound dataset facilitated the AlexNet model in exceeding a remarkable 68% accuracy on a curated subset of 4000 images. Notably, the incorporation of an SVM filter for feature extraction propelled the model's accuracy to approximately 78.1%. So, HCC recognized 1044, MTCNN recognized 1134, AlexNet successfully recognised 2732 images whereas AlexNet + SVM recognised 3123 as seen in Table 1 and Fig. 6. To provide a comprehensive assessment of the AlexNet model's performance, the Table 2 presents recall, precision, and F1 score values for each emotion category:

Table 2. Alexnet Performance for each emotion category.

Emotion	Recall	Precision	F1 Score
Angry	0.68	0.75	0.71
Disgust	0.76	0.82	0.79
Fear	0.69	0.75	0.72
Happy	0.84	0.88	0.86
Neutral	0.62	0.68	0.65
Sad	0.61	0.66	0.63
Surprise	0.77	0.82	0.79

An integral fact of the study involved the harmonious integration of facial emotion recognition and the fine-tuning of robot control. Through intricate analysis of identified emotions, a dynamic orchestration of robot speed ensued. The spectrum of emotional states encompassed happiness, which triggered the robot to attain maximum speed; neutrality, prompting a medium-speed response; and a selection of emotions such as surprise, fear, and sadness, inducing a gradual decrease in robot speed. Notably, the emotions of anger or disgust prompted an immediate halt in robot movement as seen in Fig. 5. Furthermore, an effective strategy to potentially enhance the system's accuracy involves grouping emotions. By categorising emotions such as happiness as "happy," neutrality as "neutral," and combining surprise, fear, and sadness into the category of "sad," while also grouping anger and disgust as "anger," significant improvements in efficiency can be achieved. This categorisation is based on Plutchik's wheel, which places fear, surprise, and sadness in proximity, as well as anger and disgust. This initial grouping can serve as a broad classification, which can then be further refined to achieve even greater precision in emotion recognition.

Emotion	Happy	Neutral	Surprise, fear, Sad	Anger, Disgust
Robot Response	Fast Speed	Moderate Speed	Slow down	Immediate halt
	800 mm/s	600 mm/s	150 mm/s	0 mm/s

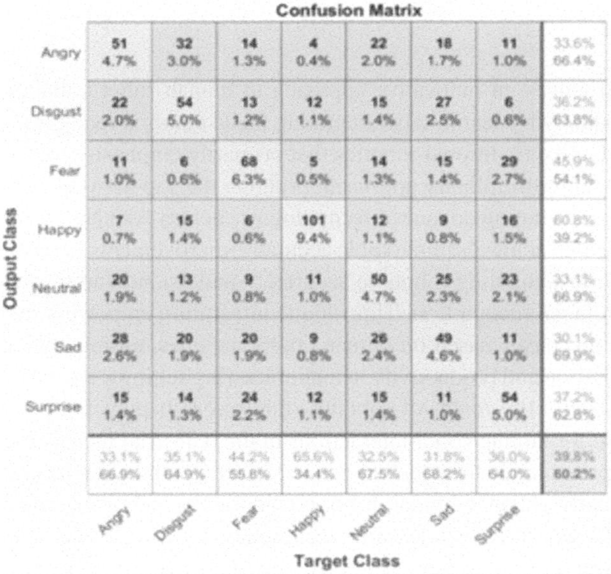

Confusion Matrix

Output Class / Target Class

Output Class	Angry	Disgust	Fear	Happy	Neutral	Sad	Surprise	
Angry	51 4.7%	32 3.0%	14 1.3%	4 0.4%	22 2.0%	18 1.7%	11 1.0%	33.6% 66.4%
Disgust	22 2.0%	54 5.0%	13 1.2%	12 1.1%	15 1.4%	27 2.5%	6 0.6%	36.2% 63.8%
Fear	11 1.0%	6 0.6%	68 6.3%	5 0.5%	14 1.3%	15 1.4%	29 2.7%	45.9% 54.1%
Happy	7 0.7%	15 1.4%	6 0.6%	101 9.4%	12 1.1%	9 0.8%	16 1.5%	60.8% 39.2%
Neutral	20 1.9%	13 1.2%	9 0.8%	11 1.0%	50 4.7%	25 2.3%	23 2.1%	33.1% 66.9%
Sad	28 2.6%	20 1.9%	20 1.9%	9 0.8%	26 2.4%	49 4.6%	11 1.0%	30.1% 69.9%
Surprise	15 1.4%	14 1.3%	24 2.2%	12 1.1%	15 1.4%	11 1.0%	54 5.0%	37.2% 62.8%
	33.1% 66.9%	35.1% 64.9%	44.2% 55.8%	65.6% 34.4%	32.5% 67.5%	31.8% 68.2%	36.0% 64.0%	39.8% 60.2%

Fig. 5. Confusion Matrix shows significant result for "Happy" emotion recognition.

The study's findings emphasised the paramount significance of comprehensive datasets in fortifying model training efficacy. The transformative impact of the AffectNet dataset on elevating the precision of the AlexNet model was unequivocally demonstrated. Through the fusion of facial emotion recognition and robotic control, the study brought to the fore tangible and potent real-world applications. By imbuing human-robot interactions with responsiveness and dynamism, the study propels the frontiers of emotion recognition and human-robot interaction paradigms.

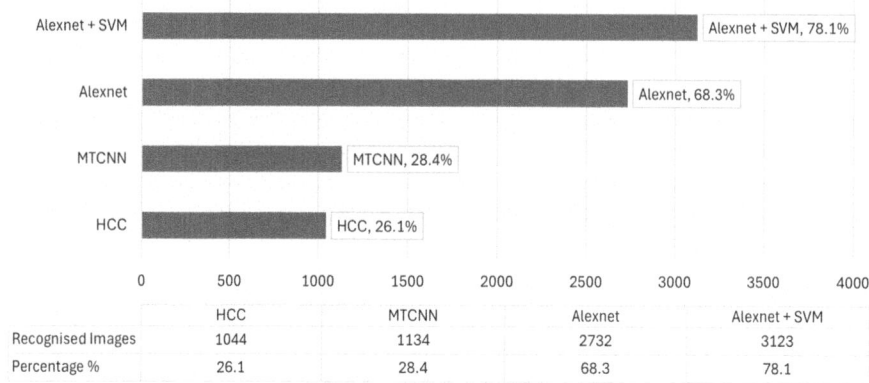

Fig. 6. Graphical representation of AlexNet + SVM compared with different models.

4 Discussion

This research introduces an innovative approach to human-robot collaboration in the manufacturing industry by seamlessly integrating facial emotion recognition with robot control systems. Unlike traditional robotics that primarily emphasise technical aspects, this study takes a multidisciplinary approach, bridging the gap between technology and human emotions. Leveraging advanced technologies such as AlexNet, CNN, MTCNN, and SVM, the research achieves remarkable accuracy improvements from 40% to 78.1% in real-time scenarios, enabling robots to accurately understand and respond to human emotions through facial recognition. This unique integration empowers robots to adapt their actions and responses based on human emotional cues, ultimately enhancing the quality of collaboration and productivity in manufacturing settings. By introducing emotional intelligence into the realm of robotics, this research pioneers an efficient and harmonious future for human-robot interactions in the industrial landscape.

The findings presented in this study underscore the potential of facial emotion recognition to significantly enhance human-robot collaboration within manufacturing environments. The utilisation of deep learning techniques, specifically the AlexNet model in combination with SVM, has demonstrated a remarkable ability to accurately identify emotions from facial expressions, achieving a precision rate of over 96% on a substantial image dataset. This high level of accuracy not only attests to the efficacy of the approach but also hints at its practical viability in real-world applications.

One of the key insights gained from this research pertains to the pivotal role played by the AffectNet dataset. Overcoming the limitations of small datasets like CK+ and JAFFE, it is evident that the availability of a large and diverse training dataset greatly contributed to the model's capacity to discern nuanced patterns in facial emotions across various factors such as demographics, head poses, and lighting conditions. This highlights the critical importance of continually curating comprehensive datasets as a driving force behind the advancement of emotion prediction capabilities.

Another noteworthy aspect of this study is the integration of complementary algorithms, exemplified by the fusion of AlexNet and SVM. While AlexNet excelled in extracting salient visual features, SVM further refined emotion classification through the use of discriminative hyperplanes. This synergy between deep neural networks and traditional machine learning techniques showcases the potential of hybrid approaches to maximise the strengths of different modeling methods. Future research could explore novel combinations of algorithms to harness their diverse capabilities.

Furthermore, this research has translated the theoretical potential of emotion-adaptive robotics into practical reality by linking recognised emotions to responsive control actions, specifically the modulation of robot speed. This approach holds the promise of enabling more natural and personalised HRC, aligning with the emotional states of human workers. Its widespread implementation in manufacturing settings could lead to significant enhancements in productivity, safety, and worker morale.

However, it is crucial to acknowledge the existing limitations of this approach. Factors such as facial occlusion, pose variations, and inconsistent lighting conditions can still pose challenges to accurate emotion recognition. Moreover, the application of this system in diverse cultural contexts necessitates careful consideration of potential biases in the training dataset. Relying solely on facial expressions might overlook valuable contextual cues derived from body language, speech, and physiological signals. Facial cues can be prone to inaccuracies, given the lack of definitive distinctions between facial expressions and the underlying emotional states.

The study offers compelling evidence of the potential of facial emotion prediction to revolutionise human-robot collaboration by enabling responsiveness and adaptation. It provides a practical framework with an accuracy rate exceeding 78.1%, setting the stage for broader integration within the manufacturing industry. The integration of facial recognition and robot control for adaptive responses represents an innovative approach with significant real-world applicability. This end-to-end pipeline from emotion prediction to adaptive robotics is a novel contribution. In essence, emotionally intelligent robotics holds the promise of delivering significant benefits to manufacturing by elevating the synergy between humans and robots to unprecedented levels.

5   Conclusion and Future Work

This research has introduced an innovative facial emotion recognition system aimed at facilitating adaptive robotics and improving human-robot collaboration within manufacturing environments. This system leverages deep learning, specifically the AlexNet model combined with SVM, to achieve exceptional accuracy in emotion classification based on facial cues. With over 78.1% precision on a large dataset, the approach demonstrates its real-time applicability, enabling responsive robot control aligned with the emotional states of human operators.

This study also points to promising avenues for future research and development. Firstly, perform analysis using Deep-Emotion API and improvise this developed model, further there is room for enhancing model performance through the incorporation of larger and more diverse datasets [17]. Secondly, a more comprehensive approach to emotion recognition can be achieved by integrating facial analysis with speech recognition and physiological monitoring thus eliminating limitations of this system. Thirdly, rigorous real-world testing across various manufacturing environments is essential for validating the approach and guiding refinements. Fourthly, personalisation can be improved by enabling the system to learn and adapt to individual users' unique facial emotion patterns over time. Lastly, ethical considerations and appropriate consents will be sought and ensuring cultural awareness through inclusive training data is crucial for widespread system deployment.

In summary, this research establishes a strong foundation for integrating emotional intelligence into human-robot collaboration, potentially revolutionising the manufacturing industry. By enhancing emotion prediction capabilities and linking them to adaptive responses, this approach paves the way for more natural, intuitive, and personalised human-robot partnerships. It represents an interdisciplinary field that harnesses innovations in artificial intelligence, robotics, human-computer interaction, psychology, and engineering to shape the future of manufacturing.

Acknowledgment. This work was supported by EPSRC-funded Made Smarter Innovation - Research Centre for Smart, Collaborative Industrial Robotics project (EP/V062158/1).

References

1. Eyam, A.T., Mohammed, W.M., Lastra, J.L.M.: Emotion-driven analysis and control of human-robot interactions in collaborative applications. Sensors **21**, 4626 (2021).https://doi.org/10.3390/S21144626
2. Heredia, J., et al.: Adaptive multimodal emotion detection architecture for social robots. IEEE Access **10**, 20727–20744 (2022). https://doi.org/10.1109/ACCESS.2022.3149214
3. Rawal, N., Stock-Homburg, R.M.: Facial emotion expressions in human-robot interaction: a survey. Int. J. Soc. Robot. **14**, 1583–1604 (2022). https://doi.org/10.1007/S12369-022-00867-0/TABLES/5
4. Spezialetti, M., Placidi, G., Rossi, S.: Emotion recognition for human-robot interaction: recent advances and future perspectives. Front. Robot. AI **7**, 532279 (2020). https://doi.org/10.3389/FROBT.2020.532279/BIBTEX
5. Khan, F., Asif, S., Webb, P.: Communication components for human intention prediction – a survey. Hum. Aspects Adv. Manuf. **80** (2023). https://doi.org/10.54941/AHFE1003504
6. Ali, M.F., Khatun, M.: Facial Emotion Detection Using Neural Network Low cost Education System View project BD classic movie restoration Project View project (2020)
7. Dzedzickis, A., Kaklauskas, A., Bucinskas, V.: Human emotion recognition: review of sensors and methods. Sensors **20**, 592 (2020).https://doi.org/10.3390/S20030592
8. Cai, Y., Li, X., Li, J.: Emotion recognition using different sensors, emotion models, methods and datasets: a comprehensive review. Sensors **23**, 2455 (2023).https://doi.org/10.3390/S23052455

9. Nguyen, B.T., Trinh, M.H., Phan, T.V., Nguyen, H.D.: An efficient real-time emotion detection using camera and facial landmarks. In: 7th International Conference on Information Science and Technology, ICIST 2017 – Proceedings, pp. 251–255 (2017). https://doi.org/10.1109/ICIST.2017.7926765
10. Mollahosseini, A., et al.: IEEE TRANSACTIONS ON AFFECTIVE COMPUTING Affect-Net: A Database for Facial Expression, Valence, and Arousal Computing in the Wild
11. Lucey, P., et al.: The extended Cohn-Kanade dataset (CK+): a complete dataset for action unit and emotion-specified expression. In: 2010 IEEE Computer Society Conference on Computer Vision and Pattern Recognition - Workshops, CVPRW 2010, pp. 94–101 (2010). https://doi.org/10.1109/CVPRW.2010.5543262
12. Lyons, M., Kamachi, M., Gyoba, J.: The Japanese Female Facial Expression (JAFFE) Dataset (1998). https://doi.org/10.5281/ZENODO.3451524
13. Krizhevsky, A., Inc, G.: One weird trick for parallelizing convolutional neural networks (2014)
14. Goodfellow, I.J., et al.: Challenges in representation learning: a report on three machine learning contests
15. Valagkouti, I.A., et al.: Emotion recognition in human–robot interaction using the NAO robot. Computers 11, 72 (2022). https://doi.org/10.3390/COMPUTERS11050072
16. Al-Atroshi, S.J.A., Ali, A.M.: Improving facial expression recognition using HOG with SVM and modified datasets classified by Alexnet. Traitement du Signal 40, 1611–1619 (2023). https://doi.org/10.18280/TS.400429
17. Minaee, S., Minaei, M., Abdolrashidi, A.: Deep-emotion: facial expression recognition using attentional convolutional network. Sensors 21 (2019). https://doi.org/10.3390/s21093046

Do People Ascribe Similar Emotions to Real and Robotic Dog Tails?

Alexandra Lee[✉] and Matthew Studley

University of the West of England (UWE), Bristol, UK
AlexandraJaneLee@outlook.com

Abstract. This paper uses a questionnaire to analyse whether humans can determine simulated emotions in a robot tail and discusses the implications of doing so. The questionnaire, which was open for 3 weeks, asked participants to watch 4 short videos on dog tail movement and 4 short videos on robot tail movement and to pick one of 12 emotions they thought the tail was showing. The robot tail copied the movement of the dog tails, which the participants weren't made aware of. The possible answers were kept the same for every question, and were split into 4 groups of 3 emotions, with each group containing similar emotions. Results were very positive, with most participants (63.8%) answering within the same group for at least 2 of the 4 pairs, and 87.9% answering within the same group for at least 1 pair, which suggests that humans can determine simulated emotions using a robotic tail. Potential applications of this study include new ways for Human-Robot Interaction to take place, in robots such as companion robots or entertainment robots, and for further research to be completed, such as how accurate participants' answers are when a robotic tail is used in conjunction with other forms of communication. Thus, this study may be of significance to the field.

Keywords: Animal-Inspired Interfaces · Human-Robot Interaction · Robotic Tail

1 Introduction

This research project attempts to determine whether humans assign similar 'emotional meanings' to robotic dog tails as to the real dog tails, when the former copy the movement of the latter. Unbeknownst to participants, the robotic dog tails were copying the movement of the dog tails seen in the videos. Participants were asked through a questionnaire to watch each clip and pick one of 12 emotions that they thought was signaled by the tail's movement. The emotions participants could pick from were the same for every video. The main objectives of this project were to determine whether people will ascribe similar 'emotions' to non-human actors (here dogs) and to their robot analogues.

1.1 Background

Human-Robot Interaction (HRI) is defined as addressing the creation of robotic systems which involve humans and robots interacting with each other [1]. It is very extensive and

© The Author(s), under exclusive license to Springer Nature Switzerland AG 2025
M. N. Huda et al. (Eds.): TAROS 2024, LNAI 15052, pp. 48–57, 2025.
https://doi.org/10.1007/978-3-031-72062-8_5

diverse, yet is neglected, with many designers choosing other ways of communication, such as human-computer interaction [2]. HRI allows for the communication of states and status between robots and humans, which can be critical to the situation. Larger robots may cause physical danger and appear threatening to users, especially ones that perform services, whereas robots used for entertainment and as a from of companion-ship may cause emotional responses without creating anxiety or fear [3]. Through the communication of states, negative emotional responses may be minimised; emotions such as fear, although may not be eradicated, may be lessened if the humans interacting with any robot can understand the intent behind the robot's actions as well as how the robot may interact with both humans, and the environment around it.

To be able to understand the most effective and efficient ways in which HRI can be implemented, it is important to understand how humans interact with other humans nonverbally and animals.

When humans interact with other humans, a variety of nonverbal signals are utilised.

These are frequently used in order to understand the participants' emotions, level of focus, and intent [3]. Non-verbal communication (NVC) is defined as any action that does not use words in order to get meaning across, including facial expressions and body language, as well as tone of voice [4]. It is considered highly important to be able to understand the meaning behind NVC, which may vary based on context or culture [4]. Gestures are used to help with comprehension and are often used alongside verbal communication to prevent ambiguity [5]. Tests with virtual agents have shown that the virtual agents are viewed more positively and more human-like when they use gestures alongside speech [5]. This suggests that using more states of communication that just speech may be beneficial in the application of robots that need to be able to communicate with humans, as humans appear to depend on visual cues as well as auditory ones. Other commonly used NVC includes eye gaze and facial expressions [3], which may be more difficult to implement in a robot as they are costly, complex or fragile [6].

When humans and animals communicate with each other, they use a range of methods. Dogs use three: auditory, visual, and olfactory [7]. Humans often attempt to use visual cues to determine an animal's body language, such as the face, the ears, and the tail [7]. Humans also often anthropomorphise animals, which is defined as the attribution of human characteristics to non-human entities, such as inanimate objects or animals [8]. Anthropomorphism is a consistent part of human culture [8]; the majority of dog owners believe their dogs feel guilt when they break a rule, even though there is no evidence of dogs feeling guilt [9]. The fact that humans anthropomorphise animals and often look for visual cues when interacting with them implies that, for HRI, the appearance of the robot is highly important, as strong and positive attachments may not be formed with a robot that looks threatening [10]; their visual cues may be negatively interpreted due to their physical appearance.

Contextual differences heavily impact how a human interacts with a robot [10]. The differences may include things such as a person's prior experiences and opinions; societal and cultural differences; and the expectations on the robot. Coeckelbergh suggests that relationships formed with robots are dependent on prior experience with robots as well as the first impression of the robot, and that information and context about the robot, such as what it can and cannot do, may change the appearance of the robot and the relationship participants may have with it, suggesting that experience of a robot and knowledge of its limitations may shape an interaction [10].

Robotic tails have been created before. Singh and Young [11] created a robotic tail to communicate different states, using two servos, some wire, and a toy to make a chain. They adapted the idea from hobbyists, who have created a wide range of tails.

2 Methodology

Like Singh and Young [12], the robot tail was created using two servos: one for horizontal movement, and one for vertical movement (Fig. 1). This allowed for a full range of movement, like a dog's tail.

The tail was made from five 3D printed vertebral segments, interlinked with ball and socket joints and nylon 'tendons' and decreasing in size with distance from the base (Fig. 2). It was important for the robot tail to looks like a real dog's tail, since the appearance of a robot heavily impacts how humans interact with it [10].

Each different tail movement was created as an individual function, which comprised of several basic functions. The basic functions set the height of the tail, the speed of the tail, and if the tail would wag fully side-to-side or only slightly wag.

Fig. 1. An example of how the servos were set up (with the breadboard and Arduino in the background).

Videos were then taken of the robotic tail and of dogs showing tail movement and uploaded to the questionnaire. All videos had the same twelve emotions in the same order for participants to pick from, which were chosen to represent four main groups - this can be seen in Table 1. The groups were carefully selected to represent four distinct types of emotions. The emotions within these categories were intentionally designed to be closely related to each other, considering the subtle nuances in how individuals perceive each specific word. As a result, participants had a wide range of answers to choose from.

Fig. 2. The version of the tail used in the questionnaire.

Table 1. The Groups and their Associated Emotions.

Emotional Groups	Associated emotions
Happy	Happy, Excited, Joy
Neutral	Neutral, Fine, Friendly
Frightened	Scared, Nervous, Tentative
Aggressive	Angry, Aggressive, Annoyed

The questionnaire was open for people to respond to for three weeks. 58 people participated in the study.

3 Results

3.1 Overall

Results were analysed in several different ways. Initially, the results of the videos were matched up with the corresponding video and compared, as can be seen in Fig. 3. The emotions were displayed in their groups to show definitive results. Figure 3 displays the group of emotion each participant picked for each pair: most participants picked from the same group of emotion for Pair 1, Pair 2, and Pair 4.

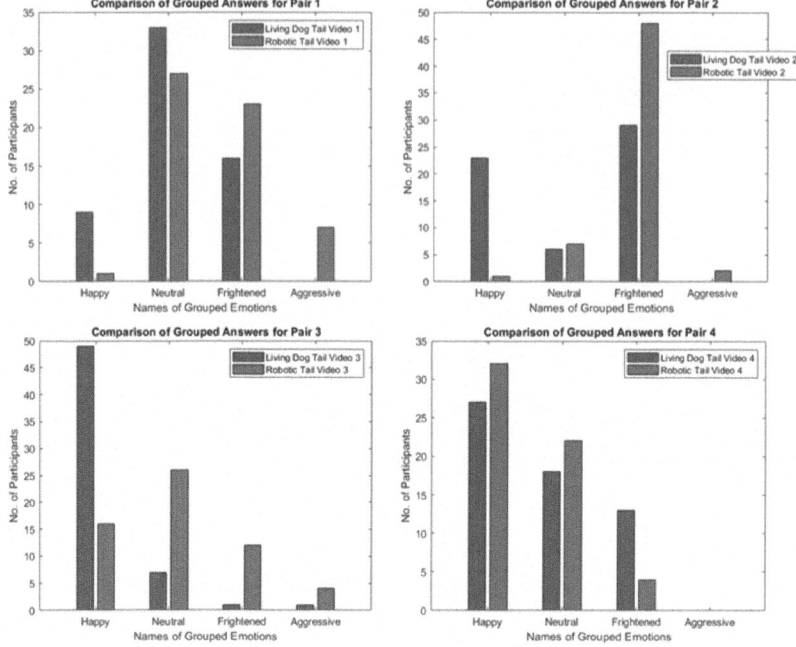

Fig.3. Overall answers for each pair.

3.2 Participant Analysis

Figure 4 displays how many participants were able to pair up the videos correctly. To determine this, each participant's answers for the correct pairs were compared to each other to see if they were in the same emotional group picked for the research.

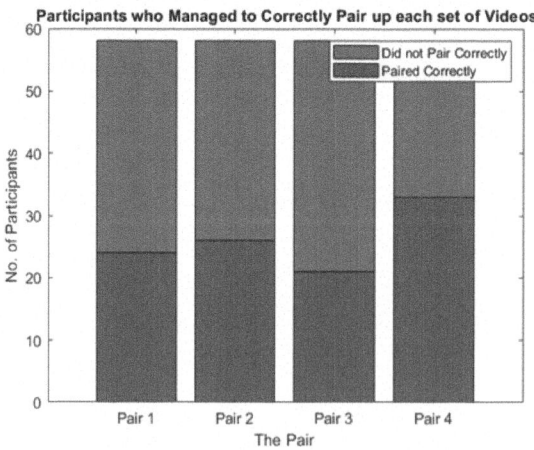

Fig. 4. Analysis Based on Individual Answers

Overall, 87.9% of participants answered within the same group for at least one pair; 63.8% answered the same for at least two pairs; and 22.4% answered the same for at least three pairs. 2 out of the 58 participants answered within the same group for all pairs.

4 Discussion

4.1 Significance

Overall results show a trend of people picking the same emotion for each tail video rather than each pair, which could show that most people think along the same lines as each other.

Figure 4 suggests that most people can tell whether a robot tail is simulating the same emotion that a dog tail can. Over 85% of people managed to get at least one pair, with almost a quarter of participants getting at least three pairs correct. This suggests that people most likely can determine the same simulated emotion in robot tails as they can in dog tails, but this could vary on context and other factors involved.

4.2 Experience with Dogs

Before participants watched any of the video clips, they were asked about their experience and opinion of dogs. Participants could pick one of three answers: "I like dogs, but have no experience with them", "I have experience owning or working with dogs", and "I don't like dogs". Participants were asked this to see whether their opinion and experience would influence the results of the study, and if so, then by how much.

Experience with dogs may have influenced the results positively or negatively. It can be assumed that having experience with dogs means that body language is better known and therefore participants who have experience with dogs would be able to recognize the body language of the dog tail and the robotic tail. However, dogs change how they wag based on how well they know the person they are with [13], which might mean that some participants assumed the wrong emotion due to being used to their dog wagging in a certain way. However, this still shouldn't have impacted the study, as the study was focused on pairing the videos up by participants putting the same emotion for the same tail movement, rather than the chosen emotions themselves.

As shown in Fig. 5, the results show that over two thirds of the participants in the survey had experience either working with or owning dogs. Whilst this could suggest that the majority of participants who correctly paired up results most likely had some form of experience working or owning dogs, the results appeared to have no connection to whether the participants had experience with dogs; 85% of participants who had experience with dogs managed to get at least one pair correct; 93.3% of participants who claimed to like dogs but have no experience working nor owning them managed to get at least one pair; and 100% of participants who claimed to not like dogs managed to make at least one pair. This suggests no correlation between opinion/experience with dogs and determining the same simulated emotion in dog tails and robot tails.

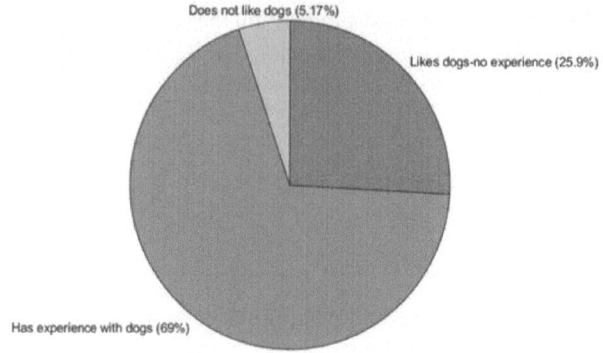

Fig. 5. A Pie Chart Showing Participants' Experience & Opinion on Dogs.

4.3 Other Bias

There may be more bias in the results than just whether people have had any experience with dogs.

The questionnaire was also open for only a limited amount of time meaning that it is possible that a lot of people missed the window in which they could answer, and thus there may be only opinions of people who happened to see the questionnaire was available to answer at that time.

Additionally, the video clips of the dogs were not always taken from the best angle. The tail was often obscured, meaning that participants may have struggled to decipher the emotion shown through the body language. As most of the dog and the background could be seen, participants may have also used that as context to help them decide which adjective to choose. This was also seen in all of the dog tail videos; context could still be taken from the background or from the body language of the dog (Fig. 6),

Fig. 6. A frame from one of the dog tail videos, showing the obscured tail & background context.

Bias around the videos of the robotic tail may include that all the videos were taken with the same background, lighting, and direction, meaning that it may make it easier for participants to see the difference of the tail movement, especially as there were minimum distractions in the video through the mostly plain background. This also means, however, that participants could not rely on any context for answering the questions to this part of the questionnaire, which, although means that the answers are only due to the movement of the tail, could mean that there is a difference in results from the dog tail videos and the robotic tail videos due to the context.

Slight blur/lack of focus was included when recording the robot tail videos, to match the varying quality of the dog tail videos so that both sections of the questionnaire were more similar in quality.

Additionally, the videos of the robot dog all had the same tail in, whereas the videos of the dog tails had different breeds of dogs in, meaning that tails varied in size & structure. This could have made it more difficult for participants to recognise the body language in the tails, as there may be bias around certain breeds [14], due to media representation, and/or due to the participants' previous experiences with the breed.

All the videos used in the questionnaire were also very short (a few seconds long), meaning that participants may have struggled to understand/properly see the tail movement. However, this was chosen so that there would be as little context as possible, and to ensure the videos were all similar in length.

5   Conclusion

5.1   Implications of This Study

This study could be used towards creating a more realistic robot dog for supporting people who are having issues with mental health or are incapable of looking after a living dog [15]. Though this could have positive benefits, as shown by Banks' study in the care home, it is also possible that using a robot dog and making it more realistic through the use of a better tail could be seen as deception. A robot dog being created for usage like this may have a good intention but could still deceive the user through emotional deception, which encourages users to believe that a robot has emotions and is able to form a relationship with users [16]. Both children and older people are more likely to form a uni-directional bond with a robot, which could cause children to miss out on opportunities to learn about how human relationships work, and older people may share things with a robot that they don't want other people to know [16]. Sharkey and Sharkey suggest these more vulnerable groups, who often rely on others to make decisions for them, need society to prevent social robots from exploiting them [16].

The robot tail could also be used as part of a robot used to entertain, which would most likely be to an audience with some knowledge in robotics [16]. Due to this, the audience may not be fooled into thinking the robot has feelings or is sentient, like they may see a magic show and know that it is an act [16]. This is a positive implication of the research, as it would not be used to exploit.

It is highly unlikely that this tail will be used in a way that could exploit humans in the near future; currently, the tail is fragile, meaning that it would be unsafe for most uses, such as for children to play with. Additionally, to exploit people, it would need a way to record what its users are saying, which it doesn't need to do in order to fulfil its function. Even as part of a robot dog, the robot wouldn't need to record anything unless it is part of its purpose. Any recordings could also be used as part of a safeguarding measure, to keep the more vulnerable groups of society safe.

5.2 Further Research

There are two directions that further research could take.

The first one involves repeating the experiment, but with some of the variables changed. This could involve randomising the order of the videos for each participant and asking participants for their experience with and opinion of robots as well as dogs, to understand if there are any bias due to participants' opinions on robots. Participants could also be asked to rank emotions from the ones they believe are most likely to be portrayed in the videos shown to the least likely, to see if this has an effect on how many participants were able to pair up the videos. Through testing these, it can be understood to what degree do humans properly understand simulated emotion from a robotic tail.

Additionally, if the experiment is repeated, the videos used should be of a higher quality, and show as little context as possible, to ensure that they are consistent with each other. This would be applied to specifically the dog videos, which often showed the body of the dog, as well as parts of the background. The pool of participants should be increased to solidify results, and participants could also be asked demographic details, such as their age range, gender, and background, so that the generalisation of the results could be understood.

The other direction would be creating a secondary study to understand the effects of the different forms of non-human NVC. This could involve using two types of NVC (such as a tail and ears) either together or separately to measure how accurately participants can determine the simulated emotions, how different forms of NVC affect it, and if using more than one form of NVC contributes to accuracy (and if so, then by how much).

Researching one or both of these directions may help further the understanding of if and how non-human NVC can contribute to HRI, as well as the effectiveness of doing so.

5.3 Conclusions

Overall, the results have supported the hypothesis that humans can determine simulated emotion from tail activity. Not only did the majority of participants answer within the same group for three out of the four pairs, 63.8% of participants paired at least two out of the four pairs, showing that humans do assign the same emotion for the same movement pattern in animals and robots.

Disclosure of Interests. The authors have no competing interests to declare that are relevant to the content of this article.

References

1. Murphy, R.R., Nomura, T., Billard, A., Burke, J.L.: Human–Robot Interaction. https://ieeexp lore.ieee.org/stamp/stamp.jsp?tp=&arnumber=5481144. Accessed 06 Apr 2024

2. Sheridan, T.B.: Human–Robot Interaction: Status and Challenges. https://journals.sagepub.com/doi/epdf/10.1177/0018720816644364?src=getftr. Accessed 06 Apr 2024
3. Kulíc, D., Croft, E.: Affective state estimation for human-robot interaction. IEEE Trans. Rob. **23**, 991–1000 (2007). https://doi.org/10.1109/TRO.2007.904899
4. Hall, J.A., Horgan, T.G., Murphy, N.A.: Nonverbal communication. Annu. Rev. Psychol. **70**, 271–294 (2019). https://doi.org/10.1146/ANNUREV-PSYCH-010418-103145/CITE/REFWORKS
5. Salem, M., Rohlfing, K., Kopp, S., Joublin, F.: A friendly gesture: investigating the effect of multimodal robot behavior in human-robot interaction. In: RO-MAN (2011)
6. Admoni, H., Scassellati, B.: Social eye gaze in human-robot interaction: a review. J. Hum. Robot. Interact. **6**, 25 (2017). https://doi.org/10.5898/JHRI.6.1.Admoni
7. Hasegawa, M., Ohtani, N., Ohta, M.: Dogs' body language relevant to learning achievement. Animals **4**, 45–58 (2014). https://doi.org/10.3390/ani4010045
8. Taylor, N.: Anthropomorphism and the animal subject. Hum.-Anim. Stud. **12**, 265–280 (2011). https://doi.org/10.1163/EJ.9789004187948.I-348.61
9. Miklósi, Á., Gácsi, M., Bugnyar, T., Vonk, J., Dolins, F.L.: On the utilization of social animals as a model for social robotics. Front. Psychol. **3**, 75 (2012). https://doi.org/10.3389/fpsyg.2012.00075
10. Coeckelbergh, M.: Humans, animals, and robots: a phenomenological approach to human-robot relations. Int. J. Soc. Robot. **3**, 197–204 (2011). https://doi.org/10.1007/s12369-010-0075-6
11. Singh, A., Young, J.E.: A dog tail for communicating robotic states. In: 2013 8th ACM/IEEE International Conference on Human-Robot Interaction (HRI) (2013)
12. Singh, A., Young, J.E.: Animal-inspired human-robot interaction: a robotic tail for communicating state. In: 2012 7th ACM/IEEE International Conference on Human-Robot Interaction (HRI) (2012)
13. Ren, W., Wei, P., Yu, S., Zhang, Y.Q.: Left-right asymmetry and attractor-like dynamics of dog's tail wagging during dog-human interactions. iScience **25**, 104747 (2022). https://doi.org/10.1016/j.isci.2022.104747
14. Caddiell, R.M.P., et al.: Veterinary education and experience shape beliefs about dog breeds. Part 2: Trust. Sci. Rep. **13**, 13847 (2023). https://doi.org/10.1038/s41598-023-40464-3
15. Banks, M.R., Willoughby, L.M., Banks, W.A.: Animal-assisted therapy and loneliness in nursing homes: use of robotic versus living dogs. JAMDA-Original Stud. **9**, 173–177 (2008). https://doi.org/10.1016/j.jamda.2007.11.007
16. Sharkey, A.: Sharkey, · Noel: we need to talk about deception in social robotics! Ethics Inf. Technol.. Technol. **23**, 309–316 (2021). https://doi.org/10.1007/s10676-020-09573-9

An Perception Enhanced Human-Robot Skill Transfer Method for Reactive Interaction

Weiyong Si[1], Jiale Dong[2], Ning Wang[3], and Chenguang Yang[4(✉)]

[1] School of Computer Science and Electronic Engineering, University of Essex, Wivenhoe Park, Colchester CO4 3SQ, UK

[2] South China University of Technology, Guangzhou 510640, China

[3] Bristol Robotics Lab, University of the West of England, Bristol BS16 1QY, UK

[4] Department of Computer Science, University of Liverpool, Liverpool L69 3BX, UK
cyang@ieee.org

Abstract. In this paper, we propose a trust human-robot skill transfer framework, interpretive and reactive dynamic system (IRDS), by investigating the behaviour tree (BT) and dynamic movement primitives (DMPs) enhanced by the feedback from perceived information for dynamic and uncertain tasks and environments. Human sensorimotor control allows interactions with various environments and accomplishes complex manipulation tasks with uncertainty; however, it is still hard for robots to own this capability. In this work, we aim to transfer these reactive skills to robots, which enable robots to interact with humans and environments under varying uncertainty. The main challenges of robot skill learning are the generalization, safety and stability issues during the skills learning and execution autonomously. BT was investigated for task planning and reactive interaction during the robot execution. The dynamic system-based model generates action for the low-level compliant controller. The convergence of the proposed IRDS framework under dynamic interaction and disturbance was proved by control theory. We conducted simulations and physical experiments on the real robots to evaluate the generalization performance and the convergence capability under uncertainty. And the results show that the reactivity, convergence and interaction performance can be guaranteed, and the learned skills can be transferred among different physical robot platforms.

Keywords: Trust human-robot skill transfer · Interpretive and reactive dynamic system (IRDS) · Compliant control · Robot learning · Human-centred environment

1 Introduction

Robot manipulators have been widely used in a range of fields, such as industrial plants, space exploration, agriculture and healthcare [1] etc. For example, collaborative robots have gained much attention for human-robot interaction and

M. N. Huda et al. (Eds.): TAROS 2024, LNAI 15052, pp. 58–69, 2025.
https://doi.org/10.1007/978-3-031-72062-8_6

collaboration, robot-assisted assembly and polishing [2], and medical examination [3,4]. However, most of these applications adopt the programmed method manually for different scenarios and tasks. The traditional robotic solutions are limited by explicit modelling and manual programming. The program designer needs to plan in advance the reactions to all possible situations that the robot may face in practice. If errors or new situations arise after the robot is deployed, the entire process has to be reprogrammed, which is inefficient and costly.

Nowadays, the learning-based approach offers better capability for robots to acquire a wider array of skills and better flexibility in adapting to new scenarios. As machine learning methods have been investigated in the robotic field, various robot learning methods have been proposed to tackle the autonomous execution of robots, such as imitation learning, also known as learning from demonstration (LfD) [5,6], reinforcement learning [7] and interactive learning [8] etc. However, the reactive capability to new tasks and dynamic environments is hard to deal with unknown scenarios and tasks, especially for safety-critical application, such as human-centred environments and human-robot collaboration [9]. Therefore, collaborative robots are expected to guarantee safe interaction and stability under various disturbances when working in human-centred tasks and environments. So far, it remains a challenge to address data-efficient learning of robot manipulation for various environments, and limited research has been conducted on data efficiency and skill adaptation.

Learning from demonstration (LfD) is one effective robot skill learning method, which relies on the human teachers demonstrations. LfD has been investigated for various manipulation tasks, such as writing character, cutting vegetables [5], composite layup [10], dough rolling [11] in our previous work. Firstly, most robot skill learning focuses on trajectory level learning, which does not consider contact tasks [12,13]. However, only trajectory planning and learning cannot work well for some contact manipulation tasks. Secondly, there have been plenty of studies on contact manipulation skill learning [14]. Force-related skill learning and generalisation have been studied in the past few years. Besides, compliant skill learning also has been investigated for collaborative robots. However, there are some challenges for the LfD. First, the demonstration quality has great influence on the LfD. Especially, when human demonstrate complex manipulation tasks, the variance or the deviation among different demonstrations can affect the training of skill model. Second, the generalisation capability is also a main challenge for LfD. Although the LfD have good generalisation capability on the demonstration space, it is hard to adapt to unseen environments and tasks.

Unfortunately, the learning-based methods cannot solve the problem of interpretation. On the other hand, the behavior trees (BTs) is a combined concepts of modularity, hierarchical structure and feedback for robot control system [15]. In [16], a BT-like robust and logical framework was proposed to conduct household manipulation task. The authors developed a collaborative robot system, integrating behavior trees and machine vision [17]. BT is combined with the RL for automated guided vehicle navigation in industrial 4.0 environment [18]. BT is also employed to design complex intelligent system or behavior of agents, such as AI game [19] (Fig. 1).

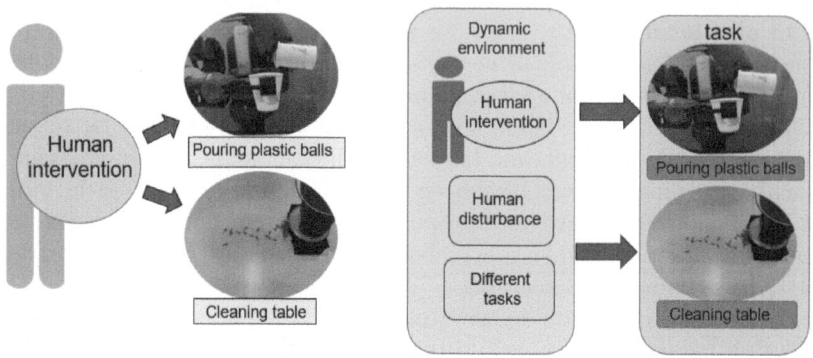

Fig. 1. Robots clean a table and pour plastic balls stably in a human-centred environment.

The challenges of robot manipulation skill learning are the generalisation capability to dynamic tasks and environments. And the safety and convergence of generalised skills for the human-centred environment and tasks are also an opening challenge. The assistive robots are able to support people with a motor disability. In this case, the robot needs to interact with the human complete compliantly and safely. For this application, the programming offline might not be enough, because unlike the isolated environments, such as industrial robots, the environment and task are dynamic. It is impossible to program for all the work environments and tasks in advance. However, these working environment and task have critical safety constraint. The pure data-driven methods may not be suitable for this kind of application, since these methods are not trustable for the users. In addition, the reactive capability is also significant for the human-centred environment under the unexpected accident and human interaction. There are some work to tackle the reactive planning and control. In our previous work, we studied the vision-based reactive planning for pizza rolling [20]. However, most of the work is based on the model-based method, lacking the generalisation capability for novel environment and tasks or requiring heavy computation capability. Therefore, an unified framework of the compliant control, reactive planning and perception are needed.

This work proposes a method combining BT and DMPs enhanced by the perception information to handle the uncertainty and dynamic tasks. The primitive skill (PS) model was built based on mapping semantic information with hierarchical labels as the basic components for constructing a database of primitive skills; it will be further generalised and enhanced for the synthetic skills library. We propose an interpretable and reactive dynamic system (IRDS) model to characterise perception-action mapping and encode motor PSs. The novel IRDS model integrates real-time perception information to adapt to the robots' behaviours reactively, dealing with unseen tasks and environmental uncertainty. The impedance controller was designed and integrated with the IRDS model to enable robots' variable impedance behaviours and robust adaptation to dynamic tasks and uncertainties. We used the behaviour tree (BT) model to synthesise multiple PSs and generalise an action chain to adapt to the requirements and constraints of a complex task. The human interaction will provide real-time feedback as input for the PS generalisations and thus generate various synthetic skills to extend the skill library according to task and environmental variations.

This work aims to propose a novel framework which could achieve reactive and interpretive planning and compliant control for human-robot interaction. Unlike the pure data-driven method, reinforcement learning, deep reinforcement learning etc., this hierarchical and hybrid method combines the advantages of the model-based method and the data-driven method. The proposed framework has the following characteristics: reactivity, convergence, safety and interpretability.

The remaining of the paper is as follows: Sect. 2 presents the preliminary knowledge of robotic dynamics and control. The proposed framework is developed in Sect. 3. The design of the experiment and the results and performance analysis are present in Sect. 4. Finally, discussions and future work are given in Sect. 5.

2 Preliminary

2.1 Robotic Dynamics and Control

For a serial robot manipulator, the dynamic equation in joint space can be given,

$$M(q)\ddot{q} + C(q,\dot{q})\dot{q} + G(q) + \tau_f(\dot{q}) + \tau_{ic} + d_u = \tau_c + J^T(q)F_e \tag{1}$$

where the $M(q)$ is the inertia matrix in Cartesian space, the $C(q,\dot{q})$ is the Coriolis term and $G(q)$ represents gravitational force. f_c is the control force and f_{ext} is the interaction force with the environment. q and $\dot{q}$ represent the joint position and velocity, respectively.

The relationship between the joint torque and the wrench can be described as,

$$\tau_{ext} = J^T(q)f_e \tag{2}$$

where $J(q)$ is the Jacobian matrix, and f_e is the interaction force with the environment.

For the teleoperation system design, the leader device and the follower robot usually have different configurations. The robot controller designed in the Cartesian space will benefit teleoperation control and telepresence. In addition, the teleoperation control in the task space will also benefit the generalisation across different platforms. Therefore, we derive the controller in the tasks space. Based on the kinematic equation of the manipulator, we can describe the position, velocity and acceleration of the end-effector as,

$$x(t) = f(q) \tag{3}$$

$$\dot{x}(t) = J(q)\dot{q} \tag{4}$$

$$\ddot{x} = \dot{J}(q)\dot{q} + J(q)\ddot{q} \tag{5}$$

where the $\dot{x}$ and $\ddot{x}$ are the velocity and acceleration of the robot end-effector in Cartesian space. $\dot{J}(q)$ is the derivate of the Jacobian matrix.

2.2   Compliant Control Methods

Impedance control, proposed by Hogan in 1985 [21], has been investigated widely for robot manipulator control, such as compliant control, human-robot interaction, etc. The impedance controller is similar to a spring-damper dynamic system, which is used to model the dynamic interaction between the robot and its environment. This control of dynamic behaviour allows human-robot and robot-environment safety interaction. The impedance control of 1-D model in Cartesian space can be represented as,

$$m_i \ddot{e}_i + d_i \dot{e}_i + k_i e_i = f_i \tag{6}$$

where the subscript i denote the ith dimension, m_i, d_i and k_i are the desired inertia, damping and stiffness, f_i is the interaction force. The impedance model is a mass-damping-spring system, which describes the relationship between the tracking errors and the contact force. To track the desired pose, we need to implement the translation impedance model and the orientation impedance model. The Eq. (6) directly describes the translation impedance model, however, the orientation tracking is complicated due to different representations. We adopt the quaternion-based orientation control for the teleoperation control. We derive the quaternion-based impedance model for orientation tracking due to the advantages of avoiding singularities.

3   Methodology

The interpretative and reactive dynamic system (IRDS) framework, as shown in Fig. 2, is introduced in this section, which includes the BT-based sequence model, DMPs-based primitive skill model and the dynamic task awareness.

3.1   Behavior Tree Based Multiple Primitive Skill Modelling

Behavior tree is an effective graph model to describe the sequence task decision and planning, which can also be integrated with the perception and interaction interface for the selection and merging of primitive actions. The behavior trees can be formulated as a combination of a planner and a metadata function used to provide feedback on the execution and external event, novel tasks etc.

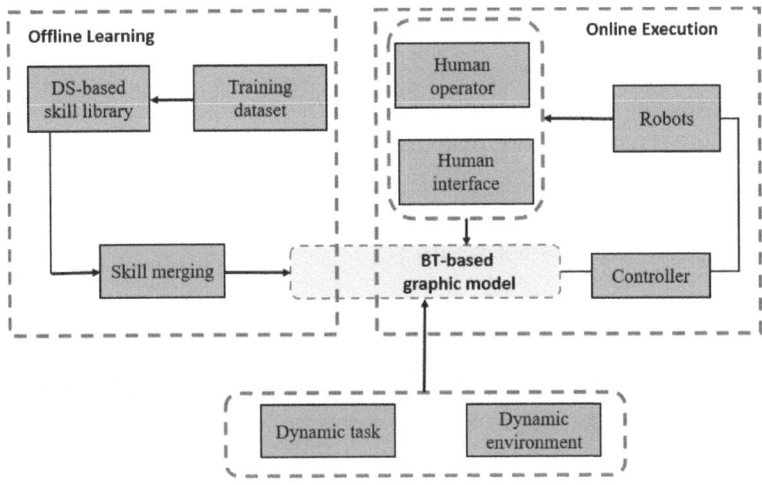

Fig. 2. The overview of the proposed interpretive and reactive dynamic system framework.

These metadata allow BT to create more complex behavior in a hierarchical tree structure.

A behavior tree (BT) is a directed tree whose leaf nodes are called execution nodes and non-leaf nodes are called control flow nodes. It selects what should be done in the current environment and state through a tree-like decision structure similar to a decision tree. The execution of BTs starts from the root node, BTs continuously generates tick signals from the root node at a fixed frequency and propagates it to the children. When the node receives the tick signal, it starts to execute and returns the result S(Success), F(Failure) or R(Running) according to the execution situation, the execution result of the child node is controlled and managed by its parent node. A node in the R state will continue to execute until it returns S or F (Figs. 4 and 6).

Nodes in BTs can be divided into two types: control flow nodes and execution nodes. There are three main control flow nodes: *Sequence*, *Fallback* and *Parallel*, and two main types of execution nodes: *Action* and *Condition* [22].

Sequence: When *Sequence* is running, it starts from the leftmost child node. When the child node returns S, the next child node starts to execute. When and only when all child nodes return S, *Sequence* returns S.

Fallback: When *Fallback* is running, it starts from the leftmost child node. If the child node returns F, the next child node starts to execute. As long as one child node returns S, *Fallback* returns S. When all child nodes return F, *Fallback* returns F.

Parallel: When *Parallel* executes, all its child nodes begin to execute at the same time. Only when all the child nodes return S, *Parallel* returns S. As long as one child node returns F, *Parallel* returns F, and in other cases, *Parallel* returns R.

Action: This is the actual operation node of the robot, such as the movement, control or state change of the robot. Returning the actual running state of the robot.

Condition: It is a judgment node. Returning S or F according to whether the condition is true, and does not return R.

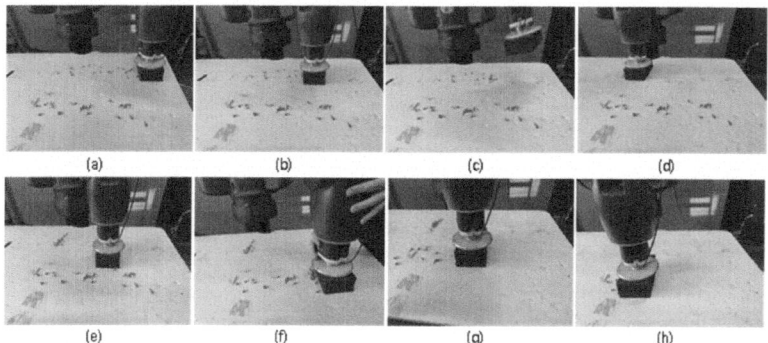

Fig. 3. In figure (a), the robot starts the wiping task, figure (b) shows that the robot is wiping the desktop, figure (c) shows that the robot deviates from the original position after receiving interference, figure (d) shows that the robot returns to the original position to continue wiping after overcoming interference, figure (e) shows that the robot is executing the second wiping track, figure (f) shows that the robot's wiping position is adjusted manually, and figure (g) shows that the robot complies with the human adjustment, Wipe the desktop from the new position, and figure (H) shows that the wiping is completed

3.2   Dynamic Movement Primitive

Dynamic movement primitives (DMPs) was proposed by Ijspeert to study motor control of humans, which was inspired by the dynamic systems and human motor control. The essential of DMPs was to encode the manipulation skills by a dynamic system. DMPs can also be employed to model multiple DoFs system, each DoF can be modelled separately, and a canonical system achieves the coupling among these DoFs. For readability, we introduce the DMPs for one degree of multiple dynamic systems,

$$\begin{aligned} \tau_s \dot{v} &= \alpha_z(\beta_z(p_g - p) - v) + F(x) \\ \tau_s \dot{p} &= v \end{aligned} \tag{7}$$

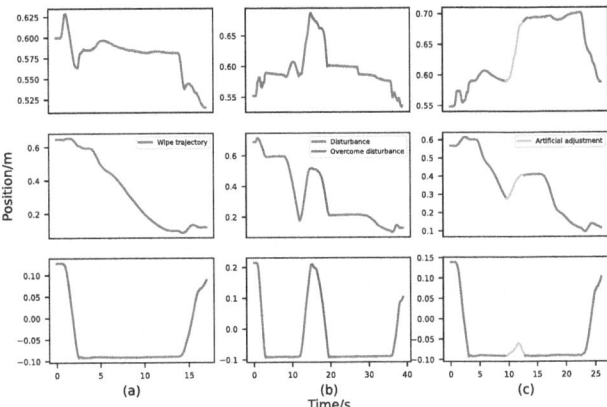

Fig. 4. Clean: Figure (a) shows the trajectories in three directions when the robot performs the wiping task, figure (b) shows the trajectories in three directions when there is disturbance in the wiping process, and figure (c) shows the position trajectories in three directions when the wiping position is manually adjusted in the wiping process.

where the p_g represents the goal position, p represents the current position; τ_s is the scaling parameter, the v is the velocity, α_z, β_z need to be designed, however, these parameters satisfying $\alpha_z = 4\beta_z$. $F(x)$ is the nonlinear forcing term, which is used to affect the trajectory. Generally, the $F(x)$ can be composed of the Gaussian functions,

$$F(x) = \frac{\sum_{i=1}^{N} \psi_i(x)w_i}{\sum_{i=1}^{N} \psi_i(x)} x(p_g - p_0) \tag{8}$$

$$\psi_i(x) = \exp(-h_i(x - c_i)^2) \tag{9}$$

A canonical system was used to determine the phase variable x, and the canonical system can coordinate different DoFs, which can be described as,

$$\tau_s \dot{x} = -\alpha_x x, \quad x \in [0, 1]; \quad x(0) = 1 \tag{10}$$

where τ_s represents the scaling parameter, α_x is a positive coefficient, and the initial value of x is $x(0) = 1$, which can converge to zero exponentially.

4   Experiment Evaluation

We evaluated the proposed framework through a typical task in life, wiping a table and pouring material. This task consists of the interaction of the robot with its environment, a task that requires both appropriate contact force control and trajectory control simultaneously. In addition, it requires the robot to adapt its behaviour to obstacle and human. A 7 degree-of-freedom (DOF) Baxter

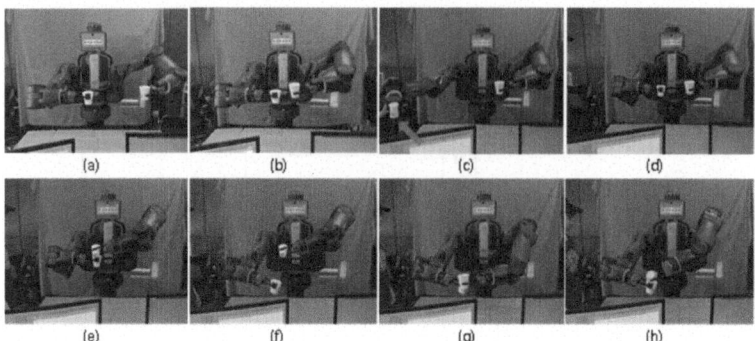

Fig. 5. Snapshot of bimanual pouring tasks. (a) starts the plastic ball pouring task, (b) shows that the left arm is approaching the right arm, (c) shows that the right arm is artificially disturbed beyond the range that the left arm can reach, (d) shows that the right arm overcomes the interference and returns to the range that the left arm can reach, (e) shows that the left arm begins to pour plastic ball, (f) shows that the right arm deviates from its original position after receiving the interference, and (g) shows that the left arm returns to its normal attitude and continues to follow the right arm, (h) shows the successful completion of the plastic ball pouring task.

collaborative robot, equipped with a Mini45 F/T sensor to measure the interacting force, is used to conduct experiments. The experimental setup is shown in Fig. 3. In the experiments, the proposed IRDS is implemented under robot operating system (ROS), whose Software Development Kit (SDK) is provided from Rethink Company, and the control rate is set as 100 Hz.

The proposed framework could generalise to different robot platforms due to the implementation in the task space. We conduct experiments on two redundant collaborative robots, Baxter robot and Franka panda, to perform several tasks, including wiping table, dough rolling and pouring plastic, are conducted to verify the performance of the proposed method.

4.1   Wiping Task by Robot

The experimental scene is shown in Fig. 3. The robot needs to clean the parts scattered on the table to one end of the table. The task is completed by three skills. Skill 1 and Skill 2 are pure displacement movements represented by DMPs. Skill 1 and 2 are used to reach the starting point of cleaning, return to the original track when encountering interference, and leave the table, and Skill 3 is composed of displacement movements and force control, which are used for wiping the desktop to ensure that the mechanical arm maintains a constant contact force with the desktop during the cleaning process. Because the desktop parts are scattered in a large range, it is impossible to clean all the parts at once, so the robot needs to clean the desktop repeatedly. The whole process will be completed by combining and generalizing the three skills through the BT model established according to the task. There are three parameters of the task, namely,

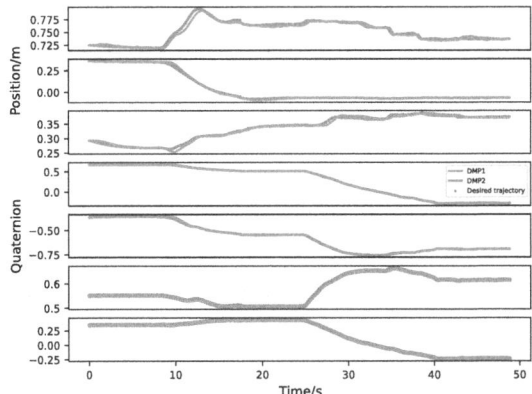

Fig. 6. pour left: This figure is the trajectory diagram of the position and direction of the robot's left arm in the plastic ball pouring task

the starting point, the end point and the direction of cleaning. We carried out three experiments in total. In the first experiment, the robot needs to sweep the scattered parts from one end of the table to the other. In the second experiment, we added artificial interference on the basis of the previous experiment to make them deviate from the original cleaning track. In the third experiment, we changed the starting point and end point of cleaning, and rotated the cleaning direction by 90°.

4.2   Pouring Task

As shown in Fig. 5, the Baxter bimanual robot is used to perform a pouring task. The two arms attempt to complete the task of pouring a plastic ball, but a human will intentionally disturb one of the arms. When human disturbance occurs, the arms must reactively re-plan their trajectory to complete the pouring task. Human intervention is detected by the contact detection function provided by the Baxter control interface.

5   Discussion and Conclusion

In this paper, we developed a robot learning system, including collaborative manipulators and force sensors etc. We propose a trust human-robot skill transfer framework, interpretive and reactive dynamic system (IRDS), by investigating the behaviour tree (BT) and dynamic movement primitives (DMPs) enhanced by the feedback from perceived information for dynamic and uncertain tasks and environments. We conducted simulations and physical experiments on the real robots to evaluate the generalization performance and the convergence capability under uncertainty. And the results show that the reactivity, convergence and interaction performance can be guaranteed, and the learned skills can be transferred among different physical robot platforms.

References

1. Si, W., Wang, N., Yang, C.: A review on manipulation skill acquisition through teleoperation-based learning from demonstration. Cogn. Comput. Syst. **3**(1), 1–16 (2021)
2. Li, Y., Ganesh, G., Jarrassé, N., Haddadin, S., Albu-Schaeffer, A., Burdet, E.: Force, impedance, and trajectory learning for contact tooling and haptic identification. IEEE Trans. Rob. **34**(5), 1170–1182 (2018)
3. Huang, Q., Lan, J., Li, X.: Robotic arm based automatic ultrasound scanning for three-dimensional imaging. IEEE Trans. Industr. Inf. **15**(2), 1173–1182 (2018)
4. Ning, G., Zhang, X., Liao, H.: Autonomic robotic ultrasound imaging system based on reinforcement learning. IEEE Trans. Biomed. Eng. **68**(9), 2787–2797 (2021)
5. Yang, C., Zeng, C., Fang, C., He, W., Li, Z.: A DMPS-based framework for robot learning and generalization of humanlike variable impedance skills. IEEE/ASME Trans. Mechatron. **23**(3), 1193–1203 (2018)
6. Yang, C., Zeng, C., Liang, P., Li, Z., Li, R., Su, C.-Y.: Interface design of a physical human-robot interaction system for human impedance adaptive skill transfer. IEEE Trans. Autom. Sci. Eng. **15**(1), 329–340 (2018)
7. Li, Z., Zhao, T., Chen, F., Hu, Y., Su, C.-Y., Fukuda, T.: Reinforcement learning of manipulation and grasping using dynamical movement primitives for a humanoid-like mobile manipulator. IEEE/ASME Trans. Mechatron. **23**(1), 121–131 (2017)
8. Franzese, G., Mészáros, A., Peternel, L., Kober, J.: ILoSA: interactive learning of stiffness and attractors. In: 2021 IEEE/RSJ International Conference on Intelligent Robots and Systems (IROS), pp. 7778–7785. IEEE (2021)
9. Liu, T., Lyu, E., Wang, J., Meng, M.Q.-H.: Unified intention inference and learning for human-robot cooperative assembly. IEEE Trans. Autom. Sci. Eng. (2021)
10. Si, W., Wang, N., Li, Q., Yang, C.: A framework for composite layup skill learning and generalizing through teleoperation. Front. Neurorobot. **16**, 840240 (2022)
11. Si, W., Guan, Y., Wang, N.: Adaptive compliant skill learning for contact-rich manipulation with human in the loop. IEEE Robot. Autom. Lett. **7**(3), 5834–5841 (2022)
12. Petrič, T., Gams, A., Žlajpah, L., Ude, A., Morimoto, J.: Online approach for altering robot behaviors based on human in the loop coaching gestures. In: 2014 IEEE International Conference on Robotics and Automation (ICRA), pp. 4770–4776. IEEE (2014)
13. Yang, C., Chen, C., He, W., Cui, R., Li, Z.: Robot learning system based on adaptive neural control and dynamic movement primitives. IEEE Trans. Neural Netw. Learn. Syst. **30**(3), 777–787 (2018)
14. Lee, M.A., et al.: Making sense of vision and touch: learning multimodal representations for contact-rich tasks. IEEE Trans. Rob. **36**(3), 582–596 (2020)
15. Ögren, P., Sprague, C.I.: Behavior trees in robot control systems. Annu. Rev. Control Robot. Auton. Syst. **5**, 81–107 (2022)
16. Paxton, C., Ratliff, N., Eppner, C., Fox, D.: Representing robot task plans as robust logical-dynamical systems. In: 2019 IEEE/RSJ International Conference on Intelligent Robots and Systems (IROS), pp. 5588–5595. IEEE (2019)

17. Paxton, C., Hundt, A., Jonathan, F., Guerin, K., Hager, G.D.: CoSTAR: instructing collaborative robots with behavior trees and vision. In: IEEE International Conference on Robotics and Automation (ICRA), pp. 564–571. IEEE (2017)
18. Hu, H., Jia, X., Liu, K., Sun, B.: Self-adaptive traffic control model with behavior trees and reinforcement learning for AGV in Industry 4.0. IEEE Trans. Industr. Inf. **17**(12), 7968–7979 (2021)
19. Sagredo-Olivenza, I., Gómez-Martín, P.P., Gómez-Martín, M.A., González-Calero, P.A.: Trained behavior trees: programming by demonstration to support AI game designers. IEEE Trans. Games **11**(1), 5–14 (2017)
20. Si, W., Guo, C., Dong, J., Lu, Z., Yang, C.: Deformation-aware contact-rich manipulation skills learning and compliant control. In: Borja, P., Della Santina, C., Peternel, L., Torta, E. (eds.) Human-Friendly Robotics 2022. HFR 2022. Springer Proceedings in Advanced Robotics, vol. 26, pp. 90–104. Springer, Cham (2022). https://doi.org/10.1007/978-3-031-22731-8_7
21. Hogan, N.: Impedance Control: An Approach to Manipulation: Part II-Implementation, vol. 107, no. 2, pp. 8–16 (1985)
22. Iovino, M., Scukins, E., Styrud, J., Ögren, P., Smith, C.: A survey of behavior trees in robotics and AI. arXiv preprint arXiv:2005.05842 (2020)

Locomotion and Manipulation

Advancing Robotic Jumping with CVT Enhanced SEA

Jingcheng Sun[1] and Chengxu Zhou[2]([✉])

[1] School of Mechanical Engineering, University of Leeds, Leeds, UK
[2] Department of Computer Science, University College London, London, UK
chengxu.zhou@ucl.ac.uk

Abstract. Series-elastic actuation (SEA) is widely employed on hopping robots due to its capability to diminish energy consumption and peak torque requirements. However, scant attention has been given in existing studies to the jump height achieved by robots utilizing SEA. To maximise jump height while retaining SEA's benefits, a solution to improve the actuator involves having a variable mechanical advantage (MA) gearbox between the elastic element and the end-effector. Continuously variable transmission (CVT) can produce such a precise MA profile with relatively straightforward structures. Although previous studies have incorporated CVT in robotics, with a predominant focus on wheeled robots, the potential advantages of using CVT on jumping robots have been overlooked. In this study, mathematical modelling is utilized to simulate and analyse the advantages of having a CVT-enhanced SEA (C-SEA) in terms of achieving higher jumps and protecting structural integrity for hopping robots. The results indicate that the C-SEA manages to reduce the peak ground reaction force by approximately 50% compared to the regular SEA. Moreover, C-SEA demonstrates superior performance in most cases with higher motor rotor mass and viscous damping coefficient when actuated by a motor with linear behaviour.

Keywords: Robotic jumping · Continuously variable transmission · Series-elastic actuators · Legged robots

1 Introduction

Series-elastic actuation (SEA) is frequently used in robots designed for continuous hopping [3,13] due to its ability to reduce energy consumption and peak torque requirements [1]. During continuous hopping, the elastic element can absorb and store the energy from the last jump as the robot lands and releases it into the next jump. However, previous studies on robots that use SEA often emphasize their agility rather than jump height. This preference stems from the common use of parallel-elastic actuators (PEA) to achieve high jumps for their ability to preload the spring during the flight phase [4]. Nonetheless, PEAs do not have the aforementioned benefits of energy conservation and torque reduction,

© The Author(s), under exclusive license to Springer Nature Switzerland AG 2025
M. N. Huda et al. (Eds.): TAROS 2024, LNAI 15052, pp. 73–84, 2025.
https://doi.org/10.1007/978-3-031-72062-8_7

making an SEA with enhanced jump capability the ideal actuator for hopping robots.

To optimise jump height while retaining the advantages of an SEA, the Salto robot provides a solution to improve the power modulation ratio (ratio between peak power output from a mechanism and that of its actuator) by having a linkage system with variable mechanical advantage (MA) as its leg mechanism [12]. However, the design process of a linkage system requires complex kinematic tuning to produce a desired MA profile. In contrast, a continuously variable transmission (CVT) can produce the same MA profile with much simpler structures. Furthermore, while the Salto robot does manage to improve the peak power output of the jumping mechanism, this enhancement does not directly contribute to the jump height, which is determined by the upward velocity of the robot body at the moment of leaving the ground. Therefore, the exact manner in how having varying MA increases the jump height of SEA-actuated hoppers needs to be further researched.

Previous studies have also incorporated CVT in robotics [5,6], but they often focus on wheeled robots instead of legged ones. Some studies utilise CVTs with compliance as the method of changing MA [8,10] but hardly any combine them with a SEA. One study investigates a CVT-SEA combination [9], mainly focusing on the control strategy for such actuators instead of specific scenarios for jumping application. Hence, the concept of CVT-enhanced SEA is novel for robotic jumping and necessitates mathematical modeling and simulation to evaluate its performance. The primary contributions of this paper are:

1. A theoretical continuously variable transmission gearbox that can be incorporated into series-elastic actuators for robotic jumping.
2. Mathematical models for the hopping robot powertrain, including the CVT gearbox and a simplified model for an electric motor powered prismatic actuator suitable for iterative simulation.
3. Comparative simulation studies to examine the performances of both the C-SEA and regular SEA in various scenarios. The advantages and limitations of the mechanism are analysed.

The structure of the paper is as follows. In Sect. 2, how to approach the analysis of a CVT-enhanced SEA is discussed; the structure of the hopping robot's powertrain is defined and the design criteria are specified. Section 3 introduces the mathematical models of the powertrain, the CVT gearbox, and the electric motor powered prismatic actuator; the constants that are later used in the simulation are also calculated from the design criteria. In Sect. 4, two simulations are done: simulation 1 examines the general behaviours of the mechanisms when being powered by an ideal actuator; simulation 2 uses the full motor model to evaluate the effects on the height performance by specific variables. Section 5 concludes the paper, and the planned future work is illustrated in this section.

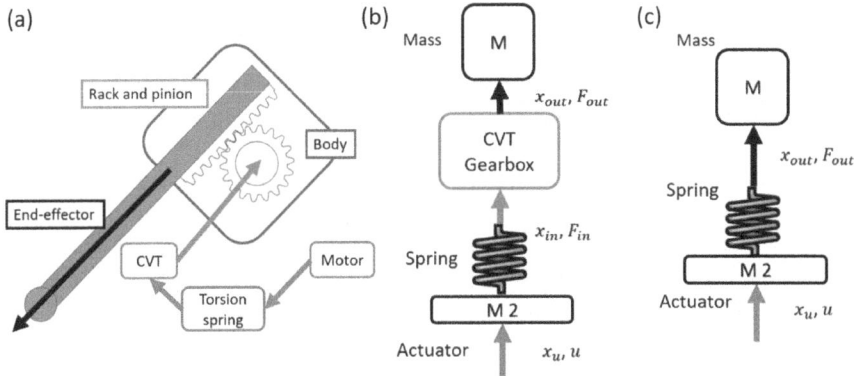

Fig. 1. A simplified structure of a hopping robot with C-SEA (a) and the isolated powertrains. Scenario (b) is the powertrain with a CVT and (c) is for a conventional SEA.

2   Methodology

To simulate a CVT-enhanced SEA robotic jumping, the advantages of having a mechanism between the elastic element and the end-effector with variable MA need to be analysed and understood. First of all, by aligning the MA profile and the spring's linear behaviour, the end-effector can produce almost constant force output, which reduces the peak ground reaction force. At the same time, having a high MA at the beginning of a stroke allows the spring to be fully compressed, thus it stores maximum energy to be released during the remainder of the stroke [12]. The first advantage improves structural integrity and reduces the material requirements for the leg mechanism, and the second one enables the robot to achieve higher jumps. This varying MA "gearbox" can be in the form of a CVT gearbox that can produce a precise MA profile that exactly matches the behaviour of the spring.

Generally in simulation, the structure of the device needs to be simplified enough so it can be mathematically defined. As shown in Fig. 1 (a), the hopper consists of a body and a leg mechanism for locomotion, with the powertrain contained inside the body. In a realistic scenario, the powertrain only provides angular motion; then, it is converted into linear motion by a rack-and-pinion mechanism. However, to avoid any complex dynamic model, the powertrain of the jumper is isolated and linearised. In Fig. 1 (b) and (c), the mass M represents the robot body and m_2 is the mass of moving components in a motor; the corresponding behaviours of each section of the powertrain are denoted by the letters on the side. In the linearised model, the motion of the robot body relative to the ground is simply represented by that of the mass M with respect to the surroundings. Since the primary objective of utilising the C-SEA is to maximise jump height, there should be a target height for the robot to reach, a fixed body mass and a maximum stroke length for the end-effector. In this case, the design criteria for the robot with a body mass of 100 g is to jump to 1 m with a stroke length (x_{max}) of 0.2 m. The parameters are averages from past light-weight jumping robots like Salto [12], EPFL jumper [7] and MSU jumper [16]; the numbers are adjusted to be mathematically convenient. The variable v_{max} is defined as the velocity of the robot body at the lift-off (the end of a stroke where the end-effector is no longer touching the ground) which directly determines the height the jump can reach. In this model, only the one dimensional vertical motion of the robot is concerned. However, balancing mechanisms such as AeroTail [15] can be easily added to allow locomotion due to its structural simplicity.

3   Mathematical Modelling

Following the assumptions made in Sect. 2, mathematical models for the power-train and the CVT gearbox can be derived. Some initial conditions that will be useful in the simulation are also provided by the models. Additionally, a model for a realistic electric motor is provided in this section.

3.1   Powertrain and CVT Models

The characteristics of CVTs compared to regular gearboxes with variable gear ratios is that a CVT can adjust gear ratio in a smooth range instead of fixed steps. Therefore, how the gear ratio changes is very flexible and open to imagination. For the simplicity of modelling, the CVT gearbox has linear inputs and outputs in the forms of displacement and force instead of angles and torques. Still using the powertrain structure in Fig. 1 (b) and assuming the spring used has perfectly linear behaviour, the input force (F_{in}) can be calculated using:

$$F_{in} = k(x_u - x_{in}) \tag{1}$$

where k is the spring constant, x_u is the displacement of the actuator and x_{in} is the input displacement into the gearbox. The gear ratio ($Gr(x_{in})$) of the CVT is set to be a linear function of the input displacement:

$$Gr(x_{in}) = G_0 - R \cdot x_{in} \tag{2}$$

where G_0 and R are constants according to the properties of the CVT gearbox. In this case, the mechanical advantage (M_A) of the gearbox is simply the inverse of the gear ratio:

$$M_A = \frac{1}{Gr(x_{in})} = \frac{1}{G_0 - R \cdot x_{in}}. \tag{3}$$

Therefore, the output displacement (x_{out}) and the output force (F_{out}) from the gearbox can also be calculated:

$$x_{out} = \int Gr(x_{in})\, dx_{in} = G_0 x_{in} - \frac{1}{2} R x_{in}^2$$

$$F_{out} = F_{in} \cdot M_A = \frac{k(x_u - x_{in})}{G_0 - R \cdot x_{in}} \tag{4}$$

and the dynamic behaviour of the powertrain can be simulated using Euler method in later sections.

3.2   Constants Calculation

Using the design criteria and assumptions made in Sect. 2, constants from equations (1) to (4) can be calculated and used as known values later on in the simulations. Given the stroke length x_{max}, target height and mass of the robot body, the ideal spring constant k, and target v_{max} can be calculated using the conservation of mechanical energy:

$$U_1 = U_2 = U_3$$
$$mgh = \frac{1}{2}k(x_{max})^2 = \frac{1}{2}m(v_{max})^2 \tag{5}$$

where U_1 is the potential energy from height, U_2 is that from the compressed spring and U_3 is the kinetic energy of the mass at the end of the stroke. In this model, drag and damping are assumed to be negligible. Thus, the ideal spring constant can be calculated as 49.05 N/m; the target v_{max} is 4.43 m/s; and the fully compressed spring produces a force of 9.81 N (denoted by F_0). To generate a constant output force, this condition has to be true:

$$\frac{dF_{out}}{dx_{in}} = 0, \tag{6}$$

and thus we get relations:

$$\frac{F_0}{G_0} = \frac{k}{R} \tag{7}$$

where F_0 and k are known. If we assume the total output stroke length is the same as the input one, the remaining two constants can be calculated: $G_0 = 2$ and $R = 10$. When the CVT consists of conical rollers, R directly reflects the slope of the cones; when it does allow the gearbox to have a mechanical advantage profile that produces constant output force, it is named the "critical gear ratio constant". For clarity and realism, the R used in the simulation is 9.9, slightly smaller than the critical gear ratio constant.

3.3   Motor Model

Various previous works have used a standard model for electric motors [1,11]. The linearised version of the equations is illustrated below:

$$V = L\frac{dI}{dt} + RI + kv$$
$$I = \frac{F + k_m v}{k} \tag{8}$$

where L is the terminal inductance, R is the winding resistance and k is the motor's back EMF constant with a unit of $N \cdot A^{-1}$ (newton over ampere); V,

I, F are the voltage input, current and the motor force output respectively. Compared to their original counterpart, several modifications have been made. Motor constant k no longer has a unit of Nm/A since it is linearised and describes the relation between the input voltage and the v (unit m/s) instead of the angular velocity ω (unit s^{-1}). At the same time, the motor torque output T_m is changed to simply force F. From (8), the expression of the motor's force output as a function of its velocity can be derived:

$$F = (\frac{k_m}{k})V - k_m v \tag{9}$$

where v is the motor velocity and k_m is a constant known as the viscous damping coefficient. It can be calculated using:

$$k_m = \frac{kI_0}{v_0} \tag{10}$$

where v_0 and I_0 are the no load velocity and current. Since the input voltage is assumed to be constant, (9) can be further simplified to:

$$F = F_0 - k_m v \tag{11}$$

where F_0 is the "stall force" of the motor (same as the maximum spring force).

4   Simulation

Two simulation studies are done in this section. Simulation 1 is to showcase the general behaviour of the C-SEA with a single end-effector stroke intending to propel the robot body to the target height. The output accelerations and velocities are compared with their SEA counterparts in specific scenarios. Simulation 2 studies the factors that can affect the v_{max} achieved by the actuators in question and the results from various input combinations are compared and analysed. The simulations in this study are exclusively done in MATLAB using the Euler method with a time step of 0.1 millisecond.

4.1   Simulation 1

An important characteristic of the C-SEA is producing reduced and almost constant output force, and this is enabled by the gear ratio being a linear function of the input displacement. To showcase this feature, a kinematic simulation of the CVT gearbox is done. The governing equation of the behaviour is based on (4) with some modifications. We assume the system is not actuated but simply powered by a compressed spring and only the analytic outputs of the gearbox are being examined (masses are ignored); thus in this case, x_u is a fixed value of x_{max}. If we plug in the constants from Sect. 3.2, the behaviour of the system in a single stroke is plotted with respect to input displacement x_{in} in Fig. 2. The

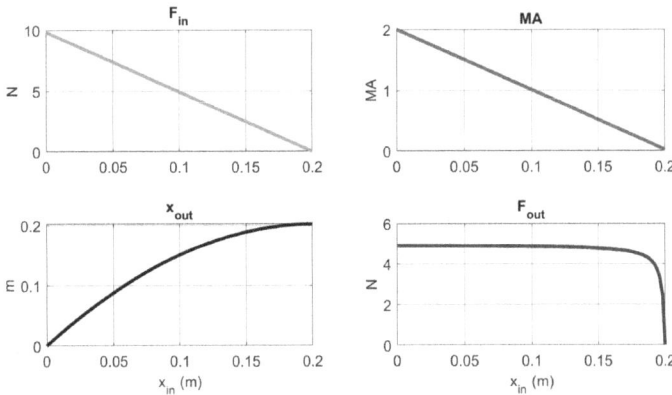

Fig. 2. The results from the kinematic simulation. The input force (F_{in}), mechanical advantage (MA), end-effector position (x_{out}) and output force (F_{out}) are plotted against the input displacement (x_{in}).

Table 1. Summary of Simulation 1 results

Scenarios	v_{max}(m/s)	Time(s)	Height(m)	Peak force(N)
Ideal scenario	4.43	–	1.000	9.81
C-SEA actuated	3.93	0.117	0.787	4.91
SEA actuated	3.97	0.091	0.803	9.81

output force is kept nearly constant and only drops at the end of the stroke with a peak value of 4.91 N, half of the peak input force of 9.81 N.

The next step is to simulate the full dynamic behaviour of both the C-SEA and the regular SEA according to the powertrain structure in Fig. 1 and the mathematical model in Sect. 3. The powertrain is placed along the vertical direction with the full effect of gravity. m_2 is set to be 10 g and the actuator is assumed to generate constant force with no limit on velocity. The maximum extent of x_u is constrained to x_{max} and the lift-off happens when the output displacement reaches x_{max}. The results are demonstrated in Fig. 3 where the x-axis is now time. The C-SEA experiences almost constant output acceleration thus linear velocity and ground reaction force throughout the stroke with a peak value around half of its SEA counterpart. The results also show that the stroke have two stages, the charging and the releasing, in a continuous manner which sees end-effector starting to displace as the spring is being compressed.

The summarised results can be found in Table 1. The system with C-SEA takes 0.117 s to reach a velocity of 3.93 m/s while the one without CVT takes 0.091 s to achieve a velocity of 3.97 m/s. C-SEA takes longer to reach the end velocity but v_{max} and thus jump height is only slightly affected (0.787 m for C-SEA and 0.803 m for SEA). Therefore, without the CVT, a regular SEA would have produced 99.8% more peak ground reaction force while reducing the time

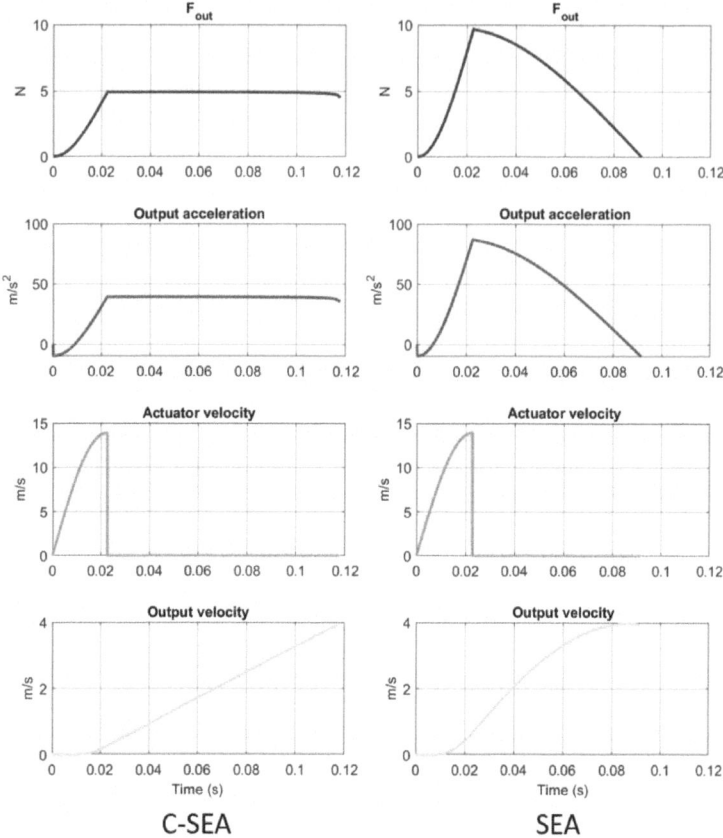

Fig. 3. The results from the dynamic simulation. The output force (F_{out}), acceleration and velocity along with the actuator velocity are plotted against time. Since the mass remains unchanged, F_{out} and the acceleration have the same shape.

duration by 22.1%. However, this simulation result still cannot reflect a realistic scenario since the actuator, if being powered by an electric motor, simply cannot generate perfectly constant force output while having no velocity limitation. The maximum velocity reached by the actuator, as shown in Fig. 3, is a whopping 13.87 m/s which is an unrealistically big number. To make the simulation more realistic, the mathematical model for the electric motor introduced in Sect. 3.3 should be included in the overall model.

4.2   Simulation 2

As discussed in Sect. 2, the second advantage arises from the initial high MA at the beginning of the stroke and a lower MA at the end, which allows for more thorough compression of the spring, thereby storing greater energy for achieving higher jumps. The rationale behind it is that prolonged charging stages can lead

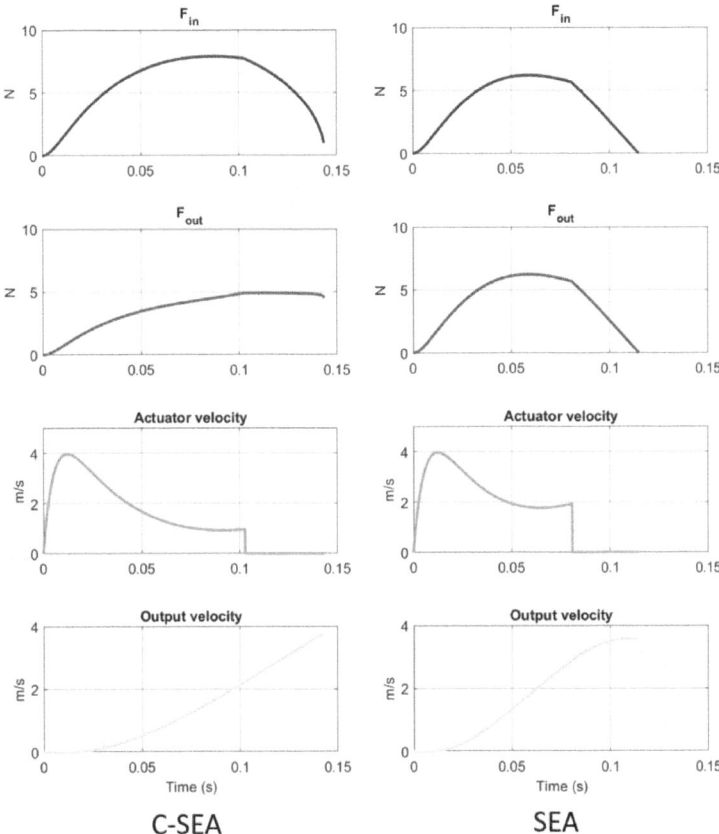

Fig. 4. The results of simulation 2. The input force from spring (F_{in}), output force (F_{out}), actuator velocity and the end-effector velocity are plotted against time. Since there is no gearbox in SEA, its (F_{in}) and (F_{out}) are identical here.

to premature expansion of the spring, resulting in incomplete compression when energy is released. However, the results from Sect. 4.1 show that the conventional SEA still outperforms in terms of jump height. This largely attributes to the assumption that the ideal actuator produces constant force with unlimited velocity, which is unrealistic and causes the charging stage to be shorter than if an electric motor is powering the actuator.

When considering an actuator powered by a linear electric motor, the time required to compress the spring is influenced by two factors: m_2 and k_m. The first factor is self-explanatory, as larger masses are harder to accelerate according to Newton's second law. The second factor determines how much the force output from the motor decreases as it speeds up. If the actuator has a linear behaviour, the charging will take longer; this consequently causes the spring to prematurely expand. When using the C-SEA, this premature expansion would be mitigated

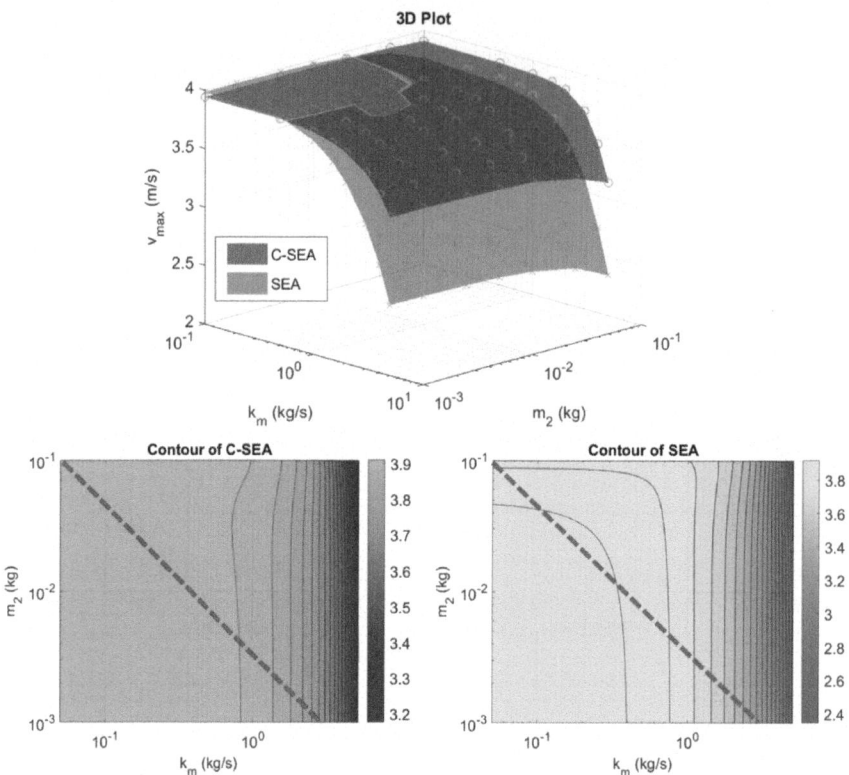

Fig. 5. The 3D plots for the results from simulation 2 are on the top and their corresponding contours are also shown on the bottom. The endpoints of the red dotted line have Cartesian coordinates $[0, 0.1]$ and $[3, 0.001]$ in the contour plots. The dotted line divides the regions according to whether C-SEA or SEA has the better performance. (Color figure online)

by the high MA at the beginning of the stroke, thus allowing the robot to reach a better height.

To prove the addition of the motor model in Sect. 3.3 does affect the performance in the manner described above, the general behaviour of the system with the new model is simulated. The k_m is set to be 2 so that when the stall force is F_0, the no-load velocity of the actuator is 4.9 m/s slightly larger than the required v_{max}; everything else remains unchanged from the setup of the dynamic Simulation 1. The results can be found in Fig. 4. As expected, the peak force output from C-SEA is still way less than the input even though the constant section is shorter compared to Simulation 1; SEA still takes less overall time to complete the stroke. A major difference from the previous simulation is that the peak force output now is only slightly higher than its C-SEA counterpart which is precisely caused by the spring in SEA not being fully compressed during the charging stage. Consequently, the v_{max} achieved by C-SEA this time is 3.81 m/s, exceeding the 3.57 m/s of the SEA. This proves the notion mentioned previously: when using a realistically modelled actuator with linear behaviour, the C-SEA still manages to compress the spring fairly well during the charging stage which is now longer, while the conventional performs poorly.

To further study the effects of m_2 and k_m on v_{max} for both C-SEA and SEA, combinations of the variables from wide ranges are used to produce a 3D plot of the results (in Fig. 5). m_2 has a range from 1 g to 100 g; the lower bound is a number small enough to represent the lowest possible value for the mass of the moving component in the actuator but still easy enough to simulate, while the upper bound is the mass of the robot body which is impossible to exceed. At the same time, k_m ranges from 0 to 5. When $k_m = 0$, $F = F_0$; this represents the scenario in simulation 1 where the actuator produces constant force. Lastly, the end-effector fails to complete a full stroke when k_m exceeds 5, thus the upper bound. The values for the variables are distributed in an exponential manner throughout the ranges, and the results are plotted in 3D. Two best-fit surfaces are generated for both the C-SEA and the SEA using the "polyfit23" function. The contours of the surfaces are also shown in Fig. 5. The red dotted line divides the contours in two regions, most m_2 and k_m combinations to the top-right of the red line, see higher v_{max} for C-SEA. This indicates that for motors with both larger m_2 and k_m, the C-SEA usually have a clear advantage over the regular SEA.

5   Conclusion and Future Works

In conclusion, the C-SEA manages to generate nearly constant and lower ground reaction force than the regular SEA sacrificing the time taken to accelerate with minimal loss on jump height. The reduced ground force is beneficial to the structural integrity and means less material requirements for the end-effector. At the same time, when actuated by a realistic motor, C-SEA has a clear advantage in most cases with higher m_2 and k_m outperforming the SEA in terms of achieving high jumps because of the C-SEA's ability to compress the spring more thoroughly even when the charging stage is extended.

For future work, the findings in the paper will facilitate the development of a physical CVT gearbox that can be integrated with an SEA for robotic jumping and a prototype of such C-SEA enabled hopping robot will be constructed. Next steps for this research include: choosing a CVT type, parameter optimization for the CVT, designing and fabricating the hopping robot as a testing platform, developing a control strategy for such a robot and finally conducting comparative experiments. Promising CVT options include Evan's variable speed counter-shaft and the ratcheting CVT; the former boasts a simple and reliable structure, while the latter excels in transmitting impressive torque with minimal loss [2]. Furthermore, a monopedal hopping robot inevitably needs flight-phase balance control when the leg mechanism has only one degree of freedom. Ultimately, combining hopping with other locomotion methods such as gliding [14] could yield a versatile robot suitable for inspecting extreme environments inaccessible to humans.

References

1. Beckerle, P., Verstraten, T., Mathijssen, G., Furnémont, R., Vanderborght, B., Lefeber, D.: Series and parallel elastic actuation: influence of operating positions on design and control. IEEE/ASME Trans. Mechatron. **22**(1), 521–529 (2016)
2. Benitez, F., Madrigal, J., Del Castillo, J.: Infinitely variable transmission of racheting drive type based on one-way clutches. J. Mech. Des. **126**(4), 673–682 (2004)
3. Csomay-Shanklin, N., Dorobantu, V.D., Ames, A.D.: Nonlinear model predictive control of a 3D hopping robot: leveraging lie group integrators for dynamically stable behaviors. In: IEEE International Conference on Robotics and Automation, pp. 12106–12112 (2023)
4. Grimmer, M., Eslamy, M., Gliech, S., Seyfarth, A.: A comparison of parallel-and series elastic elements in an actuator for mimicking human ankle joint in walking and running. In: IEEE International Conference on Robotics and Automation, pp. 2463–2470 (2012)
5. Kernbaum, A.S., Kitchell, M., Crittenden, M.: An ultra-compact infinitely variable transmission for robotics. In: IEEE International Conference on Robotics and Automation, pp. 1800–1807 (2017)
6. Kim, J., Park, F.C., Park, Y.: Design, analysis and control of a wheeled mobile robot with a nonholonomic spherical CVT. Int. J. Robot. Res. **21**(5–6), 409–426 (2002)
7. Kovač, M., Fauria, O., Zufferey, J.C., Floreano, D., et al.: The EPFL jumpglider: a hybrid jumping and gliding robot with rigid or folding wings. In: 2011 IEEE International Conference on Robotics and Biomimetics, pp. 1503–1508. IEEE (2011)
8. Matsushita, K., Shikanai, S., Yokoi, H.: Development of Drum CVT for a wire-driven robot hand. In: IEEE/RSJ International Conference on Intelligent Robots and Systems, pp. 2251–2256 (2009)
9. Mooney, L., Herr, H.: Continuously-variable series-elastic actuator. In: IEEE International Conference on Rehabilitation Robotics, pp. 1–6 (2013)
10. Nobaveh, A.A., Herder, J.L., Radaelli, G.: A compliant continuously variable transmission (CVT). Mech. Mach. Theory **184**, 105281 (2023)
11. Pentzer, J., Brennan, S.: Investigation of the effect of continuously variable transmissions on ground robot powertrain efficiency. In: American Control Conference, pp. 4245–4250 (2012)
12. Plecnik, M.M., Haldane, D.W., Yim, J.K., Fearing, R.S.: Design exploration and kinematic tuning of a power modulating jumping monopod. J. Mech. Robot. **9**(1), 011009 (2017)
13. Shin, W.D., Stewart, W., Estrada, M.A., Ijspeert, A.J., Floreano, D.: Elastic-actuation mechanism for repetitive hopping based on power modulation and cyclic trajectory generation. IEEE Trans. Rob. **39**(1), 558–571 (2022)
14. Sun, J., Yang, J., Richardson, R., Zhou, C.: Gliding mechanism with hinged aeroelastic wings. In: The 5th UK-RAS Conference, pp. 80–81 (2022)
15. Sun, J., Zhou, C.: AeroTail: a bio-inspired aerodynamic tail mechanism for robotic balancing. In: The 28th International Conference on Automation and Computing, pp. 436–441 (2023)
16. Zhao, J.: MSU jumper: a single-motor-actuated miniature steerable jumping robot. IEEE Trans. Rob. **29**(3), 602–614 (2013)

Few-Shot Transfer Learning for Deep Reinforcement Learning on Robotic Manipulation Tasks

Yuanzhi He$^{(\boxtimes)}$ ⓘ, Christopher D. Wallbridge ⓘ, Juan D. Hernndez ⓘ, and Gualtiero B. Colombo ⓘ

School of Computer Science and Informatics, Cardiff University, Cardiff, UK
{HeY65, WallbridgeC, HernandezVegaJ, ColomboG}@cardiff.ac.uk

Abstract. Robot manipulation with simulation has become a mainstream approach in the robotics field recently. It entails lower risk and cost compared to direct training a real robot. Various physics engines, such as MuJoCo, offer simulated environments tailored for robot manipulation tasks. As the robotics field rapidly grows, model complexity and training times increase exponentially to meet the demands of diverse tasks. Solving this is challenging as it requires complex models and long training times. Deep Reinforcement Learning (DRL) is the current best-performing way to solve robot manipulation problems. However, although certain algorithms utilized automated curriculum learning to tackle multi-task robot manipulation problems, the models were still too complex to be solved with one training from scratch with acceptable accuracy and reasonable training time. To address this, we introduce a novel few-shot Transfer Learning (TL) technique for DRL that applies both Forward Transfer (FT) and Reverse Transfer (RT). TL facilitates breaking down a complex problem into easier-to-solve subproblems and transferring the acquired knowledge to more complex ones. Our TL method appears able to accelerate the training process for all the MuJoCo Fetch tasks, while even improving performance by 20% and accelerating 85% for the most complex FetchSlide environment.

Keywords: Transfer Learning · Deep Reinforcement Learning · Robot Manipulation · Physics Engine

1 Introduction

In recent years, robotic experiments are often conducted in simulated environments rather than real ones due to cost and risk concerns, as well as the need for enhanced control and efficiency [1]. OpenAI Gym [2] is an open-source library which was created to simplify access to most of the latest Reinforcement Learning (RL) algorithms and soon became a significant platform for robotic simulated environments in particular with manipulation tasks (Fetch, Hand) [2,3]. Among the various physics engines introduced to build simulated environments

M. N. Huda et al. (Eds.): TAROS 2024, LNAI 15052, pp. 85–92, 2025.
https://doi.org/10.1007/978-3-031-72062-8_8

for robot manipulation tasks [1], MuJoCo presents high flexibility and accuracy for simulating robot dynamics and contact forces [4]. However, many robot training processes are still extremely time-consuming, thus often resulting in under-performing and data-dependent outcomes. Within recent Machine Learning (ML) developments in robotics, including control systems and motion planning, significant improvements have been brought about by Deep Reinforcement Learning (DRL) models, which are widely used during the training process. This involves models that are large, complex and increasingly sophisticated, leading to high resource demands and long training processes [5].

State-of-the-art DRL implementations in robotic manipulation tasks obtain good results, yet certain tasks, including FetchSlide, have only been partially solved or require excessively long times to accomplish [3,6]. To enable a more efficient and better-performing training process, we introduce a Few-shot Transfer Learning (TL) approach, utilizing a set of robotic MuJoCo tasks as testbeds. Instead of starting the learning process from scratch, TL was introduced to break down a complex problem into sub-tasks and transfer the knowledge acquired from solving simpler primary tasks to increasingly complex ones. The scope of this study is to investigate the effectiveness of TL in accelerating the training process while maintaining or improving performance across all the benchmarks considered.

2   Related Work

Transfer Learning (TL) is a Machine Learning (ML) technique that involves leveraging useful features gained from solving one task to improve performance on a related task. It relates to a number of different subcategories including Imitation Learning (IL), Multi-Task Learning (MT) and Goal-oriented Learning. Current research shows great potential for TL in the ML field [8,9]. The TL approach can be categorised into three classes: Zero-shot, Few-shot and Sample-efficient. Zero-shot TL enables the development of an agent that can be directly applied to the target domain without requiring any training interaction [10]. The Sample-efficient TL is designed to accelerate the learning process of models through the rational use of data, thereby reducing data requirements and training costs [11]. Unlike Zero-shot and Sample-efficient TL, Few-shot TL enables only a few interactions to keep more useful features and adapts more effectively to the specific characteristics of the target task [12].

To the best of our knowledge, the application of IL to robotics is mostly limited to approaches implementing zero-shot and few-shot transfer [9,13,14]. However, although IL learns high-quality behavioural policies directly from expert demonstrations, thus making it particularly suitable for tasks requiring precise execution, it is still expert-dependent and time-consuming. In addition, because of the high complexity of these types of robotic problems, traditional RL solutions are not able to achieve satisfactory results.

More recently, Deep Reinforcement Learning (DRL) has been introduced to empower robots with a greater awareness and understanding of the environment. DRL appears successful in learning adaptive strategies from interactions with the environment, which makes it applicable to various robotic tasks [15]. Current mainstream DRL algorithms have been divided into two types: on-policy and off-policy algorithms [16,17]. The latest on-policy DRL algorithms include Proximal Policy Optimization (PPO) [18], enabling better convergence and robustness of robots, especially in policy optimization. However, due to the scale of most robotic models being quite large, off-policy DRL algorithms including Soft Actor-Critic [19] and Truncated Quantile Critics (TQC) [20] were created to tackle this problem. Hindsight Experience Replay (HER) [21] was also introduced to help the challenges related to inaccuracies and large action spaces by replacing the desired goal with the achieved goal and storing their experience replay in the replay buffer [3]. To ensure a fair result comparison and make RL algorithms easy to use, the Stable-Baselines3 (SB3) and RLlib libraries were designed and assembled with most of the mainstream DRL algorithms and simulated environments, including robotic environments [22,23].

Finally, a set of simulated MuJoCo Fetch environments (Reach, Push, PickAndPlace and Stack) were implemented with alternative Goal-Oriented approaches [7,24,25]. These are often based on intrinsically motivated and curiosity-driven rewards while other proposed solutions applied forms of Multi-task Learning primarily focused on training a single model with multiple tasks with shared knowledge flow [26,27].

Although all the methods and techniques mentioned in this section were able to produce some improvements on the state of the art, as the complexity of RL algorithms grows exponentially there is still a need for enhancing the performance of the current single-environment techniques.

3   Proposed Approach

Our proposed approach aims to solve simulated robotic manipulation tasks using TL (transferring the weights and biases of the first layer to the target tasks) as an alternative to DRL model training in isolation from scratch. The experiments considered are the set of MuJoCo manipulation Fetch environments provided by OpenAI Gym [2]. These include FetchReach-v2 (reaching a target point), FetchPush-v2 (pushing a block to a target point), FetchSlide-v2 (hitting a puck to reach a target point on a long and slippery table) and FetchPickAndPlace-v2 (picking an object from a source point and placing it to a target point). RL Baselines3 Zoo is a benchmark with metrics created by SB3 [22] showing that TQC [20] currently has the best performance on success rate and rewards metrics on all of the above environments [19]. This method accepts states, actions and goals as the input of the 'critic' network to calculate the inputs to the hidden 'actor' layers while the related goals and rewards are computed applying the HER technique [3,21].

Figure 1 shows an example of the TL process with Forward (Single and Double) and Reverse Transfer with the FetchReach environment selected as the primary task. Unlike the weights and biases that need to be initialized as random values when applying a DRL model in isolation, in TL a proportion of

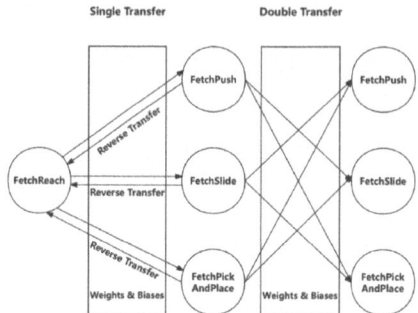

Fig. 1. Forward (Single, Double) and Reverse Transfer Learning.

the pre-trained parameters, i.e. weights and biases, are transferred between the models solving the primary and target tasks. To explain these operations in simpler terms we can observe that more complex tasks such as Pick-and-Place can be reduced to a combination of sub-tasks, one being the Reach tasks itself. Therefore if a model has already learnt through training how to complete the reaching action, then TL aims to exploit this already acquired knowledge so that the remaining training process (or indeed retraining) could only focus on the remaining sub-tasks. This is expected to require shorter training times and eventually also lead to gains in performance. It is worth noting that once some knowledge has been acquired, the related pre-trained parameters could be transferred to a number of different target tasks in parallel potentially producing further significant gains in runtime.

Existing research showing affinity to TL focuses either on trying to balance the training of the same model by using different tasks to compute the losses, similar to multi-objective optimization approaches (multi-task learning [26,27]), or on exploiting the benefits of training a model through a sequence of increasingly complex tasks (goal-oriented computing [7,24,25]). On the contrary, the principal aim of our research switches the focus to the actual mechanisms used to transfer the knowledge, namely identifying what subsets of the pre-trained parameters need to be transferred, and how these stand in relation to the different tasks and sub-tasks. When only one transfer for each training is performed this is referred to as Single Transfer. Double Transfer obtains the useful features from a trained non-primary model to another non-primary model and finally Reverse Transfer transfers the useful features from a non-primary model back to the primary one. The latter seeks to validate to what extent the features learnt during more complex tasks will remain useful to solve basic ones (in other terms if solving complex tasks could cause some basic knowledge loss).

4   Experiments and Results

The MuJoCo Fetch environments have been implemented for TL. The latest version-2 release has been adopted because of its wider MuJoCo support. Because

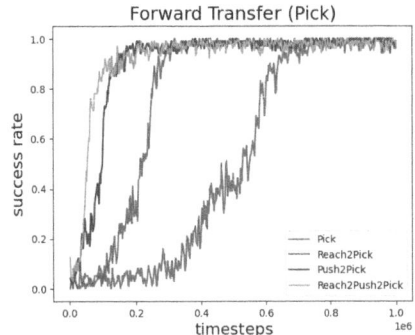

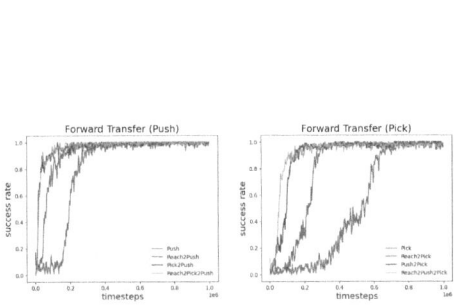

Fig. 2. Success rate vs timesteps for FetchPush-v2 with Single (Reach-> Push and Pick-> Push) and Double TL (Reach-> Pick-> Push) and without TL.

Fig. 3. Success rate vs timesteps for FetchPickAndPlace-v2 with Single (Reach-> Pick and Push-> Pick) and Double TL (Reach-> Push-> Pick) and without TL.

of its simplicity and reliability, the Stable-Baselines3 (SB3) has been utilized as a benchmark for performance comparison. To prevent uncertainty, all the experiments gathered the models with the best performance from 3 training sessions. The learning rate and batch size remain unchanged for all environments, which are 0.001 and 256 as suggested by the benchmark RL Baselines3 Zoo from SB3. We use TQC with HER as the main Deep Reinforcement Learning algorithm for environments with sparse rewards. The total training timesteps are different due to the complexity of each task. They are set to match the value of the benchmark SB3. The experiments were performed on a PC, with 64 GB RAM, an Intel Core i7-14700K 2.5 GHz CPU and an RTX 4060 GPU.

Forward Transfer. Figures 2, 3 and 4 present the results of Single and Double Transfer from FetchReach to the other Fetch environments. By analyzing Figs. 2 and 3, we can observe that the training approaches applying models pre-trained on both FetchPush and FetchPickAndPlace environments outperform those without any pre-training in runtime terms. Whether it is a Single or Double Transfer, from FetchPickAndPlace to FetchPush and vice versa, TL is always able to produce a faster convergence. Furthermore, by analyzing Fig. 4, on the FetchSlide task (the most complex of the Fetch environments), the model pre-trained with FetchPick is able to achieve the best overall performance by producing the highest success rate. It is worth noting that the pre-trained models with Double Transfer outperform those with Single Transfer, although even a simple pre-training with FetchReach produces a better performance than the model trained from scratch without any transfer applied.

Reverse Transfer. Figure 5 presents the result of Reverse Transfer from non-primary models (FetchPush, FetchSlide and FetchPickAndPlace) to FetchReach. It shows that the models pre-trained by FetchPush, FetchPickAndPlace and FetchSlide all outperform the original single-environment model by scoring success rates equal to or near to 1 compared with 0.91. However, the pre-trained

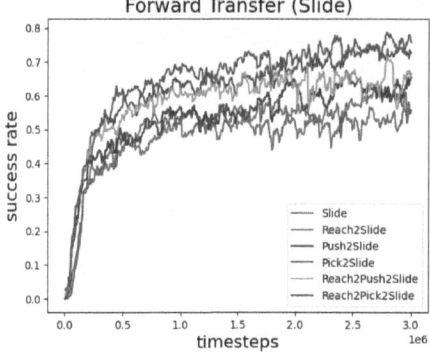 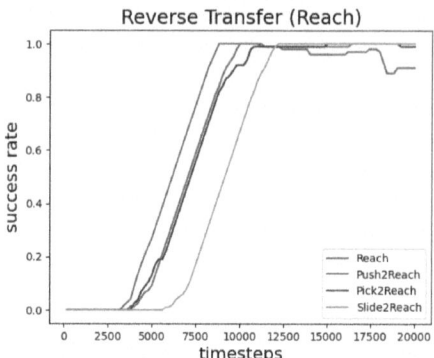

Fig. 4. Success rate vs timesteps for FetchSlide-v2 with Single (Reach-> Slide, Push-> Slide and Pick-> Slide) and Double (Reach-> Push-> Slide and Reach-> Pick-> slide) TL and without TL.

Fig. 5. Success rate vs timesteps for FetchReach-v2 with Reverse Transfer (Push-> Reach, Pick-> Reach and Slide-> Reach) and without TL.

models from FetchPickAndPlace and FetchSlide converge slower than the model trained from scratch. This is because their complexity is higher than FetchPush indicating that a potential overtraining might have happened. As a result, models starting with pre-training on more complex tasks may eventually result in affecting the converging speed, although eventually leading to higher success rates.

5   Conclusion and Future Works

This study presents some preliminary results demonstrating encouraging improvements in applying both Forward (Single, Double) and Reverse TL to the robot training process. The proposed TL methods accelerate the original training process on single environments and are able to at least equalize the current benchmarks for each of the MuJoCo Fetch environments, while also scoring higher success rates for the most complex task (FetchSlide). This suggests great potential for further TL applications to simulated Robotic tasks. Our research aims to provide more insights on the actual mechanism of transfer, by focusing on which subsets of pre-trained parameters need to be transferred between tasks and how this can be achieved.

Future work will focus on the application of the TL method to more complex robotic environments (putting objects on a shelf [28]) and alternative physical engines such as Gazebo [29]. In addition, future research will also attempt to close the gap from simulation to real environments through the so-called Sim2Real Transfer [11]. Finally, we can also consider transfer between models solving the same task, albeit implementing different forms of rewards. These could be based on different principles, such as novelty and curiosity, or on auxiliary intermediate goals, for example generated using large language models [30,31].

References

1. Erez, T., Tassa, Y., Todorov, E.: Simulation tools for model-based robotics: comparison of bullet, Havok, Mujoco, Ode and Physx. In: 2015 IEEE International Conference on Robotics and Automation (ICRA), pp. 4397–4404. IEEE (2015)
2. Brockman, G., et al.: OpenAI Gym. arXiv preprint 2016 (2016)
3. Plappert, M., et al.: Multi-goal reinforcement learning: challenging robotics environments and request for research. arXiv preprint 2018 (2018)
4. Todorov, E., Erez, T., Tassa, Y.: MuJoCo: a physics engine for model-based control. In: IEEE/RSJ International Conference on Intelligent Robots and Systems 2012, pp. 5026–5033. IEEE, Vilamoura-Algarve (2012)
5. Soori, M., Arezoo, B., Dastres, R.: Artificial intelligence, machine learning and deep learning in advanced robotics, a review. Cogn. Rob. **3**, 54–70 (2023)
6. Yang, X., Ji, Z., Wu, J., Lai, Y.-K.: An open-source multi-goal reinforcement learning environment for robotic manipulation with Pybullet. In: Fox, C., Gao, J., Ghalamzan Esfahani, A., Saaj, M., Hanheide, M., Parsons, S. (eds.) TAROS 2021. LNCS (LNAI), vol. 13054, pp. 14–24. Springer, Cham (2021). https://doi.org/10.1007/978-3-030-89177-0_2
7. Colas, C., Fournier, P., Chetouani, M., Sigaud, O., Oudeyer, P.Y.: CURIOUS: intrinsically motivated modular multi-goal reinforcement learning. In: International Conference on Machine Learning, pp. 1331–1340. PMLR (2019)
8. Pan, S.J., Qiang, Y.: A survey on transfer learning. IEEE Trans. Knowl. Data Eng. **22**, 1345–1359. IEEE (2009)
9. Jaquier, N., et al.: Transfer learning in robotics: an upcoming breakthrough? A review of promises and challenges. arXiv preprint 2023 (2023)
10. Kirk, R., Zhang, A., Grefenstette, E., Rocktäschel, T.: A survey of zero-shot generalisation in deep reinforcement learning. J. Artif. Intell. Res. **76**, 201–264 (2023)
11. Zhu, Z., Lin, K., Jain, A.K., Zhou, J.: Transfer learning in deep reinforcement learning: a survey. IEEE Trans. Pattern Anal. Mach. Intell. **45**, 13344–13362 (2023)
12. Song, Y., Wang, T., Cai, P., Mondal, S.K., Sahoo, J.P.: A comprehensive survey of few-shot learning: evolution, applications, challenges, and opportunities. ACM Comput. Surv. **55**, 1–40 (2023)
13. Hundt, A., et al.: "Good robot! Now watch this!": repurposing reinforcement learning for task-to-task transfer. In: 5th Annual Conference on Robot Learning, pp. 1564–1574. PMLR (2022)
14. Jang, E., et al.: BC-Z: zero-shot task generalization with robotic imitation learning. In: Conference on Robot Learning, pp. 991–1002. PMLR (2022)
15. Han, D., Mulyana, B., Stankovic, V., Cheng, S.: A survey on deep reinforcement learning algorithms for robotic manipulation. Sensors **23**(7), 3762 (2023)
16. Wang, H.N., et al.: Deep reinforcement learning: a survey. Front. Inf. Technol. Electron. Eng. **21**(12), 1726–1744 (2020)
17. Wang, X., et al.: Deep reinforcement learning: a survey. IEEE Trans. Neural Netw. Learn. Syst. (2022)
18. Schulman, J., Wolski, F., Dhariwal, P., Radford, A., Klimov, O.: Proximal policy optimization algorithms. arXiv preprint 2017 (2017)
19. Haarnoja, T., Zhou, A., Abbeel, P., Levine, S.: Soft actor-critic: off-policy maximum entropy deep reinforcement learning with a stochastic actor. In: International Conference on Machine Learning, pp. 1861–1870. PMLR (2018)
20. Kuznetsov, A., Shvechikov, P., Grishin, A., Vetrov, D.: Controlling overestimation bias with truncated mixture of continuous distributional quantile critics. In: International Conference on Machine Learning, pp. 5556–5566. PMLR (2020)

21. Andrychowicz, M., et al.: Hindsight experience replay. In: Advances in Neural Information Processing Systems on NeurIPS Proceedings. NIPS (2017)
22. Raffin, A., Hill, A., Gleave, A., Kanervisto, A., Ernestus, M., Dormann, N.: Stable-Baselines3: reliable reinforcement learning implementations. J. Mach. Learn. Res. **22**(268), 1–8 (2021)
23. Liang, E., et al.: RLlib: abstractions for distributed reinforcement learning. In: International Conference on Machine Learning, pp. 3053–3062. PMLR (2018)
24. Colas, C., Karch, T., Sigaud, O., Oudeyer, P.Y.: Autotelic agents with intrinsically motivated goal-conditioned reinforcement learning: a short survey. J. Artif. Intell. Res. **74**, 1159–1199 (2022)
25. Wang, R., Lehman, J., Clune, J., Stanley, K.O.: POET: open-ended coevolution of environments and their optimized solutions. In: Proceedings of the Genetic and Evolutionary Computation Conference, pp. 142–151. ACM (2019)
26. Zhang, Y., Yang, Q.: A survey on multi-task learning. IEEE Trans. Knowl. Data Eng. **34**, 5586–5609 (2021)
27. Ruder, S.: An overview of multi-task learning in deep neural networks. arXiv preprint 2017 (2017)
28. Ecoffet, A., Huizinga, J., Lehman, J., Stanley, K.O., Clune, J.: First return, then explore. Nature **590**(7847), 580–586 (2021)
29. Koenig, N., Howard, A.: Design and use paradigms for gazebo, an open-source multi-robot simulator. In: 2004 IEEE/RSJ International Conference on Intelligent Robots and Systems (IROS) (IEEE Cat. No. 04CH37566), vol. 3, pp. 2149–2154. IEEE (2004)
30. Du, Y., et al.: Guiding pretraining in reinforcement learning with large language models. In: International Conference on Machine Learning, pp. 8657–8677. PMLR (2023)
31. Colas, C., Teodorescu, L., Oudeyer, P.Y., Yuan, X., Côté, M.A.: Augmenting autotelic agents with large language models. In: Conference on Lifelong Learning Agents, pp. 205–226. PMLR (2023)

A ROS-Based Control Framework
for Simulating Locomotion of a Multi-arm
Space Assembly Robot

Sairaj R. Dillikar[1]([⊠])🆔, Cameron Leslie[2]🆔, Saurabh Upadhyay[2]🆔,
Leonard Felicetti[2]🆔, and Gilbert Tang[2]🆔

[1] Cranfield University, Cranfield, UK
`sairaj.dillikar.102@gmail.com`
[2] School of Aerospace, Transport and Manufacturing (SATM), Cranfield University,
Cranfield, UK
{`Cameron.F.Leslie,Saurabh.Upadhyay,Leonard.Felicetti,`
`g.tang`}`@cranfield.ac.uk`

Abstract. This paper proposes a ROS-based control framework for simulating the locomotion of a multi-arm space robot on a planar space structure. This framework has applications in technology demonstrations of space structure assembly, construction, and maintenance where a multi-arm robot has to traverse on a space structure building blocks known as Spatial Reticular Structures (SRS) to perform the space operation. The framework sets the desired environment and SRS structures in the first step and devises two movement primitives to achieve locomotion on the structure. A ROS-Gazebo-based simulation architecture is presented to execute the proposed control framework. The viability and limitations of the proposed control framework are discussed with simulation results.

Keywords: On-orbit operations · Manipulation in space · Space robotic locomotion · Multi-arm robot · Multi-robot cooperation/collaboration · Robot Operating System (ROS)

1 Introduction

Recent space mission studies have considered multi-arm robots to carry out the assembly tasks in orbit, e.g., the RAMST [1], MOSAR [2], PULSAR [3]. In these missions, a multi-arm robot walks or climbs on top of a spacecraft carrying a modular structure, such as a telescope mirror tile, which will be assembled onto the main structure while in orbit. Here, the space robot's locomotion is assisted by a simple mechanical coupling between the robot's end-effector and the Standard Interface (SI) on the mirror tiles. The locomotion of multi-arm space robots on the built structure is a crucial challenge to overcome [4], and these space robots must rely on a reliable control and simulation framework that can perform the coordinated multi-arm control and simulate the locomotion.

A typical coordinated control strategy for multi-arm utilises the primary arm for the specific defined aim and the secondary arm providing assistance. One such scenario is

M. N. Huda et al. (Eds.): TAROS 2024, LNAI 15052, pp. 93–105, 2025.
https://doi.org/10.1007/978-3-031-72062-8_9

capturing a moving object using coordinated control [5], where the method involves two cases: a) capturing the target when the base is free-floating and b) single-arm capturing while keeping the base's inertial pose using another arm. A free-flying dual-arm space robot utilises a coordinated motion control strategy for assembling modular parts in space, demonstrated using an air-bearing simulator [6]. They introduced two phases to accomplish the assembly task, in which coordinated motion planning is performed based on an optimisation method. The first phase (pre-assembly phase) establishes the dual-arm robots' General Relative Jacobian Matrix (GRJM, [7]) to manipulate and reach relative positions and attitudes cooperatively. For the final phase, one arm is kept at rest with respect to its base, and the other arm moves in a specified trajectory to achieve the final assembly.

The aforementioned approaches inspired the basis for the space robot to perform coordinated locomotion in this work, as the primary manipulator advances to take the next step while the secondary manipulator provides assistance or forces in the opposite direction, displacing the entire robot. Furthermore, two locomotion strategies are introduced based on bio-inspired gait mechanisms. The lateral movement is similar to the inchworm mechanism [8], while the vertical movement is adapted from the biped mechanism due to their resemblance to humans [9].

The viability of the introduced locomotion strategies will be demonstrated using the ROS-Gazebo-based simulation. In terms of ideation & developing the simulation, some insights from recently developed ROS-based projects were obtained for the locomotion strategies of lateral movement (inchworm mechanism) [10] and vertical movement (biped mechanism) [11, 12]. Moreover, a typical coordinated control demonstration utilising dual-arm configured Universal Robots (UR) manipulators can be seen from a ROS-based simulation developed by [13], which provides insights for the development of the controller setup & dual-arm configuration for multi-arm control.

Strong underlying fundamentals in multi-arm configured space robots and simulation architecture to achieve the on-orbit operations can be understood from [14], also a backbone of this research. Furthermore, the current work provides simulated validation, mainly focusing on addressing the critical challenges associated with achieving mobility and multi-arm coordination of the aforementioned project.

The paper's main contribution is to showcase the simulation of a multi-arm space robot performing locomotion on a planar space structure. Two locomotion strategies are introduced and devised for the space robot to traverse a planar space structure. Finally, a simulation based on a Robot Operating System (ROS) and Gazebo was developed to validate the locomotion strategies.

The remainder of the paper is structured as follows: Section 2 sets the background for this paper by discussing the configuration of the robot, the space structure, and the stages of implementation through the technical platforms. Section 3 explains the control architecture, discusses the basis of derived approaches for the required locomotion through multi-arm coordination, and shows the implementation carried out through the ROS-Gazebo-based simulation. Results of the developed project are discussed briefly in Sect. 4, addressing all the approaches. The paper is concluded by providing insights for the future scope in Sect. 5.

2   Problem Overview

The configuration of Multi-Arm Robot for In-Orbit Operations (MARIO) comprising of three manipulators positioned 120 degrees from each other's base (Torso, see Fig. 1a) (for more insights on robot's configuration refer [14]). Here, two robotic arms are necessary to locomote the MARIO across the structure, whilst the third arm carries a single Spatial Reticular Structures (SRS) module. The Standard Interfaces (SI) (see Fig. 1b) acts as a coupling mechanism between the robotic arm's end-effectors and the SRS modules.

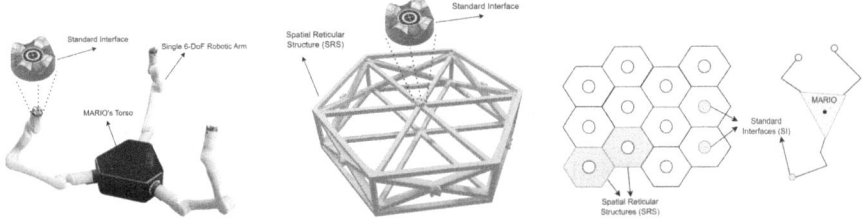

(a)   Multi-Arm   Robot   (b)   Spatial   Reticular   (c) Illustration of a 2-D Grid of
for In-Orbit Operations Structure (SRS)   SRS Structures
(MARIO)

Fig. 1. Graphical Representation of MARIO and SRS Structures

With this, consider SRS structures in a 2-D grid plane as shown in Fig. 1c; each SRS and robotic arm has an SI unit attached, allowing the robot arm to be docked on an SRS. The coordinates of every SRS and the robot torso are available at all times from the simulator, along with considering the zero gravity constant for the simulation. The development of the simulation framework considers the following two sub-problems:

1. Determine and devise appropriate locomotion strategies for the multi-arm robot.
2. Develop a ROS-Gazebo-based simulation to validate the control framework to achieve locomotion of the space robot.

3   Implementation of Control Architecture Through the ROS

This section discusses the control architecture for locomotion and its execution in ROS. The pipeline for control architecture is illustrated in Fig. 2, and it comprises three major tasks:

1. Spawning of the SRS models
2. Obtaining the coordinates of SRS models
3. Executing the locomotion sequence

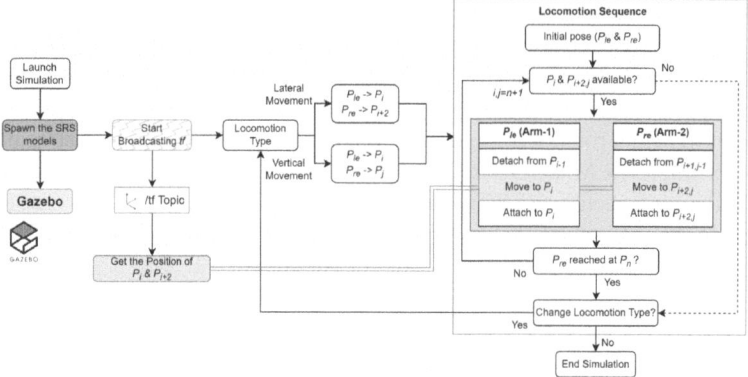

Fig. 2. Control Architecture

3.1 Spawning of the SRS Models

The first step during the simulation is to spawn the SRS models as SDF[1] type, and these models are spawned within the gazebo using */gazebo/spawn_sdf_model* service from the ROS services. The models are spawned for both lateral and vertical configurations, wherein for lateral locomotion configuration, the SRS is placed in a series manner along a single axis equidistantly as shown in the Fig. 3a. The vertical locomotion configuration from Fig. 3b requires the SRS to be placed on either side linearly to lie within the workspace of two active arms.

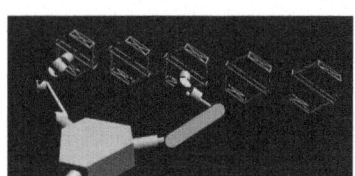

(a) Spawned SRS Models in Lateral Configuration

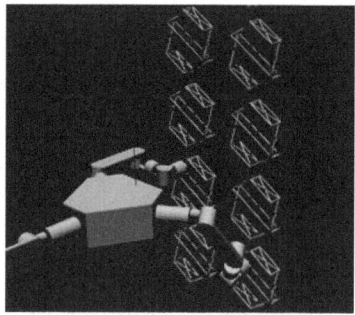

(b) Spawned SRS Models in Vertical Configuration

Fig. 3. SRS Models Spawned in Two Configurations

---

[1] SDF - Separate XML-based Description Format for gazebo.

3.2   Obtaining the Coordinates of SRS Models

The second-generation transformation frames (tf2) library[2] keeps track of multiple coordinate frames within the simulation over time and their relationship in a tree structure. Every model has its link frames[3] and the exact location of a particular model can be determined by looking up at the transformation frames tree structure. Since MARIO and all the SRS models are within the world frame, it is required such that only the locomotion is performed in terms of MARIO's torso frame (local frame of the multiarm robot (see Fig. 4a)). The relative pose and orientation of SRS coordinate frames only with respect to the MARIO's torso frame are being broadcast to the */tf* topic (see Fig. 4b & 4c for all link frames in both locomotion configuration).

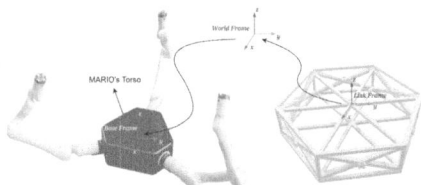

(a) Obtaining the Frame Transformation from SRS to MARIO's Torso

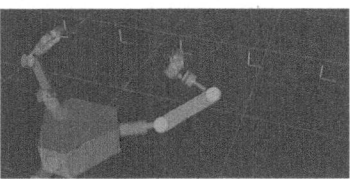

(b) Link Frames of SRS Models in Lateral Configuration

(c) Link Frames of SRS Models in Vertical Configuration

Fig. 4. Frame Transformation and Localised Link Frames

Additionally, to find a relationship between any two link frames, it is possible to look at the tree structure at any given time, which is maintained by the */tf* topic. The pose and orientation of a particular SRS model relative to the torso's frame of reference are obtained through the frame transformation from the world frame. Once the coordinates (way-point) of the SRS model are transformed, the respective manipulator will be directed towards the SRS only if it is located within the workspace.

---

[2] tf2-ros (Transformation Frames) - http://wiki.ros.org/tf2.

[3] A frame is the representation of Cartesian coordinate axes.

3.3   Locomotion Strategies and Sequence Execution

To perform locomotion on a planar structure composed of an SRS grid, we propose two locomotion primitives, namely lateral and vertical movements. A robot performing these two primitives can reach any point on the 2-D grid plane. These two primitives are discussed as follows:

Lateral Movement: The lateral movement scenario considers n SRS positioned linearly with their SI positions P_i, where $i = \{2, 3, \ldots n\}$ and the robot is moving from one SRS (left) to another SRS (right) in sequence shows a lateral movement as illustrated in Fig. 5. Two of the manipulators of MARIO perform this lateral movement, displacing its torso along the direction of the SRS sequence using an inchworm mechanism.

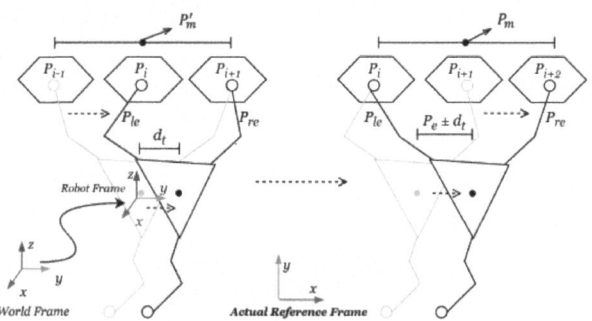

Fig. 5. Lateral Movement

For explanation, consider that the robot's left and right arms are docked on the positions P_{i-1} and P_{i+1}, respectively. The steps to achieve this movement are described as follows:

Step-1 $\rightarrow$ The left arm of the robot is moved from P_{i-1} to P_i. The left (P_{le}) and right (P_{re}) end-effector positions relative to the torso's reference frame can be calculated as,

$$P_{le} = P_i + d_s \tag{1}$$
$$P_{re} = P'_{re} - d_t \tag{2}$$

where d_s is the distance between SI position & the end-effector (due to empirical knowledge, it is a practical offset adapted within the simulation environment), d_t is the offset required to displace the torso applied to the non-moving arm, and P'_{re} is the previous position values of right end-effector.

Step-2 $\rightarrow$ The second step aligns the torso between both arms. The midpoint is calculated during each iteration, and this step is achieved initially by subtracting the mid-point of both end-effectors during the previous (P'_m) & current pose (P_m) as follows,

$$P_m = \frac{P_{le} + P_{re}}{2} \tag{3}$$

$$P_e = P_m - P'_m \tag{4}$$

During the same instance, the midpoint offset (P_e) is added to both the end-effectors because this results in the coordinated motion of both the arms and, ultimately, obtaining the net displacement required for the torso, as stated below,

$$P_{le} = P_{le} - P_e \pm d_t \tag{5}$$

$$P_{re} = P_{re} - P_e \pm d_s \tag{6}$$

However, during each locomotion cycle the P_{i+2} could be out of reachable workspace for P_{re}, and it would be only possible if P_{le} coordinates such that the link frame of the P_{i+2} lies within the workspace of the P_{re}, provided that the end-effector's pose and orientation remains the same.

Vertical Movement: The vertical movement scenario considers n SRS positioned on either side linearly with their SI positions P_i & P_j for P_{le} & P_{re} respectively, where $i, j = \{1, 2, 3, \ldots n\}$. Both arms move sequentially with their respective SRS, and the resultant displacement of the torso resembles a biped mechanism, as shown in Fig. 6.

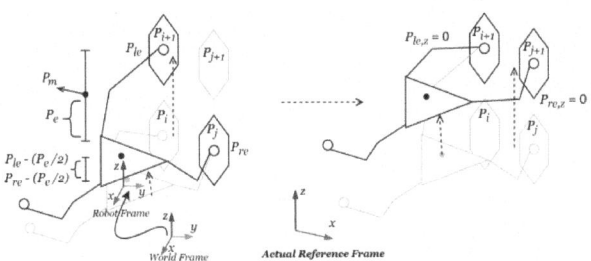

Fig. 6. Vertical Movement

Initially, consider the robot's arms P_{le} & P_{re} are docked on the SI positions $i, j = 1$. The steps to achieve this movement involve:

Step-1 $\rightarrow$ P_{le} moves to P_{i+1}, by considering left end-effector's previous (P'_{le}) & current positions (P_{le}), the mid-point (P_m) and the mid-point offset (P_e) is obtained from (8) & (9), respectively.

$$P_{le} = P_{i+1} \tag{7}$$

$$P_m = \frac{P_{le} + P'_{le}}{2} \tag{8}$$

$$P_e = P_m - P'_{le} \tag{9}$$

Now, P_e is added to both the end-effectors to achieve simultaneous motion, and can be stated with the (10) & (11) as follows,

$$P_{le} = P_{le} - \frac{P_e}{2} \tag{10}$$

$$P_{re} = P_{re} - \frac{P_e}{2} \tag{11}$$

Step-2 → During the second step, where P_{re} moves to P_{j+1}, aligning the torso by bringing both arms end-effectors to the same plane as torso (this motion can be achieved by simply equating the z-axis[4] ($P_{le,z}$ & $P_{re,z}$) of both end-effectors as zero, shown in (12) & (13)).

$$P_{le,z} = 0 \tag{12}$$

$$P_{re,z} = 0 \tag{13}$$

Note that to perform these steps in simulation, the equations from (1) to (13) for both end-effectors P_{le} and P_{re} are implemented within the scope of the programming. By sequencing these two locomotion primitives, MARIO can reach any point on a planar surface grid; as a result, combined locomotion generalises the proposed strategy to complex locomotion scenarios on a plane. A use case for this approach is discussed in Sect. 4.2.

To execute the locomotion, the sequence of standard tasks to be performed for both the arms within a single locomotion sequence cycle are mentioned as follows (refer Fig. 2 for locomotion sequence),

Step-1 → Detach from $P_{i-1,j-1}$
Step-2 → Moving the Active arm to $P_{i,j}$
Step-3 → Attach to $P_{i,j}$

At the beginning of the simulation, the active arms must move to the initial pose so that the arm's joint configuration is well suited to perform the locomotion as planned. The notation for SRS models is different for both lateral and vertical scenarios, where P_i is only considered in lateral movement because of the single linear arrangement of the SRS models, but P_j along with P_i in vertical movement for the double linear arrangement of the SRS models.

During the locomotion of MARIO across the SRS structures, it is required that the end-effectors are docked to their respective SRS, forming a tight grasp so that if the arms are manipulated, the resultant motion is affected on the MARIO's torso. PAL Robotics[5] has developed a plugin *gazebo_ros_link_attacher* [15] that attaches/detaches multiple objects as stated in Step-1 & 3 by creating/deleting an additional joint between any two links, by utilising this plugin, the coupling mechanism in simulation can be achieved.

---

[4] Axis depends on the progress of motion and orientation of the end-effectors w.r.t the torso.
[5] PAL Robotics - https://pal-robotics.com/.

As mentioned in Step-2, the manipulators must be configured with appropriate controllers to direct the arms towards the SRS within the ROS simulation. This requirement can be accomplished using *position_controller* under the *JointTrajectoryController* because this controller receives the command in the desired position as input (as the requirement is feeding the way-point of a particular SRS model within the robot's workspace) and sends the position output.

4   Simulation Architecture and Results

This section explains the simulation architecture, followed by simulation results.

4.1   Simulation Architecture

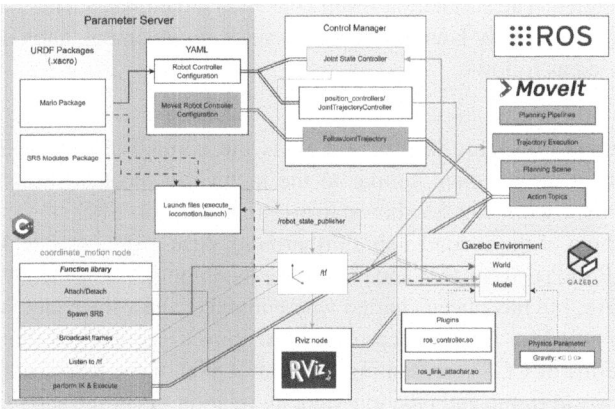

Fig. 7. ROS-Gazebo-Based Simulation Architecture

The validation of devised locomotion strategies is achieved with ROS-Gazebo-based simulation. The initial stage carried out to develop the simulation involving multi-arm-based robotic models requires the generation of necessary & working URDF packages, and the working arm groups within its semantic description file that are responsible for locomotion should be described as subgroups within a parent group so that multi-arm coordination is in effect. Furthermore, integrating the necessary plugins within the workspace ensures the appropriate working of the docking/coupling mechanisms.

The control framework was built through programming (C++) while inheriting the required functionality through essential libraries and packages such as MoveIt! and transformation frames. Since the motive for MARIO is to perform on-orbit operations, it is desired to simulate zero gravity conditions that can be established by simply zeroing the components of gravity function in the Gazebo physics simulator. Finally, Fig. 7 briefly illustrates the association of the aforementioned ROS development and connects it with the simulation environment.

The simulation was performed on a computer with Intel(R) Core™ i7-9750H CPU @2.60GHz processor paired with 16 GB RAM and an NVIDIA GeForce GTX 1050 graphics card on Ubuntu 2020.04 platform. Furthermore, the project was developed based on C++, and the simulation was validated through ROS Noetic Ninjemys 8 with Gazebo v11.11.

4.2   Simulation Results of the Locomotion Types

The data from the ROS simulation was logged with the help of rosbag[6] package into a bag file, which is recorded by subscribing to the topics: */joint_states* and */gazebo/link_states* because the joint values and the end-effector's position in terms of CCS are published in the respective topics. Additionally, the plots were generated through PlotJuggler[7] The simulation result considers three main scenarios for performing the locomotion, namely lateral movement, vertical movement, and combined locomotion, discussed as follows:

Lateral Movement: Figure 8a, 8b, and 8c is the simulation representation of the approach for lateral movement similar to the inchworm mechanism as discussed in Sect. 3.3. To begin with, Fig. 8a shows Arm-1 reaching the SRS (P_2), and Fig. 8b & 8c is the transition of the second step. Furthermore, Arm-1 (P_{le}) assists Arm-2 (P_{re}) in reaching the SRS (P_4). To further understand the locomotion performed within the simulation, plots Fig. 11c have been charted where markers show the exact point of occurrence in the timeline for these 2-steps discussed above. Furthermore, these graphs are generated in terms of CCS representing the central axis of the torso and end-effectors relative to the world frame and position-velocity relation of end-effectors.

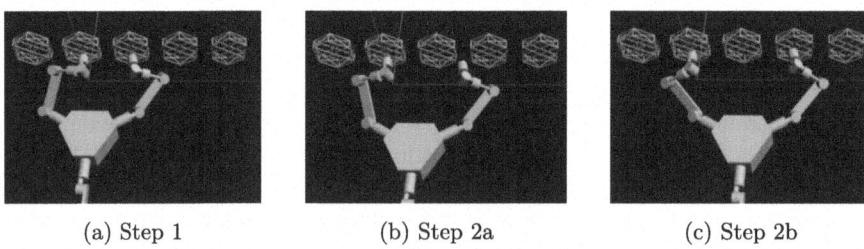

| (a) Step 1 | (b) Step 2a | (c) Step 2b |

Fig. 8. Lateral Movement

---

[6] rosbag - http://wiki.ros.org/rosbag.

[7] PlotJuggler - https://plotjuggler.io/.

Vertical Movement: The vertical movement is illustrated through Fig. 9a, 9b, & 9c, which is similar to the biped mechanism as derived in Sect. 3.3. The gradual increase in z-axis value from Fig. 11b indicates that the robot is moving along the z-axis.

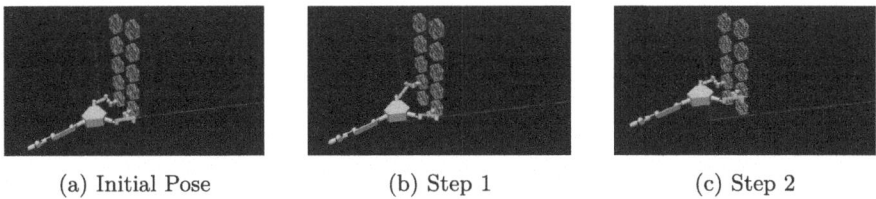

(a) Initial Pose   (b) Step 1   (c) Step 2

Fig. 9. Vertical Movement

Combined Locomotion. A combination of both locomotion types was tested and demonstrated to validate the flexibility of the locomotion control framework as mentioned in Sect. 3.3, where Fig. 10a shows the completion of the vertical movement, Fig. 10b & 10c shows the transition to lateral movement. Finally, Fig. 10d indicates that Arm-1 aims for SRS (P_2), and the lateral locomotion cycle continues until Arm-2 is reached at the last SRS model. The markers in Fig. 11c represent the switching event taking place in the timeline. The ROS-based simulation videos from a playlist can be accessed here[8] demonstrating the validation of locomotion approaches for a multi-arm robot.

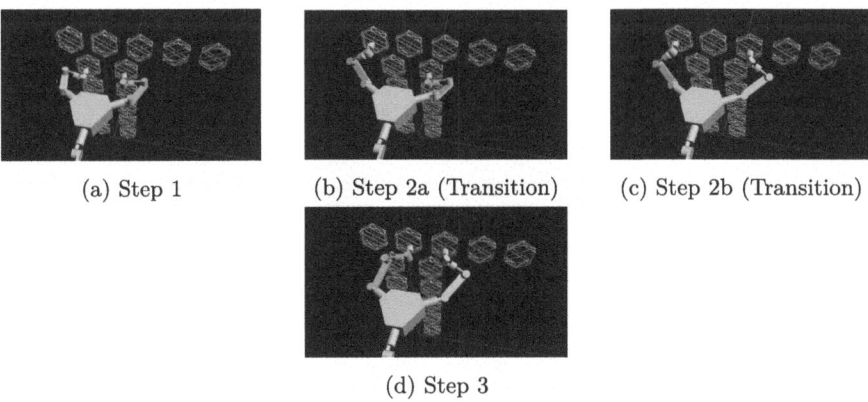

(a) Step 1   (b) Step 2a (Transition)   (c) Step 2b (Transition)

(d) Step 3

Fig. 10. Combined Locomotion

---

[8] Simulation Playlist - http://tinyurl.com/MotionPlanning-MARIO.

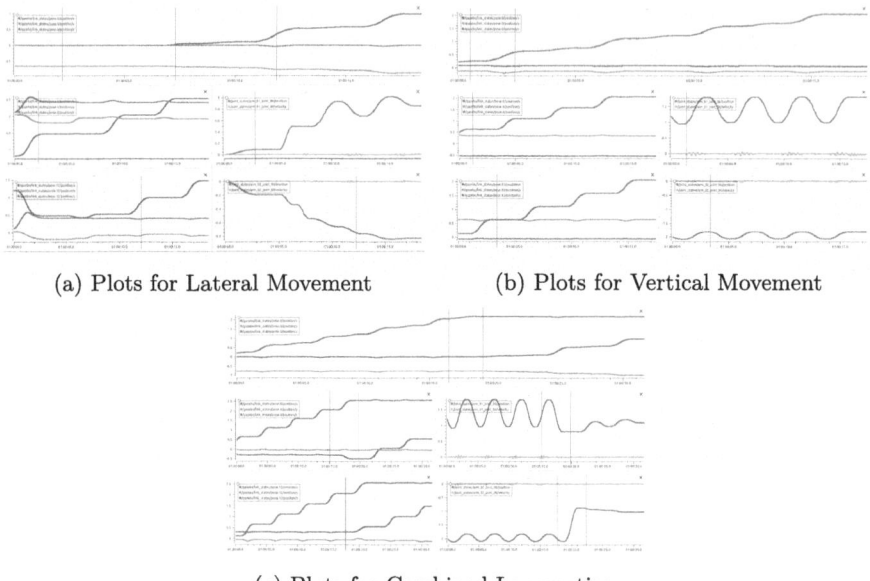

(a) Plots for Lateral Movement   (b) Plots for Vertical Movement

(c) Plots for Combined Locomotion

Fig. 11. Plots of the Locomotion Types

5   Conclusions

A simulation framework is developed for a multi-arm space robot performing locomotion on a planer space structure. The locomotion strategies are devised for lateral and vertical movements that can be combined to reach any point on a planar space structure. The locomotion is executed in three steps within a simulated environment, including the coordinated control of multiple arms. Simulation results show that the discussed ROS-based simulation architecture can generate the desired space scenario with assembly components and execute the proposed locomotion approach. Collision avoidance and improved combined locomotion approaches are immediate future works. The developed ROS-Gazebo-based simulation is available on the GitHub repository [16].

References

1. Lee, N., et al.: Architecture for in-space robotic assembly of a modular space telescope (2016). https://doi.org/10.1117/1.JATIS.2.4.041207
2. Deremetz, M.: MOSAR-WM: a relocatable robotic arm demonstrator for future on-orbit applications (2020). https://iafastro.directory/iac/paper/id/60119/summary/
3. Roa, M.A., et al.: In-space robotic assembly of large telescopes (2019). https://www.eso.org/sci/facilities/eelt/
4. Mishra, H., De Stefano, M., Ott, C.: Dynamics and control of a reconfigurable multi-arm robot for in-orbit assembly **55**, 235–240 (2022). https://repositum.tuwien.at/handle/20.500.12708/101891

5. Xu, W., Liu, Y., Xu, Y.: The coordinated motion planning of a dual-arm space robot for target capturing **30**, 755–771 (2012). https://doi.org/10.1017/S0263574711001007
6. Yang, S., Wen, H., Hu, Y., Jin, D.: Coordinated motion control of a dual-arm space robot for assembling modular parts **177**, 627–638 (2020). https://doi.org/10.1016/j.actaastro.2020.08.006
7. Yoshida, K., Kurazume, R., Umetani, Y.: Dual arm coordination in space free-flying robot **3**, 2516–2521 (1991). https://doi.org/10.1109/ROBOT.1991.132004
8. Plaut, R.H.: Mathematical model of inchworm locomotion **76**, 56–63 (2015). https://doi.org/10.1016/j.ijnonlinmec.2015.05.007
9. Xie, Y., Lou, B., Xie, A., Zhang, D.: A review: robust locomotion for biped humanoid robots **1487**(1), 012048 (2020). https://doi.org/10.1088/1742-6596/1487/1/012048
10. Chang, S., Liu, Y., Du, H., Xie, F., Zhang, J.: Modeling and gait planning simulation of tower-climbing robot based on ROS, pp. 194–200 (2021). https://doi.org/10.1109/ICNC52316.2021.9609063
11. Maciel, E.H., Henriques, R.V.B., Lages, W.F.: Control of a biped robot using the robot operating system, pp. 247–252 (2014). https://doi.org/10.1109/SBR.LARS.Robocontrol.2014.46
12. Manuele, A.: Simulation-of-walking-biped **7** (2021). https://github.com/AntoManuele/Simulation-of-Walking-Biped
13. Johannes Biermann. dual_arm_demo (2022). https://github.com/bi3ri/dual_arm_demo
14. Bhadani, S., et al.: A ROS-based simulation and control framework for in-orbit multi-arm robot assembly operations (2023). https://dspace.lib.cranfield.ac.uk/handle/1826/20622
15. PAL Robotics. Gazebo ROS link attacher (2016). https://github.com/pal-robotics/gazebo_ros_link_attacher
16. Dillikar, S., Upadhyay, S., Felicetti, L., Tang, G.: Motion planning and control of multi-arm robot for in-orbit operations (2023). https://github.com/sairajdillikar/Motion-Planning-and-Control-of-Multi-Arm-Robot-for-In-Orbit-Operations

Size Matters: Exploring the Impact of Scaling on Quadruped Robot Dynamics

Faraz Rahvar[1,2]([envelope]) [ORCID] and Abdurrahman Yilmaz[1,2] [ORCID]

[1] Department of Mechatronics Engineering, Istanbul Technical University, Istanbul, Turkey
{Rahvar22,yilmazabdurrah}@itu.edu.tr
[2] Department of Control and Automation Engineering, Istanbul Technical University, Istanbul, Turkey

Abstract. Quadruped robots are indispensable for specialized tasks, particularly in disaster scenarios like earthquakes, where their mobility surpasses that of fixed robots. However, altering their dimensions significantly impacts their mechanical requirements and control systems. This study investigates and compares the principal mechanical parameters affecting the control systems of quadruped robots during scaling, aiming to understand how scalability influences their design across different sizes. Using a standard quasi-direct driven commercial quadruped robot1(Unitree A 1) model featuring dual coaxial motor design and serial link architecture actuators, simulations are conducted at four different sizes, each expanding the robot size by 30%. These simulations analyze static and dynamic forces as the robot walks on a flat surface at a constant speed. To fortify our findings, simulations employ three trajectory scenarios, in each scenario, four robots are scaled in length, increasing by a coefficient of 0.3 from robot 1 to robot 4. The first and second scenarios feature different trajectory lengths, while the third scenario increases the trajectory height while maintaining the length of the second scenario. Specifically, we examine the required torque, energy consumption of the robot motors, reaction forces between the robot feet and terrain, and the mechanical cost of transport. These outcomes encompass critical components of the robot from a mechanical analysis perspective, including joints, providing valuable references for various actuator architecture designs. Our findings reveal a logical pattern in the increase of torques, power consumptions, and reaction forces as the robot scales. These insights lay the groundwork for developing a scalable control architecture for legged robots.

Keywords: Quadrupeds · Legged robots · Scalability

1 Introduction

Nowadays, the vital role of robots in human life is not only undeniable, but it is also essential. Quadruped robots, in particular, which mimic four-legged animals, have been significantly crucial in emergency and critical situations. In this study, we deeply focused on different mechanical parameters affecting the design of quadruped robots while considering the dimensional scaling of both robot parts and trajectory length and height,

M. N. Huda et al. (Eds.): TAROS 2024, LNAI 15052, pp. 106–117, 2025.
https://doi.org/10.1007/978-3-031-72062-8_10

thereby potentially leading us to achieve a scalable control architecture for Quadruped Robots (QRs). The scalability of QRs can significantly enhance their capabilities. If a QR can adjust its size, it can quickly conceal itself under debris to observe enemy operations or navigate through narrow pathways under earthquake rubble. Achieving this ability necessitates a scalable mechanical design, which in turn requires a scalable control architecture. Such an architecture relies heavily on mechanical parameters. To realize this scalable control architecture, it is imperative to meticulously monitor the behavior of these parameters and establish relationships among them. Our research continues previous studies by providing a new perspective on scalable control architecture in quadruped robots. Specifically, our contributions include:

- Analysis of Mechanical Parameters: We examine the required torque, energy consumption of the robot motors, reaction forces between the robot feet and terrain, and the mechanical cost of transport across different robot sizes.
- Simulation Scenarios: We conduct simulations using a standard quasi-direct driven commercial quadruped robot model at four different sizes, analyzing static and dynamic forces as the robot walks on a flat surface at a constant speed.
- Trajectory Scenarios: Our study includes three trajectory scenarios to robustly assess the mechanical behavior of the robots, varying trajectory length and height.
- Scalable Control Insights: Our findings provide insights into the development of scalable control architectures by understanding how torques, power consumptions, and reaction forces change with robot scaling.

2   Literature Review

Legged robots have emerged as promising platforms for high-speed locomotion, with a focus on achieving low duty factor and high stride frequency [1]. These requirements entail swift leg extensions during the swing phase, necessitating significant accelerations and decelerations. To meet these demands, enhancing actuator capacity and reducing leg inertia are crucial, achieved through methods like positioning actuators closer to the body and employing compliant joints. Integrating tendon into prototypes like MIT Robotic Cheetah reduces leg stress and enhances performance [2]. Quasi-Direct Drive (QDD) actuators offer high backdrivability and torque control bandwidth, surpassing Series Elastic Actuators (SEAs). Medium-sized quadrupeds like the MIT Cheetah series demonstrate exceptional maneuverability [3]. Small-sized quadrupeds utilize modular QDD actuators for agility and cost-effectiveness [4, 5]. Motion imitation provides a comprehensive approach for executing diverse behaviors [6]. However, its application in legged robots has focused mainly on upper-body movements [7]. Integrating motion imitation with reinforcement learning shows promise but faces challenges in replicating simulated capabilities in real-world scenarios [8]. Recent advancements in learning-based controllers have extended the capabilities of legged robots in navigating challenging terrains [9]. Integrating perception enhances dynamic locomotion by providing accurate positioning of feet [10]. Trajectory optimization techniques optimize torso and leg movements, albeit with computational demands. Addressing these challenges requires reliable initial guesses for feasible motions [11]. Strategies like deconstructing terrain complexities into separate touch transitions simplify planning [12]. Real-time feasible approaches are proposed using dynamic models [13]. The utilization of elevation maps remains crucial in legged robotics for perceptual locomotion control [14]. Energy efficiency is paramount, with strategies like Total Cost of Transport (TCoT) evaluation and morphological computation reducing power usage [15, 16]. MIT Cheetah's design principles prioritize energy efficiency, leading to superior performance comparable to natural running animals [17]. Recent research in quadruped robot design highlights the critical roles of mechanical flexibility and scalability. Incorporating foldable mechanisms greatly improves these robots' adaptability and performance on challenging terrains by allowing them to adjust their dimensions for better navigation in varied environments. Liu et al. describe a bio-inspired quadruped robot with a scalable torso, which enhances its ability to overcome obstacles and increases its efficiency in high-speed movement [18]. Semini et al. emphasize the need for optimized actuators to achieve robust, dynamic motion, which is vital for fast and energy-efficient locomotion [19]. Fadini et al. introduce a co-design framework that optimizes both hardware and control strategies, resulting in dynamic and energy-efficient behaviors in quadruped robots [20]. These advancements are crucial for creating highly versatile, dynamic robots capable of operating effectively in complex and diverse terrains [18–20]. Despite valuable contributions to quadruped robots, there is a lack of exploration into how all the mechanical principles alter with scaling. Our study aims to address this gap by examining variations in mechanical factors during scaling, offering insights into adaptable control structures in quadruped robots.

3   Scalable Mechanical Design

While numerous valuable studies have explored the optimization of legged robots, our understanding of their behavior during scaling remains limited. Despite the wealth of research in this field, there is a notable gap concerning the effects of scaling on legged robot dynamics. Comprehending the variations in mechanical parameters such as joint torques, power consumption, MCOT (Mechanical Cost of Transport), and reaction forces during scaling is essential for advancing the optimization of legged robots. This study aims to fill this gap by analyzing a standard Unitree A1 quadruped robot at various sizes, thereby elucidating the changing behaviors of mechanical parameters that influence the robot's dynamics. This comprehensive analysis contributes to the broader understanding of legged robot optimization and provides insights that can inform future advancements in this field.

3.1   General Hypotheses

The Unitree A1, a sophisticated quadruped robot employing QDDs (Sombolestan, 2021), serves as our standard model for scaling. Due to its structural attributes, this robot offers a conducive framework for gathering valuable data relevant to various categories of legged robots, including series elastic and QDDs. As precise dimensional drawings of the Unitree A1 quadruped robot were unavailable, we embarked on the task of redesigning and assembling its components to approximate the dimensions of the commercially available Unitree A1 model. Figure 1 depicts the reference model utilized in this study. Our investigation assumes that the robots walk at a consistent speed on level terrain, disregarding the influence of fluid mechanics such as air resistance during locomotion. Additionally, the robot is assumed to lack back drivability, while ground acceleration and reaction forces between the end effectors and terrain are accounted for. The study delves into the behavior of mechanical principles affecting the robot, irrespective of motor type, motor placement architecture, or the presence of elastic elements. This simplified model facilitates the provision of foundational and fundamental data pertinent to various types of legged robots. By comprehending the behavior of a robot during scaling and understanding the requisite mechanical parameters such as hip torque, designers can make informed decisions regarding actuator type and placement or the integration of elastic elements.

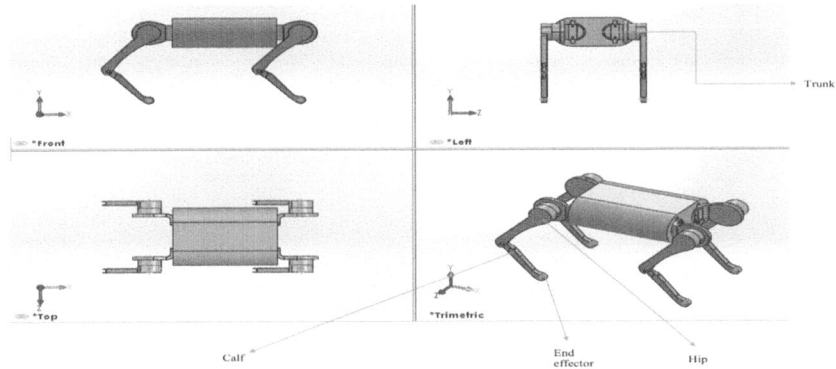

Fig. 1. The reference model based on Unitree A1 quadruped robot

3.2   Scaling Plans

Utilizing CATIA V5 R21, we crafted and assembled the various components of the robot, including the Torso, Hips, and Calves. Employing the SolidWorks motion analysis environment, the robot underwent scaling and simulation across three distinct scenarios. Throughout these scenarios, the robots traversed even, level terrain at a consistent pace, maintaining identical starting positions. Each simulation spanned five seconds, commencing with the robots being released from proximity above the terrain. Primarily, within the initial 0.8 s of each simulation, an adjustment phase was observed. The trajectory of the end effector is delineated in Fig. 2, while Table 1 outlines pertinent details such as the robots' dimensions, mass, material composition, walking speed, as well as the length and height of each step of the end effector. In the first scenario, spanning from the reference robot (designated as "robot one") to the fourth iteration, there exists a 0.3 increment in the length of the Torso, arms, and Calves. The second scenario involves an increase in the length of the step size. In the third scenario, the step length from the second scenario is maintained, but there is an increase in the step height, analogous to the second scenario.

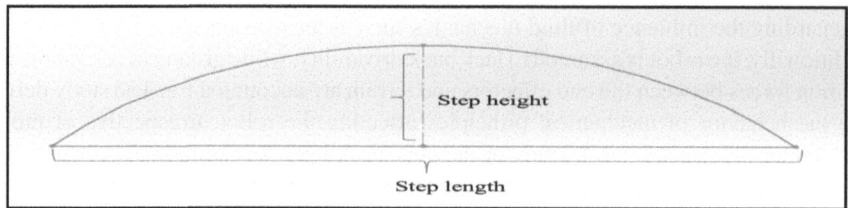

Fig. 2. The trajectory of the end effector

Table 1. Scaled robots' general data

Scenario	Robot	Arm length (mm)	Calf length (mm)	Torso length (mm)	Torso width (mm)	Torso height (mm)	Total mass (kg)	Material	Step length (mm)	Step height (mm)	Lineer speed (mm/s)
1, 2, 3	1	160	140	250	200	100	15.3	2018 Alloy	120	27	250
									165	27	
									165	40	
	2	208	182	325	200	100	18.7	2018 Alloy	120	27	250
									165	27	
									165	40	
	3	256	224	400	200	100	22.6	2018 Alloy	120	27	250
									165	27	
									165	40	
	4	304	266	475	200	100	28.7	2018 Alloy	120	27	250
									165	27	
									165	40	

4   Comparison and Results

In this section, we undertake a comparative analysis of the outcomes derived from our simulations pertaining to hip torque, calf torque, hip power consumption, calf power consumption, MCOT (Mean Center of Torque), and the reaction forces exerted between the end effector and the terrain. The determination of these parameters involved the utilization of eight motors strategically positioned at the center of mass of each joint, comprising four motors within the hip assembly and an additional four within the knee assembly. The trajectory was subdivided into eight discrete points, facilitating the subsequent measurement of arm and calf positions. Notably, the hip motors were programmed to follow distinct displacement patterns aimed at attaining a consistent linear velocity across all scenarios and robot configurations. Consequently, both angular displacement and angular velocity exhibited variations across the spectrum of scenarios and robot designs. In this study using Solidworks motion analysis, robots were dropped onto terrain and navigated a specified path. Initial positioning was achieved with parallel and centering mates, which were then suppressed to allow free movement. The only active mate was a contact mate between the robot and the terrain. A video[1] showcases the first, second, and third robot of scenario one, providing an overview before detailed results and plots are presented in subsequent sections.

4.1 Scenario One, Two, and Three

In this investigation, the initial scenario delineates the trajectory's dimensions as 120 mm in length and 27 mm in height. Subsequently, in the second scenario, an augmentation of the trajectory length to 165 mm was implemented to scrutinize the ramifications of such elongation. Transitioning to the third scenario, the trajectory length remained fixed at 165 mm, while the trajectory height was elevated to 40 mm. across all scenarios, the four robots showcased analogous trends in hip torque, calf torque, hip power consumption, calf power consumption, and reaction forces between the end effector and the terrain over time. Initially, minimal torque, nearing zero, was observed, gradually escalating with temporal progression, accompanied by fluctuations throughout the movement cycle. despite the 30% increase in robot length and alterations in trajectory dimensions across the three scenarios, the fundamental shape and magnitude of hip torque (illustrated in Fig. 3(a)), calf torque (depicted in Fig. 3(b)), hip power consumption (presented in Fig. 4(a)), calf power consumption (shown in Fig. 4(b)), and reaction forces (portrayed in Fig. 5) remained consistent. These modifications did not substantially deviate from the established hip torque pattern; instead, they appeared to proportionally scale the values either upwards or downwards. The persistent conformity in the configuration of these mechanical principles' curves across varied robot scales and scenario adjustments implies that the envisaged alterations and scaling paradigms exerted negligible influence on the dynamical behavior of the scaled robots. in each scenario, although the overall pattern is consistent, there are noticeable differences in the magnitude of values between the robots, generally increasing as the robot's length increases. nonetheless, it is crucial to note that this correlation does not adhere strictly to a linear pattern. throughout the data, there are occurrences of abrupt spikes or declines in values across all robots. these fluctuations signify transient events or adaptations within the walking motion, likely aligning with distinct phases of the walking cycle. Conversely, the magnitude of the values for the respective robots remains relatively constant. The primary disparity lies in the angular displacement of the motors. For instance, Fig. 6 illustrates the angular displacement of robot three across scenarios one, two, and three.

In other words, if the motors have adjustable capabilities in terms of motor angular speed and possess logical power and torque, the robot can be scaled using the aforementioned scaling patterns and scenarios.

---

[1] https://www.youtube.com/watch?v=5fq2vdHWJ7Q.

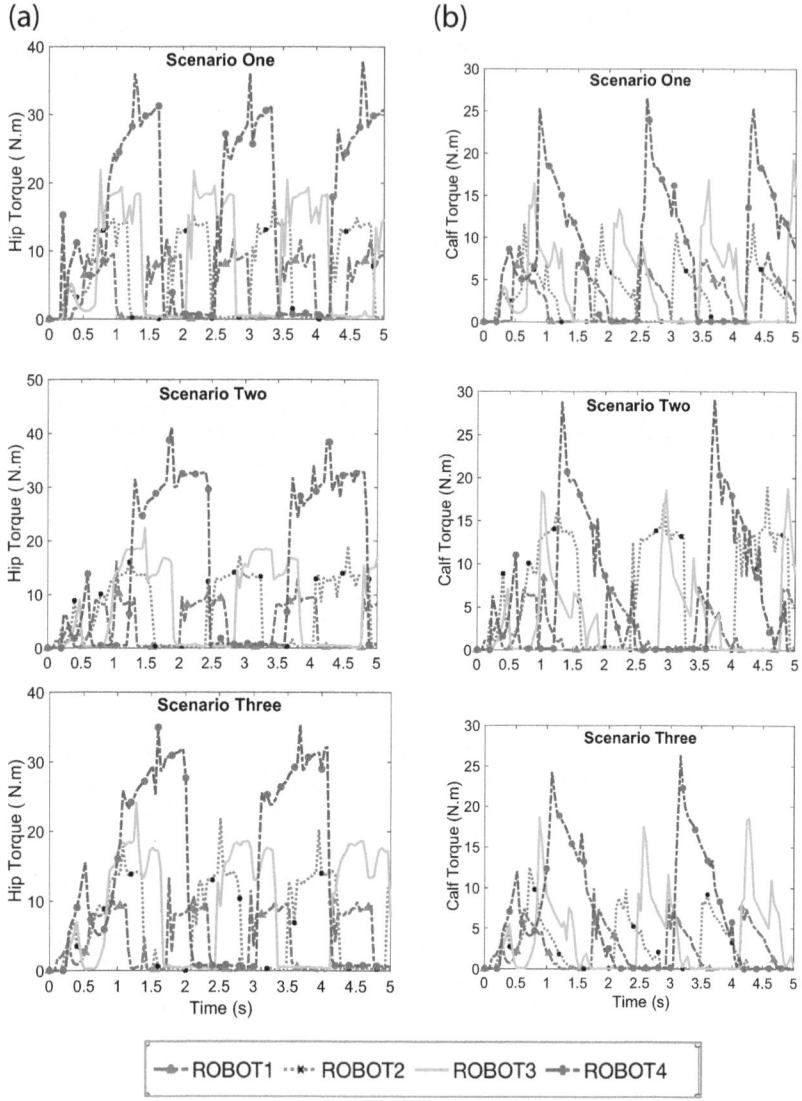

Fig. 3. (a) Similar patterns of hip torque (b) Similar patterns of calf torque

4.2 Mechanical Cost of Transport

The mechanical cost of transport can be calculated as seen in Eq. 1.

$$\text{MCOT} = \frac{\sum P}{w \times d} \tag{1}$$

where P represents power consumption, w is the mass being transported, typically measured in kilograms (kg), and d denotes the distance traveled during locomotion, usually

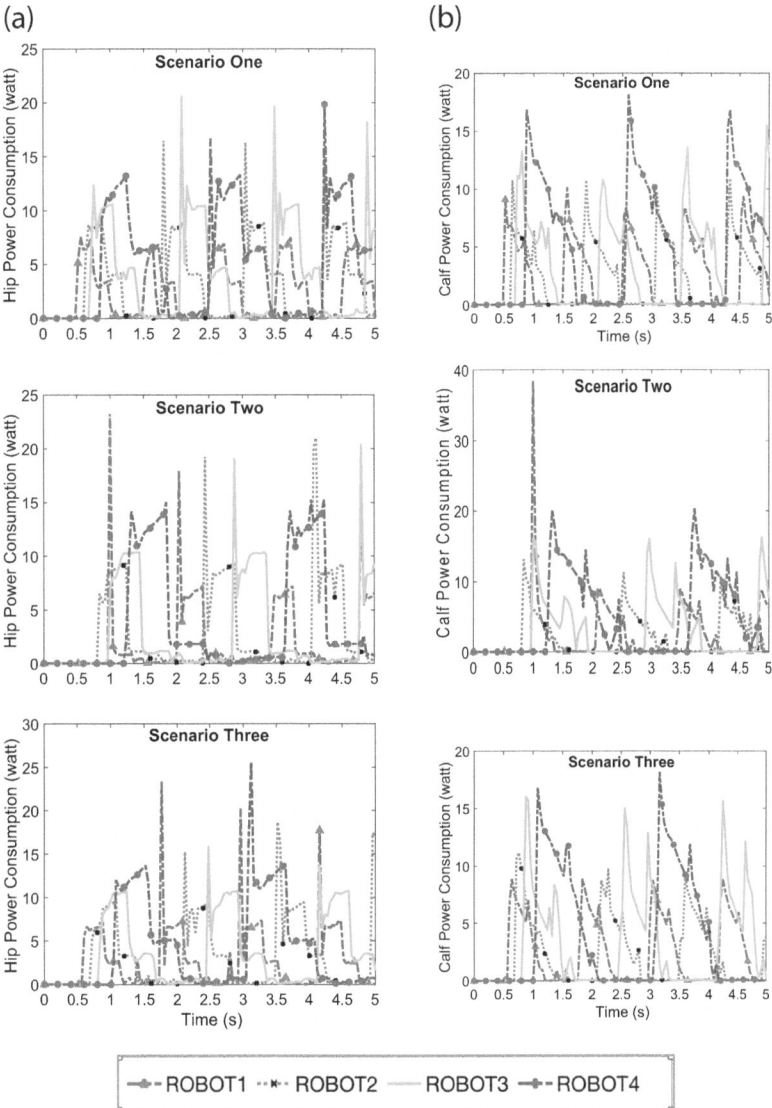

Fig. 4. (a) Similar patterns of hip power (b) Similar patterns of calf power

measured in meters (m). The total power consumption, usually measured in watts (W), is calculated through the total mechanical power expended during locomotion over the earned time interval. Since the power consumption data was provided in intervals of 0.04 s, we used the trapezoidal rule to approximate the area under the power consumption curve and then multiplied it by the time interval. The MCOT of the robots through the three scenarios with two example gear ratios are shown in Table 2.

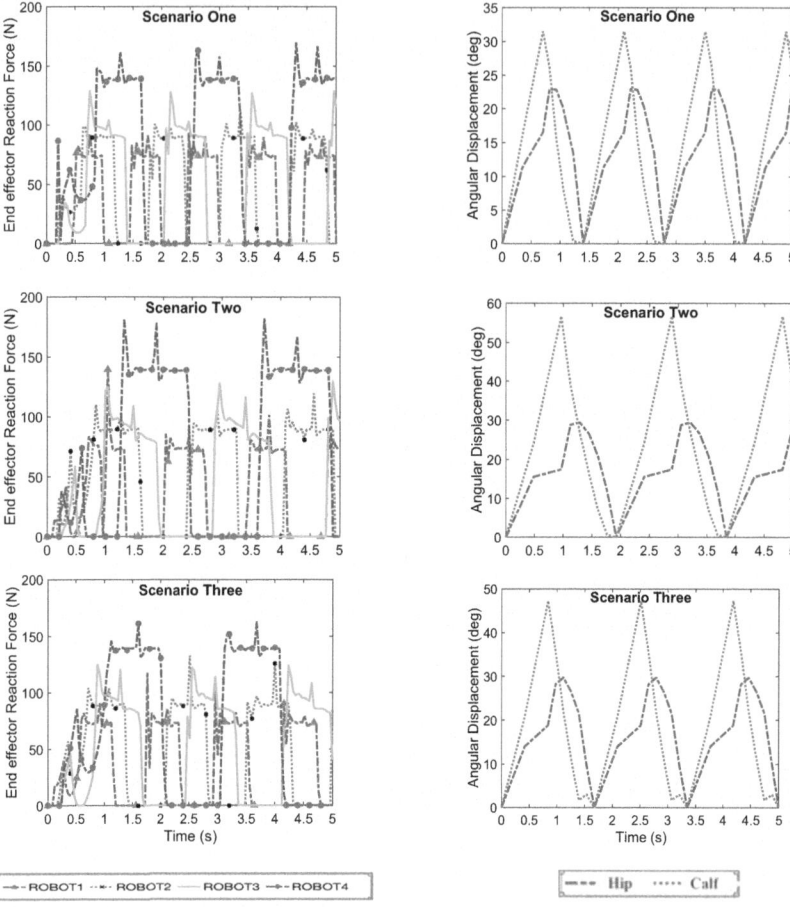

Fig. 5. Reaction forces similar patterns   **Fig. 6.** Angular displacement of Robot 3

The MCOT of the robots through the three scenarios with two example gear ratios are shown in Table 2.

Table 2. MCOT comparison in all scenarios

Gear ratio	Scenario	Robot 1	Robot 2	Robot 3	Robot 4
8	1	0.83	0.935	0.72	0.82
	2	0.58	0.65	0.62	0.68
	3	0.76	0.7	0.75	0.66
6	1	1.1	1.24	0.96	1.1
	2	0.77	0.86	0.82	0.93
	3	1.01	0.93	1	0.88

4.3 The Comparison of the MCOT Through the Three Scenarios

TCOT is the most common tool for measuring the energy efficiency of legged robots (Bhounsule, 2012), while MCOT does not consider the energy wastage within the electric motor. In this context, we focus on MCOT due to our mechanical design principles. Upon examining Table 2, we noted that MCOT rises as the gear ratio decreases. This is expected because a lower gear ratio results in higher total power consumption, thereby increasing MCOT. The influence of gear ratio on MCOT has been discussed in previous studies (Seok, 2012) and is not the primary focus of this investigation. The subsequent comparisons are applicable for both given gear ratios.

Upon scrutinizing Table 2, we note a consistent MCOT level across all robots within each scenario, albeit with minor discrepancies. Given the potential impact of even slight variations in MCOT on energy efficiency, it is imperative to give due consideration to these findings. We initially expected to observe an increase in MCOT from robots one to four across scenarios one, two, and three due to the escalation in weight and dimensions. However, contrary to our expectations, this trend did not materialize. We propose that this deviation can be ascribed to the influence of mechanical design on MCOT. For instance, in scenario one, robot two demonstrates an MCOT of 0.935, surpassing that of robot one but falling short of robot four, despite the latter's larger size and weight compared to robot two. Hence, it is evident that mechanical design plays a significant role in shaping energy efficiency during scaling. Simulation outcomes further support this notion, revealing that elongating the trajectory's length and height leads to a decrease in MCOT, with the reduction being more pronounced as the trajectory extends.

5 Conclusion and Future Works

Quadruped robots demonstrate invaluable capabilities for specialized tasks, especially on catastrophic scenarios, where their mobility often outperforms that of static-bodied robots. However, modifying the dimensions of these robots significantly impacts their mechanical requirements and control systems. This study aimed to investigate and compare key mechanical parameters, including hip and calf torque, hip and calf power consumption, reaction forces between the end effector and the ground, and the mechanical cost of transport of scaled robots across three different scenarios. Our objective was to observe their effects on the dynamics of quadruped robots during scaling. Utilizing the motion analysis tool within SolidWorks, we conducted simulations involving a standard Unitree A1 quadruped robot walking on an even flat surface at a constant linear velocity of 250 mm per second. We observed that the behavioral patterns of the robots during scaling with various trajectory dimensions remained relatively consistent. However, the magnitude of these mechanical principles increased in each scenario due to the elongation of the torso, arms, and legs. Significant variations were noted in the angular velocity and displacement of the arms and legs, which directly correlated with the motors' performance. Thus, successful scaling of the robots during operations hinges on the motors' ability to meet maximum torque and power consumption requirements, while adapting to the necessary angular velocity. Our observations revealed a consistent cost of transport across scenarios, albeit with a decrease in the mechanical cost of transport (MCOT) as trajectory length and height increased. This finding underscores the impact of even minor differences in mechanical parameters on energy efficiency. These outcomes provide comprehensive insights into critical mechanical components, including joints, offering valuable references for various actuator architecture designs such as series elastics or QQDs. Importantly, these references are not limited to a singular type of actuator, thereby enhancing their applicability. Our findings elucidate a discernible pattern of torques, power consumptions, and reaction forces as the robot scales in size. By investigating the behavior of these mechanical principles at different constant speeds in future research endeavors, we intend to leverage these fundamental data to develop a scalable control architecture with a focus on machine learning integration.

Acknowledgments. This study is supported by ITU Bilimsel Arastirma Projeleri Koordinasyon Birimi under grant number MYL-2023-44659.

Disclosure of Interests. The authors have no competing interests to declare that are relevant to the content of this article.

References

1. Maes, L.D.: Steady locomotion in dogs: temporal and associated spatial coordination patterns and the effect of speed. J. Exp. Biol., 138–149 (2008)
2. Ananthanarayanan, A.: Towards a bio-inspired leg design for high-speed running. Bioinspiration Biomimetics, 046005 (2012)
3. Hooks, J.: Alphred: A multi-modal operations quadruped robot for package delivery applications. IEEE Robot. Autom. Lett., 5409–5416 (2020)
4. Sombolestan, M.: Adaptive force-based control for legged robots. In: 2021 IEEE/RSJ International Conference on Intelligent Robots and Systems (IROS), pp. 7440–7447. içinde IEEE (2021)
5. Kim, J.: Design and control of a open-source, low cost, 3D printed dynamic quadruped robot. Appl. Sci., 3762 (2021)
6. Yamane, K.: Controlling humanoid robots with human motion data: experimental validation. In: 2010 10th IEEE-RAS International Conference on Humanoid Robots, pp. 504–510. içinde IEEE (2010)
7. Koenemann, J.: Real-time imitation of human whole-body motions by humanoids. In: 2014 IEEE International Conference on Robotics and Automation (ICRA), pp. 2806–2812. içinde IEEE (2014)
8. Liu, L.: Learning basketball dribbling skills using trajectory optimization and deep reinforcement learning. ACM Trans. Graph. (TOG), 1–14 (2018)
9. Miki, T.: Learning robust perceptive locomotion for quadrupedal robots in the wild. Sci. Robot., eabk2822 (2022)
10. Jenelten, F.: Perceptive locomotion in rough terrain–online foothold optimization. IEEE Robot. Autom. Lett., 5370–5376 (2020)
11. Melon, O.: Reliable trajectories for dynamic quadrupeds using analytical costs and learned initializations. In: 2020 IEEE International Conference on Robotics and Automation (ICRA), pp. 1410–1416. içinde IEEE (2020)
12. Kalakrishnan, M.: Fast, robust quadruped locomotion over challenging terrain. In: 2010 IEEE International Conference on Robotics and Automation, pp. 2665–2670. içinde IEEE (2010)
13. Mastalli, C.: Motion planning for quadrupedal locomotion: coupled planning, terrain mapping, and whole-body control. IEEE Trans. Robot. 1635–1648

14. Kweon, I.-S.: Terrain mapping for a roving planetary explorer. In: IEEE International Conference on Robotics and Automation, pp. 997–1002. içinde IEEE (1989)

15. Bhounsule, P.A.: Design and control of ranger: an energy-efficient, dynamic walking robot. Adapt. Mob. Robot., 441–448. World Scientific (2012)

16. Tedrake, R.: Actuating a simple 3D passive dynamic walker. In: 2004 IEEE International Conference on Robotics and Automation. Proceedings, ICRA 2004, pp. 4656–4661. IEEE (2004)

17. Seok, S.: Design principles for energy-efficient legged locomotion and implementation on the MIT cheetah robot. IEEE/ASME Trans. Mechatr., 1117–1129 (2014)

18. Liu, Y., Bi, Q., Li, Y.: Design of a bio-inspired quadruped robot with scalable torso. In: IEEE 17th International Conference on Automation Science and Engineering (CASE)

19. Semini, C., et al.: Design and scaling of versatile quadruped robots. Istituto Italiano di Tecnologia (2012)

20. Fadini, G., Rigatos, G., Tsaramirsis, G.: Co-design framework for optimization of hardware and control policies in dynamic quadruped robots. Robot. Autonom. Syst. J.

Learning Bipedal Walking on a Quadruped Robot via Adversarial Motion Priors

Tianhu Peng[1], Lingfan Bao[1], Joseph Humphreys[1],
Andromachi Maria Delfaki[2], Dimitrios Kanoulas[2], and Chengxu Zhou[2(✉)]

[1] School of Mechanical Engineering, University of Leeds, Leeds, UK
[2] Department of Computer Science, University College London, London, UK
chengxu.zhou@ucl.ac.uk

Abstract. Previous studies have successfully demonstrated agile and robust locomotion in challenging terrains for quadrupedal robots. However, the bipedal locomotion mode for quadruped robots remains unverified. This paper explores the adaptation of a learning framework originally designed for quadrupedal robots to operate blind locomotion in biped mode. We leverage a framework that incorporates Adversarial Motion Priors with a teacher-student policy to enable imitation of a reference trajectory and navigation on tough terrain. Our work involves transferring and evaluating a similar learning framework on a quadruped robot in biped mode, aiming to achieve stable walking on both flat and complicated terrains. Our simulation results demonstrate that the trained policy enables the quadruped robot to navigate both flat and challenging terrains, including stairs and uneven surfaces.

Keywords: Legged Robots · Bipedal Locomotion · Deep Reinforcement Learning · Adversarial Motion Priors

1 Introduction

Legged robots exhibit superior terrain adaptability compared to their wheeled and tracked counterparts. Although quadrupedal robots are known for their stability and agility, bipedal robots offer greater flexibility by freeing the upper body for complex tasks. This flexibility suggests the potential for quadrupedal robots to walk in a bipedal gait, using the rear legs for walking and the front legs for manipulation.

The primary challenge in adapting quadruped robots for bipedal locomotion stems from their mechanical design constraints. First, unlike typical bipedal robots that have firm, flat feet, quadruped robots often feature soft, point-contact feet that inherently lack stability. Second, the rear legs of quadruped robots are

This work was supported by the Royal Society [grant number RG\R2\232409] and the UKRI Future Leaders Fellowship [grant number MR/V025333/1]. For a visual overview of our framework and results, please refer to the video at https://youtu.be/JYD1RlrQRWM.

M. N. Huda et al. (Eds.): TAROS 2024, LNAI 15052, pp. 118–129, 2025.
https://doi.org/10.1007/978-3-031-72062-8_11

not specifically designed for bipedal walking, their limited range of motion and underactuation contribute to unnatural and unstable bipedal gaits. This design mismatch explains why quadruped robots struggle with bipedal walking modes. This leads to high requirements for locomotion controllers during bipedal modes.

To achieve bipedal walking for quadruped robots, there are primarily two approaches: the model-based method and the learning-based method [1]. Model-based methods are based on highly accurate mathematical models, which have proven to be effective in executing highly dynamic motions in both quadruped and bipedal robots. However, these methods lack robustness and generalization in unseen scenarios, largely due to the difficulty of accurately modeling ground interactions and contact dynamics. In contrast learning-based methods, reinforcement learning (RL), provides a more adaptable solution by enabling the exploration of the robots' full dynamics and interactions with the environment, thus offering greater flexibility in controlling complex locomotive behaviors. Early research in RL on legged robots primarily utilized unrealistic models within physical simulators [2–4]. In transitioning to a practical bipedal robot and learning natural and robust gaits, previous studies have primarily designed a reference-free learning framework by designing periodic composition reward [5] or mimicking predefined references [6–8].

Reference-free methods explore various gait patterns efficiently, while reference-based methods leverage prior knowledge to accelerate learning, resulting in efficient policy exploration and robust locomotion skills. These methods incorporate expert information and predefined reference trajectories from motion capture data or trajectory optimization (TO). Generative Adversarial Imitation Learning [2] and Adversarial motion priors (AMP) [8] predict state transitions and evaluate the similarity between reference and agent data, promoting stable gait maintenance. AMP was implemented with a study of human reference behavior in biped robots [9,10], combining it with periodic rewards to promote stable gait maintenance.

To generate appropriate predefined references, several methods are utilized. Motion capture technology is commonly employed to produce reference data for various types of legged robots [8,10,11]. This technology captures comprehensive kinematic data from real-world scenarios, enabling versatile data collection that is not confined to a specific robot model. However, adapting this data to different robotic platforms often requires a re-targeting process, which increases both complexity and manual labor. On the other hand, TO in reduced-order [12] and full dynamics models [7], has been employed. This approach reduces complexity and eliminates the need for further re-targeting, making it a more efficient solution for generating references. Additionally, compared to full dynamics models, reduced-order models can decrease the computational resources required for optimization and offer greater generalization across all quadruped robots. Regarding predesigned references from the optimization method, various models are utilized. Besides HZD-based full dynamics reference [7], reduced-order dynamics such as Single Rigid Body Dynamics (SRBD) [13] are also utilized in the training

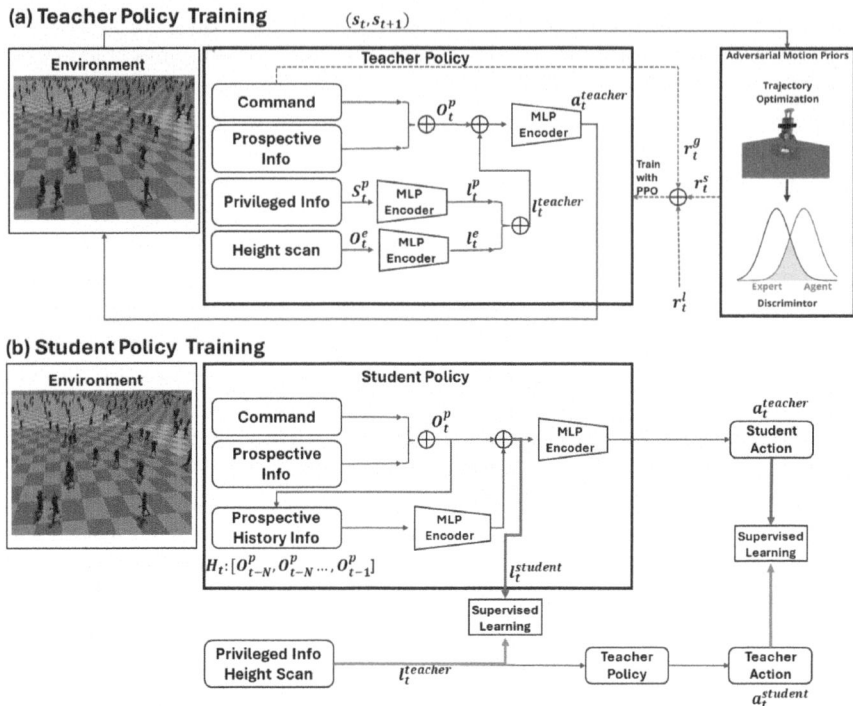

Fig. 1. Overview of the teacher-student learning framework. (a) The teacher policy, which leverages privileged data S_t^p, terrain information o_t^e and prospective data o_t^p through RL, aims to maximize a total reward r_t comprising command task reward r_t^g, style reward r_t^g based on AMP, and regulation reward r_t^g for ensuring safety and smooth motion. (b) The student policy, trained via supervised learning, seeks to imitate the teacher's actions $a_t^{teacher}$ and reconstruct the teacher's latent states $l_t^{teacher}$ from historical and prospective observations $H_t : [O_{t-N}, O_{t-N-1}, ..., O_{t-1}]$

procedure. Based on the reduced-order model, task-space learning focuses on the foot setpoints and based velocity [14,15].

Another significant challenge exists in bridging the sim-to-real gap. To extend the robustness of locomotion and overcome the sim-to-real gap, frameworks using the privileged learning paradigm have been introduced [16,17]. By combining the strengths of AMP in reference-based learning and privileged learning, there is potential to enable quadrupedal robots to adopt bipedal walking modes. Similar frameworks have been introduced [9,11], but none have been validated on bipedal robots or quadruped robots for bipedal mode.

Our objective is to train a policy that enables a quadrupedal robot to achieve bipedal locomotion using only its two rear legs, thus freeing its front legs for more complex tasks. This capability aims to enhance the robot's versatility and functionality in various practical applications. This paper presents a novel framework that enables quadrupedal robots to achieve robust and agile blind bipedal locomotion on flat terrain. We adopt a teacher-student policy framework, where privileged information that the robot cannot directly access is encoded. The student policy uses historical observation information to infer this privileged information, thereby enhancing robustness. Additionally, we integrate the AMP training framework to learn and imitate the style behaviors of reference data generated through TO based on a SRBD model. Different from previous work [18] with assistant devices, this comprehensive training framework equips the policy to support agile bipedal motions in quadrupedal robots.

In summary, the primary contribution of this paper is to develop a novel framework (shown in Fig. 1) that allows quadrupedal robots to perform robust and agile bipedal blind locomotion on flat terrain using only their rear legs. This bipedal mode frees the front legs for more complex tasks, significantly enhancing the robot's versatility and functionality. Besides, We evaluate our model on the A1 quadrupedal robot using a biped gait model in the Isaac Gym simulation environment, demonstrating agile and robust movements.

2 Methodology

2.1 Reinforcement Learning on Legged Robots

The task of learning legged locomotion poses significant challenges due to the complex environment and limitations in sensor data. To address this, a partially observable Markov decision process (POMDP) framework was adopted, denoted as $(s_t, a_t, P, r_t, p_0, \gamma)$, where s_t represents the state at time step t, a_t is the action taken by the agent, $P(s_{t+1}|s_t, a_t)$ describes the system dynamics, predicting the next state based on the current state and action, $r_t(s_t, a_t, s_{t+1})$ is the reward function, quantifying the immediate benefit of taking action a_t in state s_t leading to s_{t+1}, p_0 denotes the initial state distribution, and γ^t is the discount factor, determining the importance of future rewards in the decision-making process. The objective of RL in this context is to identify an optimal policy π_θ parameterized by θ, that maximizes the expected discounted return over future trajectories. This is formalized by the objective function $J(\theta)$:

$$J(\theta) = \mathbb{E}_{\pi_\theta}[\sum_{t=0}^{\infty} \gamma^t r_t] \quad (1)$$

In the state space, the state $s_t^{teacher}$ includes proprioceptive observation $o_t^p \in \mathbb{R}^{48}$, privileged state $s_t^p \in \mathbb{R}^{45}$ and terrain information $o_t^e \in \mathbb{R}^{187}$. The proprioceptive observation $o_t^p \in \mathbb{R}^{48}$ encompasses critical data such as the orientation of the gravity vector, base linear and angular velocity in the robot's frame, joint positions and velocities, the previous action $a_{t-1} \in \mathbb{R}^{12}$ executed by the current policy. The privileged state s_t^p contains the information include ground friction coefficients, ground restitution coefficients, contact forces, external forces within positions on the robot, and collision state. The privileged state $s_t^p \in \mathbb{R}^{45}$ includes additional key details that the physical robot cannot directly access in the real-world environment. This information encompasses ground friction and restitution coefficients, contact and external forces at specific robot positions, collision state information. The terrain information $o_t^e \in \mathbb{R}^{187}$ contains the 187 height measurement sampled from grid around robot base to the ground. In contrast, the student policy state utilizes a sequence history of proprioceptive observations $H_t : [O_{t-N}, O_{t-N-1}, ..., O_t]$ to approximate the privileged information. By learning from this historical data, the student policy aims to imitate the inaccessible privileged state, enhancing its decision-making capabilities in the absence of direct access to certain environmental variables. Regarding the action space, the policy action a_t is a 12 dimensional target joint position offset added to the time-invariant nominal joint position. This specification guides the joint PD controller, utilizing fixed gains to compute torque commands effectively for motor position control.

2.2   Adversarial Motion Priors and Rewards Design

The AMP framework utilizes adversarial learning to train two neural networks—a generator and a discriminator—in a competitive setup. The generator produces motion predictions for the robot, while the discriminator evaluates their quality and realism. Style rewards are used to measure the similarity between the demonstrator's behavior and the robot's, with higher similarity yielding more rewards.

Using the reference dataset D, the AMP-based style reward function encourages the robot to replicate the same gait style. According to [19], a neural network-based discriminator D_φ predicts whether a state transition (S_t, S_{t+1}) is from the dataset D or generated by the agent A. Each state $S_t^{AMP} \in \mathbb{R}^{31}$ includes joint positions, point velocities, base linear velocity, base angular velocity, and base height relative to the terrain. To avoid mode collapse, the dataset D contains only trot gait motion clips.

The discriminator's training objective includes a gradient penalty to enforce smoothness and is defined as:

$$
\arg\min_{\varphi} \mathbb{E}_{(s_t,s_{t+1}) \sim D}[(D_\varphi(s_t, s_{t+1}) - 1)^2]
$$

$$
+ \mathbb{E}_{(s_t,s_{t+1}) \sim A}[(D_\varphi(s_t, s_{t+1}) - 1)^2] \tag{2}
$$

$$
+ \frac{\alpha^{qp}}{2} \mathbb{E}_{(s_t,s_{t+1}) \sim D}[\|\nabla_\varphi D_\varphi(s_t, s_{t+1})\|_2^2],
$$

where α^{qp} is a manually specified coefficient ($\alpha^{qp} = 10$).

The style reward is defined by:

$$
r_t^s[(s_t, s_{t+1}) \sim A] = \max[0, 1 - 0.25(d_t^{score} - 1)^2], \tag{3}
$$

where $d_t^{score} = D_\varphi(s_t, s_{t+1})$ and is scaled to the range $[0, 1]$.

The overall reward function is:

$$
r_t = r_t^g + r_t^s + r_t^l, \tag{4}
$$

where r_t^g is the task reward, r_t^s is the style reward, and r_t^l is the regularization reward. Task rewards typically include tracking base linear and angular velocities:

$$r_t^g = \omega^v \exp\left(-|\hat{v}_t^{xy} - v_t^{xy}|\right) + \omega^\omega \exp\left(-|\hat{\omega}_t^z - \omega_t^z|\right), \qquad (5)$$

where ω^v and ω^ω are coefficients, and $\hat{v}_t^{xy}$ and $\hat{\omega}_t^z$ are the velocity commands.

Regularization rewards promote safe, smooth motion and minimize energy costs, enhancing the adaptability and efficiency of learned behaviors for real-world applications. This component contributes to the robustness and effectiveness of the overall motion.

2.3   Reference Generation

In our research, we refine the imitation and learning of reference data for precise bipedal locomotion. Using a single TO formulation from previous work [12], we generate walking and running gaits for the A1 biped robot, focusing on its back legs' trajectories.

We streamline this process with TOWR (Trajectory Optimization for Walking Robots) [20], which eliminates the need for manual tuning. TOWR generates dynamically feasible, energy-efficient motions by optimizing smooth and stable trajectories. To improve imitation fidelity, we integrate inverse kinematics into TOWR, producing joint space data that closely mimics reference trajectories.

Our generated trajectories encompass various locomotion patterns, including forward walking and two distinct running gaits, each lasting 2.4 s. Utilizing TO for motion dataset generation offers several advantages. Firstly, it enables precise matching of the state space between the simulated agent and demonstrator, leveraging kinematic dynamics models to refine trajectory suitability. Moreover, this approach circumvents complexities associated with other motion re-targeting techniques, ensuring a more seamless and accurate replication of desired motions.

3   Framework and Training

3.1   Learning Framework

Teacher Policy Architecture. The teacher policy $\pi_\theta^{\text{teacher}}$ is trained using Proximal Policy Optimization [21] with the total reward r_t as specified in Sect. 2.2. During training, the teacher performs a rollout in the environment to generate a state transition $(s_t^{\text{AMP}}, s_{t+1}^{\text{AMP}})$. This state transition is then fed into the discriminator described in Sect. 2.2.

The teacher policy consists of three Multilayer Perceptron (MLP) networks. Two of these MLPs encode low-dimensional latent representations: $l_t^e \in \mathbb{R}^{16}$ for terrain data and $l_t^p \in \mathbb{R}^8$ for privileged data. Using two separate MLPs for encoding helps mitigate information loss that often occurs during the compression process, thereby preserving crucial and necessary information.

The preservation of essential data significantly aids the student policy in reconstructing the latent representations, facilitating a more efficient and accurate learning process. The third MLP acts as a low-level network, utilizing the proprioceptive observation state along with the two encoded latent representations to generate the teacher's action in the environment. The learning framework is shown in Fig. 1.

Student Policy Architecture. The student policy is designed to emulate the teacher policy, replicating actions without relying on privileged state and terrain information. Throughout the student training process, a supervised approach is employed, minimizing two key losses: imitation loss and reconstruction loss. The imitation loss ensures that the student policy closely mimic the action $a_t^{teacher} \in \mathbb{R}^{12}$ dictated by the teacher's policy. Simultaneously, the reconstruction loss encourages the memory encoder within the student's policy to faithfully reproduce the latent representation $l_t^{teacher} \in \mathbb{R}^{24}$ consists of the terrain latent $l_o^t \in \mathbb{R}^{16}$ and privileged latent $l_o^t \in \mathbb{R}^8$ employed by the teacher.

The overarching architecture comprises a memory encoder and a low-level MLP [22] that maintains an identical structure to the teacher's low-level network. Memory encoders are implemented by stacking a sequence of 45 historical observations information $H_t : [O_{t-N}, O_{t-N-1}, ..., O_{t-1}]$ into the input of an MLP.

3.2   Training and Implementation Details

Termination. The robot's locomotion training is governed by specific termination conditions designed to ensure safety and effectiveness. Episodes are terminated upon detecting collisions involving the trunk, upper limbs, thighs, and calves, prioritizing the study's focus on bipedal locomotion with only two-foot ground contact.

Domain Randomization. To facilitate the transfer of learned behaviors from simulation to the real world, domain randomization has been employed. This approach involves randomizing various parameters crucial for robot locomotion, such as terrain friction, base mass, joint PD controller gains, ground friction, restitution, and perturbations to the robot's base velocity. During training, sampled velocity vectors are added to the robot's current base velocity at random intervals. The specific randomization variables and their corresponding uniform distribution ranges are detailed in Table 1, enabling robust policy adaptation and testing in diverse real-world environments.

Simulation Setup. We trained 500 parallel agents on different types of terrains with increasing difficulties using the Isaac Gym simulator [23] in overall 26000 iterations and cost 15.88 h in total. Figure 2 showed the simulation in different terrain type. Each RL episode lasts for a maximum of 1000 steps, and terminates

Table 1. Randomized Simulation Parameters

Parameters	Range	Unit
Link Mass	[0.8, 1.2]	kg
Payload Mass	[0, 3]	kg
Payload Range	[−3]	-
Ground Friction	[0.05, 2.75]	-
Ground Restitution	[0.0, 1.0]	-
Joint K_p	[0.8, 1.2] × 20	-
Joint K_d	[0.8, 1.2] × 0.5	-

early if it reaches the termination criteria. The control frequency of the policy is 50 Hz in the simulation. All training's were performed on a single NVIDIA RTX 4070 GPU. In A1 locomotion scenarios, actions are represented by $a_t \in R^{12}$ a 12-dimensional vector specifying the desired positional adjustments for each actuated joint as dictated by the Proportional-Derivative (PD) controller.

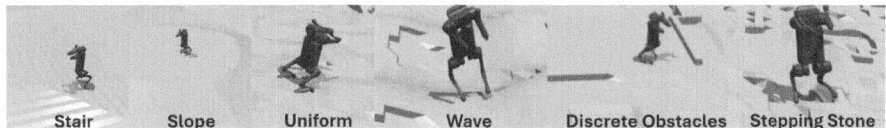

Fig. 2. Terrains in Isaac Gym Simulations.

4 Results

In our experiments within the Isaac Gym simulation environment, we evaluated the performance of the updated policy for a quadrupedal robot in biped locomotion. The results, depicted in Table 2, show varying success rates and tracking accuracy across different terrains and speeds.

The robot generally performs well at lower speeds, with higher success rates on less challenging terrains. However, success rates and tracking accuracy decrease as speed and terrain difficulty increase, notably on sloped terrains and obstacles due to mechanical design limitations.

The uniform terrain and discrete obstacle terrain were specifically chosen for detailed analysis as they are highly representative of uneven terrains. The uniform terrain has uneven features, while the discrete obstacle terrain combines elements of both flat and stepped obstacles, akin to stairs, providing a comprehensive challenge for evaluating the robot's locomotion policy.

Figure 3 presents the base linear velocity in the x direction and the base angular velocity in yaw for the robot on both uniform and discrete obstacle terrains.

Table 2. Comparison of Tracking Accuracy and Success Rate Across Various Terrains and Speeds. The success rate and tracking accuracy were evaluated over 10 trials for each terrain type.

Terrain Types		Uniform	Wave	Stepping Stones	Sloped	Stairs	Obstacles
0.5 m/s	**Acc (%)**	82.76	80.84	80.06	81.37	77.67	80.88
	Succ (%)	92.3	91.76	91.6	84.21	90.09	89.55
1.0 m/s	**Acc (%)**	81.29	79.64	77.73	80.01	71.57	73.51
	Succ (%)	90	91.59	90.01	82.14	86.14	86.59
1.5 m/s	**Acc (%)**	76.60	77.19	73.46	78.43	64.57	71.47
	Succ (%)	86.96	85.06	88.73	79.5	84.91	84.55
2.0 m/s	**Acc (%)**	63.73	66.28	64.69	64.68	53.38	61.52
	Succ (%)	79.55	84.22	78.55	76.23	81.53	78.1

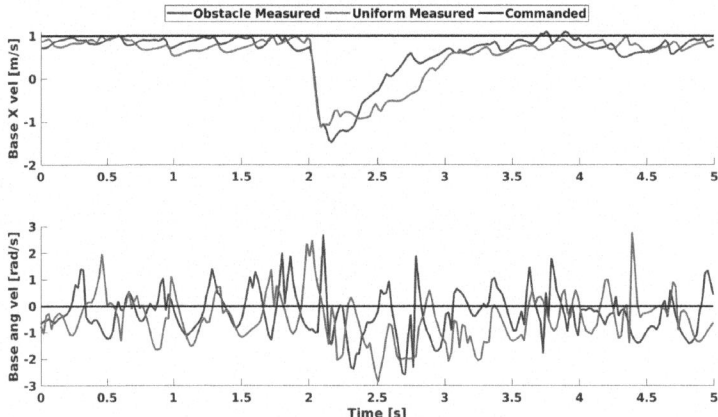

Fig. 3. Base linear velocity in the x direction and base angular velocity in yaw for the robot on both uniform and discrete obstacle terrains.

During the push test, noticeable changes in these velocities occur, reflecting the robot's dynamic response to the external force. The robot's base linear velocity tracking shows acceptable performance, with the measured velocities closely following the commanded values. The base angular velocity tracking, while not as accurate as the linear velocity, still demonstrates the robot's ability to follow the commanded trajectory to a reasonable extent. The large spike observed around the 2-s mark is attributed to the external push force. Despite this disturbance, the robot quickly regains stability, indicating robust recovery capabilities of the locomotion policy. This ability to reject disturbances is crucial for maintaining stable locomotion in unpredictable environments.

The contact force data, shown in Fig. 4, provides additional insights into the interaction between the robot's feet and the terrain. The vertical ground reaction forces F_z were measured for both the rear-left (RL) and rear-right (RR) legs on discrete obstacles and uniform terrains. The data reveals asymmetry in the contact forces, indicating that the learned policy does not exhibit a perfectly symmetrical gait. To address this issue and enhance the robot's performance, future work could incorporate a symmetrical reward function or a reward based on the orientation of base tracking into the learning algorithm.

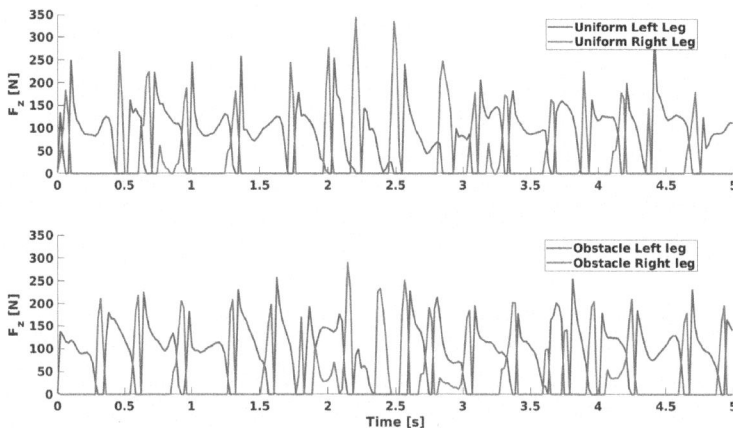

Fig. 4. Robot's feet contact forces on both uniform and discrete obstacle terrains.

The push test results and performance analysis provide additional insights into the robot's capabilities and resilience. As illustrated in Fig. 5, the robot's response to a 100N push applied along the x-axis over a duration of 0.1 s is captured, showcasing the policy's effectiveness in handling external disturbances. Despite the significant push, the robot manages to stabilize itself quickly, demonstrating the robustness and resilience of the developed locomotion policy.

Fig. 5. Snapshots illustrating the response of a robot subjected to a 100N push force (indicated by a red arrow) applied along the x-axis over a duration of 0.1 s. The sequence shows the robot's movement and stabilization process. (Color figure online)

The results indicate that the legged robot's locomotion policy exhibits satisfactory performance in both base linear velocity tracking and recovery from external disturbances. The ability to maintain stability and follow commanded trajectories on different terrains, as well as the efficient recovery from disturbances, underscores the robustness of the developed policy.

Overall, the experimental outcomes validate the effectiveness of our legged locomotion policy in achieving stable, efficient, and resilient movement across various terrains, even under external disturbances. These findings contribute valuable knowledge towards the development of energy-efficient and robust legged robots capable of operating in diverse and challenging environments.

5   Conclusion and Future Work

In this paper, we proposed a novel framework for learning robust, agile, and natural bipedal locomotion skills for quadruped robots in simulation. Utilizing a teacher-student learning framework with privileged and terrain information, we enhanced the robustness of the learned policy and helped bridge the sim-to-real gap. By integrating adversarial motion imitation, the learned gait mimics the style and behavior of a TO reference gait. Our results demonstrate high-performance blind locomotion in a quadruped robot in biped mode.

Overall, our findings highlight the potential of imitation learning and TO in achieving agile and robust locomotion across diverse robotic platforms. Future work will focus on developing more robust biped motion capabilities on uneven terrain, transferring these capabilities to physical robots, and refining the transition from quadrupedal to biped mode to enhance legged robots' versatility.

References

1. Bao, L., et al.: Deep reinforcement learning for bipedal locomotion: a brief survey (2024). arXiv: 2404.17070 [cs.RO]
2. Ho, J., Ermon, S.: Generative adversarial imitation learning. In: International Conference on Neural Information Processing Systems, pp. 4572–4580 (2016)
3. Peng, X., et al.: DeepLoco: dynamic locomotion skills using hierarchical deep reinforcement learning. ACM Trans. Graph. **36**, 1–13 (2017)
4. Xie, Z., et al.: ALLSTEPS: curriculum-driven learning of stepping stone skills. Comput. Graph. Forum **39**, 213–224 (2020)
5. Siekmann, J., et al.: Sim-to-real learning of all common bipedal gaits via periodic reward composition. In: IEEE International Conference on Robotics and Automation, pp. 7309–7315 (2021)
6. Xie, Z., et al.: Learning locomotion skills for Cassie: iterative design and sim-to-real. In: Conference on Robot Learning, pp. 317–329 (2020)
7. Li, Z., et al.: Reinforcement learning for robust parameterized locomotion control of bipedal robots. In: IEEE International Conference on Robotics and Automation, pp. 2811–2817 (2021)
8. Zhang, Q., et al.: Whole-body humanoid robot locomotion with human reference. arXiv preprint arXiv:2402.18294 (2024)
9. Wu, J., et al.: Learning robust and agile legged locomotion using adversarial motion priors. IEEE Rob. Autom. Lett. **8**(8), 4975–4982 (2023)
10. Escontrela, A., et al.: Adversarial motion priors make good substitutes for complex reward functions. In: IEEE/RSJ International Conference on Intelligent Robots and Systems, pp. 25–32 (2022)

11. Wang, Y., Jiang, Z., Chen, J.: Learning robust, agile, natural legged locomotion skills in the wild. In: RoboLetics: Workshop on Robot Learning in Athletics@ CoRL (2023)
12. Winkler, A., et al.: Gait and trajectory optimization for legged systems through phase-based end-effector parameterization. IEEE Rob. Autom. Lett. **3**(3), 1560–1567 (2018)
13. Yu, F., et al.: Dynamic bipedal turning through sim-to-real reinforcement learning. In: IEEE-RAS International Conference on Humanoid Robots, pp. 903–910 (2022)
14. Duan, H., et al.: Learning task space actions for bipedal locomotion. In: IEEE International Conference on Robotics and Automation, pp. 1276–1282 (2021)
15. Castillo, G.A., et al.: Template model inspired task space learning for robust bipedal locomotion. In: IEEE/RSJ International Conference on Intelligent Robots and Systems, pp. 8582–8589 (2023)
16. Lee, J., et al.: Learning quadrupedal locomotion over challenging terrain. Sci. Rob. **5**(47), eabc5986 (2020)
17. Kumar, A., et al.: RMA: rapid motor adaptation for legged robots. arXiv preprint arXiv:2107.04034 (2021)
18. Yu, C., Rosendo, A.: Multi-modal legged locomotion framework with automated residual reinforcement learning. IEEE Rob. Autom. Lett. **7**(4), 10312–10319 (2022)
19. Peng, X.B., et al.: AMP: adversarial motion priors for stylized physics based character control. ACM Trans. Graph. **40**(4), 1–20 (2021)
20. Winkler, A.W.: TOWR-an open-source trajectory optimizer for legged robots in C (2018). https://github.com/ethz-adrl/towr
21. Schulman, J., et al.: Proximal policy optimization algorithms. arXiv preprint arXiv:1707.06347 (2017)
22. Margolis, G.B., et al.: Rapid locomotion via reinforcement learning. arXiv preprint arXiv:2205.02824 (2022)
23. Rudin, N., et al.: Learning to walk in minutes using massively parallel deep reinforcement learning. In: Conference on Robot Learning, pp. 91–100. PMLR (2022)

Ground-Truth Based Calibration for a Two Wheel Differential Drive Robot

Youssef G. Alboraei$^{(\boxtimes)}$ ⓘ, Chayada Thidrasamee ⓘ, and Paul O'Dowd ⓘ

University of Bristol, Bristol BS8 1QU, UK
youssef.alboraei@bristol.ac.uk

Abstract. This study builds upon existing odometry calibration formulation for two-wheeled differential drive robots. The proposed approach increases the potential for autonomous calibration by utilising closed-loop control where the mobile robot is programmed to follow a line marked out on a surface. The proposed approach negates the need for external measurements made by a human operator within the calibration process. This study makes a comparative analysis and evaluation between the presented approach against an established prior approach by implementing both on a real (not simulated) mobile robot and capturing performance data.

Keywords: Odometry · Pose errors · Calibration · Wheel robot

1 Introduction

This study investigates a novel calibration approach to correct the kinematic parameters in two-wheel differential drive mobile robots. The objective of this technique is to enhance odometry performance while removing the necessity for final pose measurement procedures by a human. This is achieved by using a line-following behaviour programmed into the mobile robot, where a line placed onto the surface of the task environment represents a ground-truth source of simple localisation information perceivable to the robot.

Robot localisation is one of the main challenges in the area of autonomous mobile robots [7]. Odometry is the capability of a mobile robot to compute and estimate its change in position over time. Errors in odometry can be categorised into two types: *systematic* and *non-systematic*.

Non-systematic errors are unpredictable or have extrinsic cause, such as a loss of traction. It is possible to address non-systematic errors through the inclusion of complementary information from sensors such as inertia measurement units [3], or with advanced localisation technologies like Simultaneous Localisation and Mapping (SLAM) [13], Monte Carlo Localisation (MCL) [14], or Visual Odometry [11]. These complementary approaches to odometry are outside the scope of our study, where addressing systematic errors underpinning odometry is the focus.

© The Author(s), under exclusive license to Springer Nature Switzerland AG 2025
M. N. Huda et al. (Eds.): TAROS 2024, LNAI 15052, pp. 130–142, 2025.
https://doi.org/10.1007/978-3-031-72062-8_12

Systematic errors are deterministic and can therefore be mitigated through calibration. Systematic errors often relate to properties of the robot geometry and how the robot senses displacement. For example, unequal wheel sizes, irregular distance between wheels, quantisation error when detecting the robot displacement, or loss through approximations within the kinematic model. Systematic errors are significant because they compound over time whilst the robot is operating [9].

Whilst non-systematic errors are important to address, it stands to reason that mitigating systematic errors within odometry will improve its utility in all other contingent areas. It is therefore advantageous to develop a technique to resolve systematic errors efficiently prior to a full system deployment.

Importantly, the majority of prior odometry calibration techniques require a human operator to take external measurements of a mobile robot's final pose, or at specific intervals of operation. Furthermore, in prior approaches utilising open-loop control there can be a high variability in the final robot pose, and the external measurement is required as a point of reference against the robot odometry estimation. In our approach, by following a line, the robot finishes closed-loop motion in approximately the same final pose near the start point, providing a convenient point of reference to the robot. Our results show that calibration error can then be calculated effectively from internal odometry estimation, negating external human made measurements.

Prior open-loop approaches create a fundamental limitation to the autonomy of the approach. Junior & Todd [6] also note in conclusion that the capacity of a final external measurement of a robot pose will have an impact on the accuracy and precision of a calibration method. In our approach, we have experimentally validated our proposed technique by calculating and applying the calibration feedback offline. However, the approach is suitable for an autonomous implementation. Along these lines, our study has a greater focus on a *data acquisition* approach, which builds upon prior work that has refined the calibration formulation. Specifically, we make a novel application of the *Experimental Heading Errors* (EHE) formulated by Jung & Chung [10] (see Sect. 2 for more detail) increasing the potential for autonomy. Because we have built upon EHE, we also evaluate the original EHE approach by the same measures to serve as comparison to our proposed calibration.

Exceptions to external manual measurements include [4] who utilise a mobile robot fitted with a laser range-finder so that the robot can detect its displacement within a rectangular environment, and [8] utilise an indoor positioning system to inform the calibration process. Both approaches are relatively expensive and require a specific environment. We present a line following approach where a line shape could be situated much more conveniently within a task environment (such as a sticker/decal on a surface) and at very low-cost.

The structure of this paper is organised as follows. Subsection 1.1 delineates the calibration problem and offers a broad perspective of our methodology. Following this, Sect. 2 presents an in-depth calibration technique devised. Section 3 then details the experimental procedures employed. Ultimately, the results are presented and examined in Sects. 4 and 5, respectively.

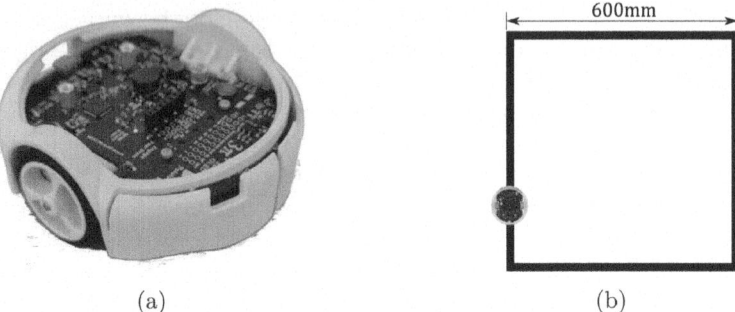

Fig. 1. (a) Pololu 3PI+ robot used [5] (b) the line track used and 3Pi+ robot drawn to scale.

1.1 Problem Definition

Similar to [1,2,10], we adopt a problem definition of addressing two types of systematic error in odometry calibration. These are *Type A* and *Type B* errors [9]. A detailed review of type A and type B errors is provided in [9] and the combined effect of both types is investigated in [10]. Type A error occurs when the actual distance between the robot's wheels deviates from the nominal value defined in the kinematic model. This introduces an error in the actual angle of rotation calculated by odometry. For instance, a trajectory with a Type A error may lead to a turning motion of 85° instead of the intended 90° due to an inaccurately estimated wheelbase. Type B errors stem from variations in wheel diameter, either wheels of unequal sizes, or deviating from the nominal value defined in the kinematic model. Type B error introduces error in the actual translation of the robot calculated by odometry.

To address Type A and Type B errors, the calibration formulation of the EHE technique [10] accounts for the coupled effects between type A and type B errors, as well as avoids the approximated equations of position errors, addressing the limitations of the UMBmark technique [2]. Similar to [2,9], and [10], we adopt the same premise of a two wheeled mobile robot that transits a square track. However, we present a new approach where the robot will actually follow a square track defined as a line on the surface (illustrated in Fig. 1b), and utilise the detection of this line as a source of absolute pose information to autonomously calibrate its odometry.

Our proposed technique eliminates the need to collect measurements of the robot's position or heading for calibration. In contrast to other methods that depend on a square line and require a comparison of the robot's final position [1,2] or heading [10] with its odometry-estimated values to identify errors, our technique ensures that the robot returns near its initial absolute pose. As a result, errors can be directly computed from the odometry data alone, which is readily available to the robot.

This technique can be integrated with the calibration equations of any calibration technique that employs square motion for calibration. In this work, we have adopted the calibration equations proposed by Jung and Chung in [10], which is referred to as the *Experimental Heading Error* (EHE) technique. Our proposed technique is referred to as the *Experimental Heading Error with Line Following* (EHE-LF) technique.

2 Implementation

This section outlines the practical steps involved in implementing the EHE-LF calibration technique to correct the parameters of the 3pi+ robot. This technique is implemented through a calibration routine in which the robot follows a square track on the ground and collects the calibration data. Notably, this procedure can be implemented as an autonomous calibration routine for online calibration, as it does not require access to external pose information. However, in this study, we collected odometry errors from the robot and performed corrections offline to concentrate on evaluating the viability of the calibration technique.

2.1 Overview of EHE-LF Calibration Procedure

The calibration procedure requires the robot to follow a square track at a 90-degree turn for each corner in both clockwise (CW) and counterclockwise (CCW) directions. This is repeated five times to enhance data robustness by minimising the potential effects of non-systematic errors. The procedure followed to implement the EHE-LF technique is summarised as follows:

1. The robot begins on the centre of one of the square track corners and facing the direction of CW rotation.
2. Start the run; the robot should:
   (a) Follow one edge of the square.
   (b) Detect the next corner and stop at its centre.
   (c) Performs a (90°) rotation in CW.
   (d) Repeat substeps a–c four more times.
   (e) At the end of the last turn, record the odometry-estimated heading θ_{odom}
3. Repeat step 2 four more times.
4. Repeat steps 1–3 in the CCW direction.
5. Use Eqs. 16–17 to calculate the corrected parameters, r_{left_actual}, r_{right_actual}, and b_{actual}.

As detailed in the procedure above, executing this technique requires programming the robot to adeptly follow a straight line on the ground, identify corners, and execute 90° turns. The smoothness of the line following is crucial; aggressive line following can introduce slippage and impede the accuracy of calibration. Similarly, accurately detecting corners and performing 90° rotations guided by the line is necessary, especially at the end of each turn, where data is collected for the final heading. The extent to which this procedure needs to be accurate can depend on the robot parameters as well as the track size (see [1, 10]).

Subsections 2.2–2.4 describe the implementation of EHE-LF on a Pololu 3pi+ robot [5], while Subsects. 2.5–2.6 describe the odometry parameters correction process.

2.2   Line Following and Turning

The 3pi+ robot is equipped with five infrared (IR) sensors (DN1 to DN5) on its front, facing the ground. These sensors undergo a calibration process to normalise their readings between $[0, 1]$, which facilitates accurate sensor output across their sensitive range. Sensors DN2 and DN4 are primarily used for line following, while DN1, DN3, and DN5 are used for detecting corners and executing precise turning motions.

The formula for the normalised reading N_i of any sensor i is:

$$N_i = \frac{R_i - R_{\min,i}}{R_{\max,i} - R_{\min,i}} \tag{1}$$

where R_i, $R_{\max,i}$, and $R_{\min,i}$ are the raw, maximum, and minimum readings of sensor i, respectively.

For line following, the normalised readings from sensors DN2 and DN4 (N_2 and N_4) are used to calculate the weighted error $\epsilon_{\text{heading}}$, representing the robot's heading deviation from the line:

$$\epsilon_{\text{heading}} = \frac{N_2 - N_4}{N_2 + N_4} \tag{2}$$

This normalised error $\epsilon_{\text{heading}}$ ranges between $[-1, 1]$ and feeds into the heading Proportional-Integral-Derivative (PID) controller, which adjusts the turning values for each wheel's PID speed controller, ensuring smooth line following and precise turning.

At each corner of the predefined square track, the turning mechanism is initiated when sensors DN2, DN3, and DN4 detect the endpoint of a straight segment. The robot progresses until sensors DN1 or DN5 re-encounter the line, which sets the turning direction by assigning a turn angle of either $90°$ or $-90°$. This setup enables the robot to autonomously determine its rotation direction as either clockwise (CW) or counterclockwise (CCW). The robot continues to advance until it loses the line, positioning itself precisely for the turn. This step can differ for different robots depending on their dimensions and sensor locations, but adjustments should be straightforward. The robot then executes the rotation according to the predetermined turn angle and continues until sensor DN3 re-detects the line, marking the completion of the turn. This process allows for precise and controlled navigation through each corner, facilitating smooth transitions along the square track.

2.3   Robot Kinematics

It is essential to employ the kinematic model of the robot for accurate odometry calculation and speed control. The 3pi+ robot uses a two-wheel differential drive model, detailed in [12].

The 3pi+ robot is equipped with dual quadrature rotary encoders [5], each connected to a motor. The kinematic equations are updated every 15 ms, and wheel speed is estimated via encoder counts using a complementary filter (Eq. 3), which mitigates noise from quantisation errors:

$$\phi_{k_w} = \alpha\phi_{(k-1)_w} + (1-\alpha)\frac{\Delta d_w}{t_{cycle}} \tag{3}$$

where ϕ_{k_w} is the angular velocity of wheel w at time k, t_{cycle} is the time interval in milliseconds, and α is the complementary filter coefficient, tuned from 0 to 1.

2.4 System Parameters

Following the above description of the implementation, the final and fixed parameters used to configure the system are presented in Table. 1:

Table 1. 3pi+ Control System Parameters

Heading PID	$K_P = 95.0$	$K_I = 0.75$	$K_D = 0.1$
Left Wheel PID	$K_P = 0.14$	$K_I = 0.013$	$K_D = 0.001$
Right Wheel PID	$K_P = 0.14$	$K_I = 0.013$	$K_D = 0.001$
Cycle Interval (ms)	15		
Wheel Speed (encoder counts/s)	200		
Complementary Filter Coefficient α	0.92		

2.5 Data Collection

The EHE-LF technique requires an estimate of the heading error after completing the square track specifically after performing the final turn. The heading error (θ) is computed as the difference between the absolute heading (θ_{abs}) and the odometry-estimated heading (θ_{odom}), as shown in Eq. (4).

Although the position error is not needed for calibration, it was also collected for the purpose of the evaluation and comparison of the odometry accuracy. Similar to the heading error, the position error values (x and y) are calculated from the absolute position (x_{abs} and y_{abs}) and the odometry-estimated position (x_{odom} and y_{odom}), as shown in Eq. 4.

$$\theta = \theta_{abs} - \theta_{odom}, \quad x = x_{abs} - x_{odom}, \quad y = y_{abs} - y_{odom} \tag{4}$$

Since the robot is programmed to finish the track at approximately the same starting pose, the absolute pose of the robot after finishing the track is defined as in Eqs. 5, 6.

$$\theta_{abs,cw} = -2\pi, \quad x_{abs,cw} = 0, \quad y_{abs,cw} = 0 \tag{5}$$

$$\theta_{abs,ccw} = 2\pi, \quad x_{abs,ccw} = 0, \quad y_{abs,ccw} = 0 \tag{6}$$

Thus, the pose error can be determined solely from the robot's odometry-estimated pose as shown in Eqs. 7, 8.

$$\theta_{cw} = -2\pi - \theta_{odom,cw}, \quad \theta_{ccw} = 2\pi - \theta_{odom,ccw}, \tag{7}$$

$$x_{cw/ccw} = -x_{odom,cw/ccw}, \quad y_{cw/ccw} = -y_{odom,cw/ccw} \tag{8}$$

By following the EEH-LF procedure outlined in Subsect. 2.1 and applying Eqs. 7–8, one can derive five sets of pose error values each for clockwise (CW) and counter-clockwise (CCW) directions (total of 10 sets, 3 values each). The centre of gravity (c.g.) for both CW and CCW is determined by averaging the error values in each respective direction, as illustrated in Eqs. 9–11.

$$\theta_{cg,cw/ccw} = \frac{1}{n} \sum_{i=1}^{n} \theta_{i,cw/ccw} \tag{9}$$

$$x_{cg,cw/ccw} = \frac{1}{n} \sum_{i=1}^{n} x_{cg,cw/ccw} \tag{10}$$

$$y_{cg,cw/ccw} = \frac{1}{n} \sum_{i=1}^{n} y_{cg,cw/ccw} \tag{11}$$

where $n = 5$ is the total number of runs in each direction.

Values from Eq. 9 are used for calibration. While all values from Eqs. 9–11 are used for evaluation and comparison.

2.6   EHE-LF Calibration Process

The EHE-LF calibration technique employs the same calibration equations as the EHE technique, detailed in [10].

The combined impact of Type A and Type B errors is demonstrated in [10] through Eq. 12. These equations show that both the error stemming from 90° rotations (α) and the error resulting from linear motions (β) are influenced by inaccuracies in wheelbase (ε_b) and wheel diameter (ε_d).

$$\theta_{error,90°turn} : \alpha = \alpha_{\varepsilon b} + \alpha_{\varepsilon d}, \quad \theta_{error,straightline} : \beta = \beta_{\varepsilon b} + \beta_{\varepsilon d} \tag{12}$$

The heading error values $\theta_{cg,cw}$ and $\theta_{cg,ccw}$ from Eq. 9 are used to calculate the parameters $\alpha_{\varepsilon b}$ and $\beta_{\varepsilon d}$ as follows in Eq. 13.

$$\alpha_{\varepsilon b} = \frac{\theta_{cg,cw} - \theta_{cg,ccw}}{8}, \quad \beta_{\varepsilon d} = \frac{\theta_{cg,cw} + \theta_{cg,ccw}}{8} \tag{13}$$

The correction parameters E_b and E_d are determined using Eq. 14. Here, $\theta_{nominal}$ represents the angle turned by the robot at each corner, and R is the radius of the curved track executed by the robot when instructed to move straight, reflecting odometry errors. In the EHE-LF procedure, the turning angle is consistently 90°. Importantly, although the robot in EHE-LF appears to follow a straight line, the value R signifies the radius of the curved track of the *odometry-estimated* trajectory due to odometry errors. The derivation of the term R is explained in [9].

$$E_b = \frac{\theta_{nominal}}{\theta_{nominal} - \alpha}, \quad E_d = \frac{R + (b_{actual}/2)}{R - (b_{actual}/2)} \tag{14}$$

The corrected kinematic parameters, r_{left_actual}, r_{right_actual}, and b_{actual} can be calculated following the equations proposed in [10] as in Eqs. 15–17.

$$r_c = \left(\frac{r_{left\_nominal} + r_{right\_nominal}}{2} \right) \tag{15}$$

$$r_{left\_actual} = \frac{2}{E_d + 1} r_c, \quad r_{right\_actual} = \frac{2}{(\frac{1}{E_d}) + 1} r_c \tag{16}$$

$$b_{actual} = E_b b_{nominal} \tag{17}$$

3   Experimental Methodology

Two experiments were designed to assess the efficacy of the proposed EHE-LF technique. The first experiment compares the EHE-LF technique with the EHE technique to validate the applicability of the proposed method. The second experiment tests the EHE-LF technique across different wheelbase values for the robot to evaluate its robustness.

Section 2 provides a detailed implementation of the EHE-LF technique. The EHE technique implementation follows the procedure described in [10], with the only variation being the selection of the parameter L to match the value used in EHE-LF, as discussed in Subsect. 3.1.

3.1   Discussion of Variables

Controlled Variables:

– **Length of the Square Edge** ($L = 600\,\text{mm}$): Selected to be sufficiently large to produce observable errors, particularly of Type B, while allowing for convenient execution within a reasonable timeframe (see [1] for a discussion on track size selection).
– **Type of Ground (Flat Surface)**: Chosen to minimise the impact of potential nonsystematic errors on the calibration procedures.
– **Width of the Black Track Line** ($W = 20\,\text{mm}$) in **EHE-LF**: Ensures efficient line following by maintaining close proximity of the line-following sensors DN2 and DN4 to the line edges, minimising oscillations and enhancing calibration results.

Independent Variables:

– In the calibration techniques comparison, the independent variable is the **track following method**: a programmed square for EHE versus an actual square on the ground for EHE-LF.

– In the wheelbase variation experiment, the independent variable is the robot's **wheelbase size**. The wheelbase was adjusted to three sizes: 86 mm, 91 mm, and 100 mm. Calibration was done for each size, with a fixed odometry nominal value of 86 mm.

Dependent Variables: The difference between the odometry-estimated pose and the final actual pose is observed to evaluate the performance of calibration Further details are provided in Subsect. 3.2 below.

3.2 Discussion of Metrics

The objective of performing calibration techniques is to minimise the difference between the absolute pose $(x_{abs}, y_{abs}, \theta_{abs})$ of the robot and the odometry-estimated pose $(x_{odom}, y_{odom}, \theta_{odom})$. Thus, the dependent variables for this experiment were determined to be this pose error. To analyse this error, it was separated into heading error (θ) and position error (x, y).

In [9], it was proposed to use the root mean square error of position to reflect the accuracy of the odometry. This value was determined as the maximum between the two root mean squares of the position errors in CW and CCW.

Following a similar approach, the metric for position accuracy $(r_{x,y})$ was calculated as the position root mean square error as in Eq. 18. However, the metric for heading accuracy (r_θ) was selected by directly referring to the heading error θ as in Eq. 19.

$$r_{x,y} = \sqrt{x^2 + y^2} \tag{18}$$
$$r_\theta = \theta \tag{19}$$

4 Results

4.1 Results of Varying the Calibration Technique

Table 2 displays the corrected kinematic parameters, r_{left}, r_{right}, and b, obtained using EHE and EHE-LF techniques. It is worth noting that the direction of change is consistent across all parameters, indicating that both techniques had a similar effect on the parameters to varying degrees. Figure 3 illustrate the resulting position and heading errors, using the metrics $r_{x,y}$, and r_θ introduced by Eqs. 18 and 19, for both EHE and EHE-LF techniques.

Table 2. Robot parameters before and after calibration with EHE and EHE-LF

Parameter	b	r_{left}	r_{right}
Nominal	86.0000	16.4000	16.4000
EHE	86.0629	16.3881	16.4119
EHE-LF	86.1838	16.3869	16.4131
(in mm)			

The error distribution presented by Fig. 2 indicates that during CCW runs, both EHE and EHE-LF exhibited comparable accuracy, clustering around the origin. However, in CW runs, EHE-LF demonstrated enhanced accuracy. These findings suggest that EHE-LF may be less prone to the effects of nonsystematic errors when compared to EHE. When considering both CW and CCW runs, the results of the EHE technique exhibit notable variability in errors, as depicted in Figs. 3a and 3b, with an average position error of 14.1 mm. This indicates a broad dispersion of errors across the runs, ranging from 5.5 mm to 25 mm. In contrast, the EHE-LF produced errors within a narrower range, as shown in Fig. 3b, ranging from 2.5 mm to 13.3 mm, with an average of about 8.7 mm. The clustered data implies a more consistent and uniform performance. The decreased disparity suggests that EHE-LF is highly performant at minimising calibration errors and delivering more precise measurements.

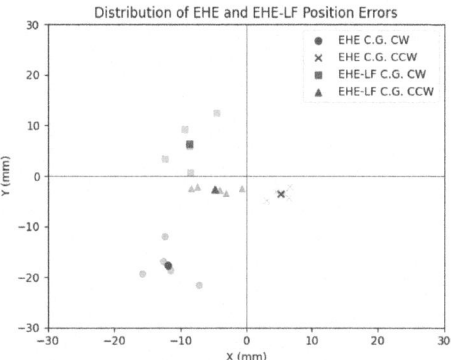

Fig. 2. Distribution of position error of EHE and EHE-LF after calibration

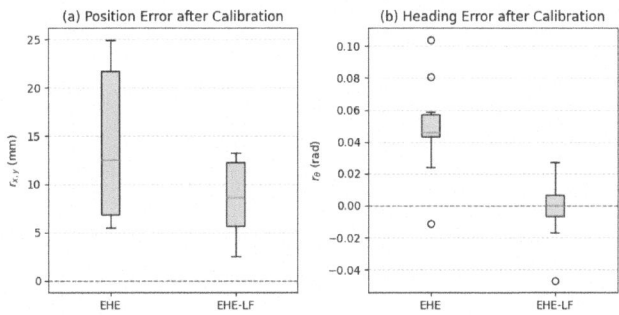

Fig. 3. Errors in (a) position and (b) heading for 10 runs (5 CW and 5 CCW)

In Fig. 3b, EHE-LF displays an average angular deviation of -0.003 rad, indicating minimal discrepancies in measurements. This represents a significant enhancement in comparison to EHE which exhibits an average deviation of 0.049 rad with a non-uniform distribution of errors.

4.2   Results of Varying the Wheelbase

Table 3 displays the estimated kinematic parameters resulting from the application of the EHE-LF calibration method for three distinct wheelbase lengths: 86 mm, 91 mm, and 100 mm. Notably, for all three wheelbase sizes, calibration manages to estimate a value close to the actual size. Additionally, the estimated values for wheel radii has similar direction of change. This underscores the robustness of our approach in calibrating the parameters.

Figure 4 presents the position and heading errors for robots with 86 mm, 91 mm, and 100 mm wheelbases, measured before and after the calibration process. The calibration has notably decreased both the range and median values of errors for all wheelbase measurements, demonstrating the effectiveness of the calibration process. Specifically, from Table 3 for the 86 mm wheelbase, there was an 88.48% reduction in positional error and an 88.31% percent reduction in heading error, indicating a substantial enhancement in precision. This pattern of improvement is mirrored in the 91 mm and 100 mm wheelbases, with positional errors diminishing by 77.40% and 85.07%, respectively, and heading errors by 93.36% and 96.40%. These results underscore the calibration's consistent impact on accuracy across different wheelbase lengths.

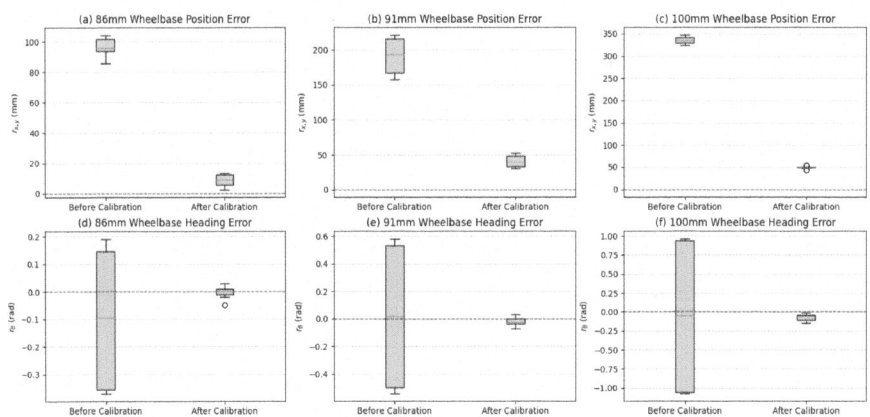

Fig. 4. Resulting errors after EHE-LF calibration for different wheelbases

Table 3. Calibration results with EHE-LF for different wheelbase values.

$b_{measured}$	$b_{nominal}$		r_{left}		r_{right}		Accuracy improvement	
	Before	After	Before	After	Before	After	Heading	Position
86.0000	86.0000	86.1838	16.4000	16.3869	16.4000	16.4131	88.31%	88.48%
91.0000	86.0000	93.7423	16.4000	16.3876	16.4000	16.4124	93.36%	77.40%
100.0000	86.0000	100.3247	16.4000	16.3962	16.4000	16.4038	96.40%	85.07%

(parameters in mm)

5   Discussion and Conclusion

The proposed EHE-LF calibration technique demonstrates significant effectiveness in achieving the expected changes across all odometry parameters, with reference to the established EHE technique that utilises the same equations. This effectiveness is evidenced by the consistent direction of change in the kinematic parameters corrected by both the EHE and EHE-LF techniques (see Table 2). Such consistency is crucial to evaluate and show the predictability and reliability of the technique. Furthermore, the technique exhibits robust adaptability, as demonstrated by its successful application across varying wheelbases values for the robot. This underscores the EHE-LF's potential for broad use across diverse systems of mobile robots.

More significantly, EHE-LF not only matches but surpasses the performance of the established EHE technique. This improvement can be attributed to its ability to reduce wheel slippage, since following a straight line on the ground makes the robot's wheels rotate at relatively similar velocities, which minimises slippage. This is in contrast to following a programmed line, where odometry errors often cause the robot to move in a curved line, leading to different wheel velocities and introducing a source of non-systematic errors. By isolating errors to exist only in odometry estimation, non-systematic errors are minimised, and the observed errors more accurately reflect the inaccuracies in the nominal kinematic parameters. Utilising ground truth data makes the experiments more applicable in real-world scenarios, allowing the enhancements observed under controlled conditions to be evaluated across various environments.

References

1. Accurate calibration of kinematic parameters for two wheel differential mobile robots (2011). https://doi.org/10.1007/S12206-011-0334-Y
2. Borenstein, J., Feng, L.: UMBmark: a benchmark test for measuring odometry errors in mobile robots. In: Mobile Robots X, vol. 2591, pp. 113–124. SPIE (1995)
3. Cantelli, L., Ligama, S., Muscato, G., Spina, D.: Auto-calibration methods of kinematic parameters and magnetometer offset for the localization of a tracked mobile robot. Robotics **5**(4) (2016).https://doi.org/10.3390/robotics5040023. https://www.mdpi.com/2218-6581/5/4/23
4. Censi, A., Franchi, A., Marchionni, L., Oriolo, G.: Simultaneous calibration of odometry and sensor parameters for mobile robots. IEEE Trans. Rob. **29**(2), 475–492 (2013). https://doi.org/10.1109/TRO.2012.2226380
5. Corporation: Pololu 3pi+ 32u4 user's guide. https://www.pololu.com/docs/0J83/all. https://www.pololu.com/docs/0J83/all
6. D.L Tomasi, Todt, T.: Rotational odometry calibration for differential robot platforms. IEEE **12**(6), 869 – 880 (2017). https://doi.org/10.1109/SBR-LARS-R.2017.8215315
7. Desai, A., Ghagare, N., Donde, S.: Optimal robot localisation techniques for real world scenarios. In: 2018 Fourth International Conference on Computing Communication Control and Automation (ICCUBEA), pp. 1–9. IEEE (2018)

8. Goronzy, G., Hellbrueck, H.: Weighted online calibration for odometry of mobile robots. In: 2017 IEEE International Conference on Communications Workshops (ICC Workshops), pp. 1036–1042 (2017). https://doi.org/10.1109/ICCW.2017. 7962795

9. Borenstein, J., Feng, L.: Measurement and correction of systematic odometry errors in mobile robots. IEEE **12**(6), 869– 80 (1996).https://doi.org/10.1109/70.544770

10. Jung, C., Chung, W.: Accurate calibration of two wheel differential mobile robots by using experimental heading errors. In: 2012 IEEE International Conference on Robotics and Automation, pp. 4533–4538, May 2012. https://doi. org/10.1109/ICRA.2012.6224660. https://ieeexplore.ieee.org/abstract/document/ 6224660. ISSN: 1050-4729

11. Scaramuzza, D., Fraundorfer, F.: Visual odometry [tutorial]. IEEE Rob. Autom. Mag. **18**(4), 80–92 (2011)

12. Siegwart, R., Nourbakhsh, I.R., Scaramuzza, D.: Introduction to autonomous mobile robots. In: Intelligent Robotics and Autonomous Agents, 2nd edn. MIT Press, Cambridge, Mass (2011). oCLC: ocn649700153

13. Stachniss, C., Leonard, J.J., Thrun, S.: Simultaneous localization and mapping. In: Siciliano, B., Khatib, O. (eds.) Springer Handbook of Robotics, pp. 1153–1176. Springer, Cham (2016). https://doi.org/10.1007/978-3-319-32552-1_46

14. Thrun, S., Fox, D., Burgard, W., Dellaert, F.: Robust Monte Carlo localization for mobile robots. Artif. Intell. **128**(1–2), 99–141 (2001)

Control Strategies Using the Extended Cooperative Dual Task-Space

Seyonne Leslie-Dalley$^{(\boxtimes)}$ ⓘ, Bruno Vilhena Adorno ⓘ, and Keir Groves ⓘ

Manchester Centre for Robotics and AI, University of Manchester, Oxford Road,
Manchester M13 9PL, UK
{seyonne.leslie-dalley,bruno.adorno,keir.groves}@manchester.ac.uk

Abstract. An existing extension of the *Cooperative Task Space* is reformulated in the dual quaternion domain to form the *Extended Cooperative Dual Task Space (ECDTS)*.This representation allows for the description of both uncoordinated and coordinated cooperative manipulation tasks into a unified formulation free of representational singularities. Additionally, a general framework is proposed to describe cooperative manipulation tasks considering the required manipulation mode in the ECDTS, the desired control objective, and additional task constraints. Vector Field Inequalities (VFIs) and geometric primitives such as Plücker lines and planes are used to define such constraints. Simulations of a task from each manipulation mode are carried out and show the generality of the proposed framework and applicability of the ECDTS formulation.

Keywords: Cooperative Manipulation · Collision Avoidance · Robot Control

1 Introduction

Cooperative manipulation provides capabilities that exceed those of a single manipulator such as greater payload capacity and the handling of large objects, allowing for the completion of a wider range of tasks [5,15]. Nonetheless, it also poses a new set of challenges that may further complicate the control problem; for instance, the additional task-space constraints and greater kinematic complexity of the system [1,5]. Consequently, research efforts have focused into methods of parametrising the cooperative system to attempt to reduce this complexity and allow for the adequate definition of a general cooperative manipulation task.

The framework proposed in [3] reformulated the Cooperative Task Space (CTS) [6] in the dual quaternion domain, resulting in a compact representation of cooperative manipulation tasks, named Cooperative Dual Task Space

This project was funded by EPSRC through the CRADLE programme (EP/X02489X/1)and by the Royal Academy of Engineering under the Research Chairs and Senior Research Fellowships programme.

M. N. Huda et al. (Eds.): TAROS 2024, LNAI 15052, pp. 143–154, 2025.
https://doi.org/10.1007/978-3-031-72062-8_13

(CDTS). That work has been applied to applications beyond two manipulators, such as whole-body manipulation and physical human-robot-interaction (pHRI), [1,13]. However, one drawback of that formalism is that it only considers the parallel coordination mode [10] where both manipulators share a common absolute and relative goal. Thus, when considering tasks that require different degrees of coordination, switching may be required between task representations, limiting the general applicability of that method.

Park and Lee [10] proposed an extension of the CTS describing both uncoordinated and coordinated manipulation tasks in a unified formulation. Although more general, that work lacks the advantages provided by dual quaternion algebra, such as the singularity-free representation preventing inconsistencies in the task execution [3], and the expressiveness of geometrical primitives that are naturally embedded in its algebraic structure [2].

Additionally, a crucial requirement is to ensure the safety of the entire system throughout the execution of a constrained cooperative manipulation task. This requires avoiding undesirable configurations that could damage a manipulated object or the system itself (e.g., hitting joint limits, self-collisions), and also explicitly considering the interaction of the system with the surrounding environment. Constraints related to collision avoidance in the workspace are not always known in advance. Therefore, reactive methods (i.e., sensor-based closed-loop control) must be implemented. The problem of singularity and joint-limit avoidance for cooperative manipulation has been addressed in [7] by using a task-priority framework, where tasks are split into sub-tasks with different priorities. However, their work has not considered any constraints due to the environment, or task-specific constraints. Alternatively, Quiroz-Omaña and Adorno [13] have coupled Vector Field Inequalities (VFIs) [9] with the CDTS framework to define constraints preventing the violation of joint limits and undesired end-effector orientations for bimanual manipulation tasks, but has not considered collision avoidance with obstacles in the environment.

1.1   Contributions

In this paper, we first merge the extension proposed by [10] of the CTS with the CDTS proposed by [3] through the use of dual quaternion algebra, resulting in the Extended Cooperative Dual Task Space (ECDTS). This allows for the representation of both coordinated and uncoordinated bimanual manipulation tasks in a single, compact, geometrically meaningful, and singularity-free formulation. Similar to the work in [13], we relax the control task, thus releasing degrees of freedom (DoFs) to enforce constraints defined through VFIs and geometric primitives, such as planes and Plücker lines. In addition, we introduce a framework to allow for the intuitive description of a range of different cooperative manipulation tasks, and consider not only intrinsic constraints (e.g., joint limits, actuation saturation) and environmental constraints, but also constraints specific to a manipulation task, such as undesired end-effector poses or maintaining a specific relative position. As a proof of concept, we simulate several examples of manipulation tasks using the ECDTS and the primitives described in the proposed framework.

2   Mathematical Background

2.1   Dual Quaternion Representation

Quaternions can be seen as an extension of the complex numbers, where the unit quaternion $r = \cos(\phi/2) + \sin(\phi/2)n$ represents a rotation of angle ϕ around the axis $\boldsymbol{n} = \hat{\imath}n_x + \hat{\jmath}n_y + \hat{k}n_z$, with $n_x, n_y, n_z \in \mathbb{R}$ and $\hat{\imath}^2 = \hat{\jmath}^2 = \hat{k}^2 = \hat{\imath}\hat{\jmath}\hat{k} = -1$. The unit dual quaternion $\boldsymbol{\underline{x}} = r + (1/2)pr$ represents a rigid motion in $\mathrm{Spin}(3) \ltimes \mathbb{R}^3$ that double covers $\mathrm{SE}(3)$, where r is the unit quaternion that represents the rotation, $\boldsymbol{p} = \hat{\imath}x + \hat{\jmath}y + \hat{k}z$ is a pure quaternion that represents the translation, and is the nilpotent dual unit; that is $\neq 0, ^2 = 0$. A general dual quaternion belongs to set $\mathcal{H}$ and has the form $\boldsymbol{\underline{h}} = h_1 + h_2\hat{\imath} + h_3\hat{\jmath} + h_4\hat{k} + \varepsilon(h_5 + h_6\hat{\imath} + h_7\hat{\jmath} + h_8\hat{k})$, where $\mathcal{P}(\boldsymbol{\underline{h}}) = h_1 + h_2\hat{\imath} + h_3\hat{\jmath} + h_4\hat{k}$ and $\mathcal{D}(\boldsymbol{\underline{h}}) = h_5 + h_6\hat{\imath} + h_7\hat{\jmath} + h_8\hat{k}$ extract the primary and dual parts, respectively. The vec_8 operator performs the bijective mapping $\mathrm{vec}_8 : \mathcal{H} \to \mathbb{R}^8$. Therefore, given $\boldsymbol{\underline{x}}, \boldsymbol{\underline{y}}, \boldsymbol{\underline{z}} \in \mathcal{H}$ such that $\boldsymbol{\underline{z}} = \boldsymbol{\underline{x}}\boldsymbol{\underline{y}}$, the Hamilton operators, $\overset{+}{\boldsymbol{H}}$ and $\overset{-}{\boldsymbol{H}}$, are matrices that satisfy $\mathrm{vec}_8\,\boldsymbol{\underline{z}} = \overset{+}{\boldsymbol{H}}_8(\boldsymbol{\underline{x}})\,\mathrm{vec}_8\,\boldsymbol{\underline{y}} = \overset{-}{\boldsymbol{H}}_8(\boldsymbol{\underline{y}})\,\mathrm{vec}_8(\boldsymbol{\underline{x}})$. In addition, by the definition of the dual quaternion conjugate, $\boldsymbol{\underline{h}}^* \triangleq h_1 - h_2\hat{\imath} - h_3\hat{\jmath} - h_4\hat{k} + \varepsilon(h_5 - h_6\hat{\imath} - h_7\hat{\jmath} - h_8\hat{k})$, we define $\boldsymbol{C}_8 \triangleq \mathrm{diag}(1, -1, -1, -1, 1, -1, -1, -1)$ such that $\mathrm{vec}_8\,\boldsymbol{\underline{x}}^* = \boldsymbol{C}_8\,\mathrm{vec}_8\,\boldsymbol{\underline{x}}$. For a more comprehensive overview of dual quaternion algebra, refer to [2].

2.2   Constrained Kinematic Controller

The closed-loop kinematic controller used in this work implements a linearly-constrained quadratic optimization problem where hard constraints, such as obstacle avoidance, joint limits, and task-specific constraints, are enforced as differential inequalities in the control inputs. Considering a desired task variable $\boldsymbol{x}_d \in \mathbb{R}^m$ and the task error $\tilde{\boldsymbol{x}} \triangleq \boldsymbol{x} - \boldsymbol{x}_d$, the control input $\boldsymbol{u}$ is obtained from

$$\boldsymbol{u} \in \underset{\dot{\boldsymbol{q}}}{\arg\min} \qquad \|\boldsymbol{J}\dot{\boldsymbol{q}} + \eta\tilde{\boldsymbol{x}}\|_2^2 + \lambda^2 \|\dot{\boldsymbol{q}}\|_2^2$$
$$\text{subject to} \quad \boldsymbol{W}\dot{\boldsymbol{q}} \preceq \boldsymbol{w}, \tag{1}$$

where $\boldsymbol{J} \in \mathbb{R}^{m \times n}$ is the task Jacobian, $\boldsymbol{q} \in \mathbb{R}^n$ is the joint configuration vector, $\eta \in (0, \infty)$ is the proportional gain that determines the convergence rate of the closed-loop error, and $\lambda \in (0, \infty)$ is the damping factor used to penalise high joint velocities. In addition, $\boldsymbol{w} \triangleq \boldsymbol{w}(\boldsymbol{q}) \in \mathbb{R}^p$ enforce p linear constraints on the control inputs, which can be designed to avoid joint limits, collision with obstacles using VFIs, etc. [9].

3   Extended Cooperative Dual Task Space (ECDTS)

Different bimanual manipulation tasks may require different types of end-effector coordination. For instance, carrying a box requires a parallel coordination, but

the motion of pouring and handing over a drink requires a leader-follower movement and the original pick-up of the objects for the pouring task requires no
explicit coordination between the manipulators. Park and Lee [10] defined these
different levels of manipulation as:

Parallel Coordination: symmetric coordination motion, creating a closed kinematic chain where the manipulators share a common absolute and relative variable.

Serial Coordination: asymmetric coordination motion, alternatively represented as a leader-follower coordination, where one hand is seen as the reference
for the other hand.

Orthogonal Coordination: uncoordinated manipulation in which tasks are
executed independent of each other.

We reformulate Park and Lee's extension of the CTS into the dual quaternion
domain to form the ECDTS by introducing the balance coefficient α and the
coordination coefficient β, as proposed in [10].

Let us first introduce the CDTS. Consider a two-arm system, as shown in
Fig. 1, where $\underline{\boldsymbol{x}}_1$ and $\underline{\boldsymbol{x}}_2$ represent the pose of the left and right end-effectors,
respectively. The *cooperative variables* $\underline{\boldsymbol{x}}_r$ and $\underline{\boldsymbol{x}}_a$ are used to represent the
CDTS, where the relative pose $\underline{\boldsymbol{x}}_r$ describes the pose of one end-effector with
respect to the other, whereas the absolute pose $\underline{\boldsymbol{x}}_a$ describes the pose of a frame
located between the two end-effectors with respect to a common frame.

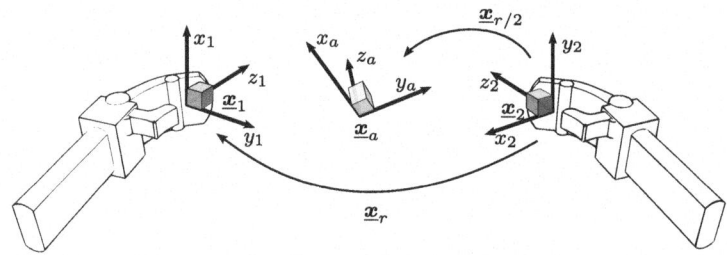

Fig. 1. Cooperative dual task-space representation: a cooperative manipulation task
can be completely described using the cooperative variables, the absolute pose $\underline{\boldsymbol{x}}_a$ and
relative pose $\underline{\boldsymbol{x}}_r$ [3].

Definition 1. *The relative and absolute poses are defined as*

$$\underline{\boldsymbol{x}}_r \triangleq \underline{\boldsymbol{x}}_2^* \underline{\boldsymbol{x}}_1 \text{ and } \underline{\boldsymbol{x}}_a \triangleq \underline{\boldsymbol{x}}_2 \underline{\boldsymbol{x}}_r^{\{1/2\}}, \tag{2}$$

*where $\underline{\boldsymbol{x}}_r^{\{1/2\}}$ is the transformation that corresponds to half of the angle ϕ_r around
the axis $\boldsymbol{n}_r = \hat{\imath} n_x + \hat{\jmath} n_y + \hat{k} n_z$ of the unit quaternion $\mathcal{P}(\underline{\boldsymbol{x}}_r)$, and half of the
translation between the two end-effectors [3].* These two cooperative variables
define a parallel coordination, where both arms share a common absolute and
relative pose.

Herein, we introduce the ECDTS.

Definition 2. *The ECDTS is described by the (extended) relative and absolute poses, defined as*

$$\underline{\boldsymbol{x}}_r \triangleq \underline{\boldsymbol{x}}_2^{*\{\beta\}} \underline{\boldsymbol{x}}_1 \ \text{and} \ \underline{\boldsymbol{x}}_a \triangleq \underline{\boldsymbol{x}}_2 \underline{\boldsymbol{x}}_r^{\{\alpha\beta\}}, \tag{3}$$

where $\alpha \in [0,1]$ is the balance coefficient and $\beta \in \{0,1\}$ is the coordination coefficient. As

$$\underline{\boldsymbol{x}}^{\{\lambda\}} = \boldsymbol{r}^{\{\lambda\}} + \varepsilon \frac{1}{2}\lambda \boldsymbol{pr}^{\{\lambda\}}, \tag{4}$$

where $r^{\{\lambda\}} = \cos(\lambda\phi/2) + n\sin(\lambda\phi/2)$ [2, Section 2.3], the CDTS turns out to be a particular case of ECDTS, where $\beta = 1$ and $\alpha = 0.5$, as shown in Fig. 2.

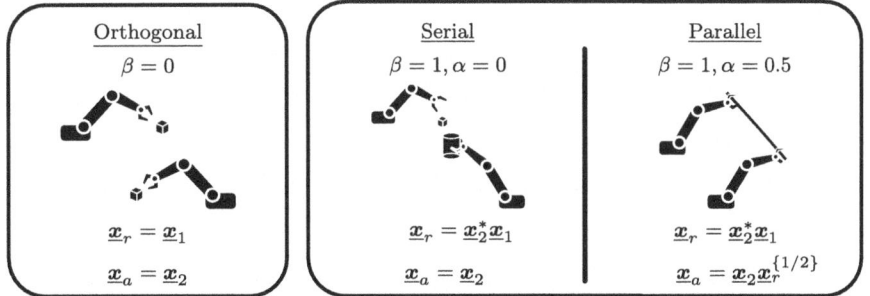

Fig. 2. Orthogonal, serial, and parallel manipulation modes in ECDTS.

To find the relative Jacobian $\boldsymbol{J}_{\underline{\boldsymbol{x}}_r} \in \mathbb{R}^{8\times n}$ that satisfies $\text{vec}_8 \, \underline{\dot{\boldsymbol{x}}}_r = \boldsymbol{J}_{\underline{\boldsymbol{x}}_r} \dot{\boldsymbol{q}}$, where $\boldsymbol{q} = [\boldsymbol{q}_1^T \ \boldsymbol{q}_2^T]^T$, with $\boldsymbol{q}_1 \in \mathbb{R}^{n_1}$ and $\boldsymbol{q}_2 \in \mathbb{R}^{n_2}$ being the configuration vectors of the first and second manipulators, respectively,

let us consider first the case when $\beta = 0$, in which $\underline{\boldsymbol{x}}_r$ in (3) simplifies to $\underline{\boldsymbol{x}}_r = \underline{\boldsymbol{x}}_1$. By taking the first time derivative and using the commutativity properties of the Hamilton operators [3], we obtain

$$\text{vec}_8 \, \underline{\dot{\boldsymbol{x}}}_r = \text{vec}_8 \, \underline{\dot{\boldsymbol{x}}}_1 = \boldsymbol{J}_{\underline{\boldsymbol{x}}_1} \dot{\boldsymbol{q}}_1 = \underbrace{\left[\boldsymbol{J}_{\underline{\boldsymbol{x}}_1} \ \boldsymbol{0}_{8\times n_2} \right]}_{\boldsymbol{J}_{\underline{\boldsymbol{x}}_{1\text{ext}}}} \begin{bmatrix} \dot{\boldsymbol{q}}_1 \\ \dot{\boldsymbol{q}}_2 \end{bmatrix},$$

where $\boldsymbol{J}_{\underline{\boldsymbol{x}}_1}$ is the Jacobian matrix of the first manipulator satisfying $\text{vec}_8 \, \underline{\dot{\boldsymbol{x}}}_1 = \boldsymbol{J}_{\underline{\boldsymbol{x}}_1} \dot{\boldsymbol{q}}_1$. When $\beta = 1$, then $\underline{\boldsymbol{x}}_r = \underline{\boldsymbol{x}}_2^* \underline{\boldsymbol{x}}_1$, and the relative Jacobian is the same as in the CDTS [3]. Taking the first time derivative of the relative pose gives $\underline{\dot{\boldsymbol{x}}}_r = \underline{\dot{\boldsymbol{x}}}_2^* \underline{\boldsymbol{x}}_1 + \underline{\boldsymbol{x}}_2^* \underline{\dot{\boldsymbol{x}}}_1$. Therefore,

$$\text{vec}_8 \, \underline{\dot{\boldsymbol{x}}}_r = \overline{\boldsymbol{H}}_8(\underline{\boldsymbol{x}}_1)\text{vec}_8 \, \underline{\dot{\boldsymbol{x}}}_2^* + \overset{+}{\boldsymbol{H}}_8(\underline{\boldsymbol{x}}_2^*)\text{vec}_8 \, \underline{\dot{\boldsymbol{x}}}_1 = \left[\overset{+}{\boldsymbol{H}}_8(\underline{\boldsymbol{x}}_2^*) \boldsymbol{J}_{\underline{\boldsymbol{x}}_1} \ \ \overline{\boldsymbol{H}}_8(\underline{\boldsymbol{x}}_1) \boldsymbol{C}_8 \boldsymbol{J}_{\underline{\boldsymbol{x}}_2} \right] \begin{bmatrix} \dot{\boldsymbol{q}}_1 \\ \dot{\boldsymbol{q}}_2 \end{bmatrix}.$$

Hence for $\beta \in \{0, 1\}$, the relative Jacobian $\boldsymbol{J}_{\underline{\boldsymbol{x}}_r}$ is given by

$$\boldsymbol{J}_{\underline{\boldsymbol{x}}_r} = \begin{cases} \boldsymbol{J}_{\underline{\boldsymbol{x}}_{1\text{ext}}} & \text{if } \beta = 0, \\ \left[\overset{+}{\boldsymbol{H}}_8\left(\underline{\boldsymbol{x}}_2^*\right)\boldsymbol{J}_{\underline{\boldsymbol{x}}_1} \ \ \overset{-}{\boldsymbol{H}}_8\left(\underline{\boldsymbol{x}}_1\right)C_8\boldsymbol{J}_{\underline{\boldsymbol{x}}_2} \right] & \text{otherwise.} \end{cases}$$

Analogously, the absolute Jacobian $\boldsymbol{J}_{\underline{\boldsymbol{x}}_a} \in \mathbb{R}^{8 \times n}$ satisfies $\text{vec}_8\, \dot{\underline{\boldsymbol{x}}}_a = \boldsymbol{J}_{\underline{\boldsymbol{x}}_a}\dot{\boldsymbol{q}}$. When $\beta = 0$, the absolute pose in (3) simplifies to $\underline{\boldsymbol{x}}_a = \underline{\boldsymbol{x}}_2$. Therefore,

$$\text{vec}_8\, \dot{\underline{\boldsymbol{x}}}_a = \text{vec}_8\, \dot{\underline{\boldsymbol{x}}}_2 = \boldsymbol{J}_{\underline{\boldsymbol{x}}_2}\dot{\boldsymbol{q}}_2 = \underbrace{\left[\boldsymbol{0}_{8 \times n_1}\ \ \boldsymbol{J}_{\underline{\boldsymbol{x}}_2} \right]}_{\boldsymbol{J}_{\underline{\boldsymbol{x}}_{2\text{ext}}}} \begin{bmatrix} \dot{\boldsymbol{q}}_1 \\ \dot{\boldsymbol{q}}_2 \end{bmatrix}. \tag{5}$$

When $\beta = 1$, then $\underline{\boldsymbol{x}}_a = \underline{\boldsymbol{x}}_2\underline{\boldsymbol{x}}_r^{\{\alpha\}}$. Taking the first time derivative and applying the Hamilton operators yields

$$\text{vec}_8\, \dot{\underline{\boldsymbol{x}}}_a = \overset{-}{\boldsymbol{H}}_8\left(\underline{\boldsymbol{x}}_r^{\{\alpha\}}\right)\text{vec}_8\, \dot{\underline{\boldsymbol{x}}}_2 + \overset{+}{\boldsymbol{H}}_8\left(\underline{\boldsymbol{x}}_2\right)\text{vec}_8\left(\frac{d\underline{\boldsymbol{x}}_r^{\{\alpha\}}}{dt}\right). \tag{6}$$

Using (4) to expand $d\underline{\boldsymbol{x}}_r^{\{\alpha\}}/dt$ gives

$$\text{vec}_8\left(\frac{d\underline{\boldsymbol{x}}_r^{\{\alpha\}}}{dt}\right) = \begin{bmatrix} \text{vec}_4\left(\frac{dr_r^{\{\alpha\}}}{dt}\right) \\ \frac{\alpha}{2}\left(\overset{-}{\boldsymbol{H}}_4\left(r_r^{\{\alpha\}}\right)\text{vec}_4\, \dot{p}_r + \overset{+}{\boldsymbol{H}}_4\left(p_r\right)\text{vec}_4\left(\frac{dr_r^{\{\alpha\}}}{dt}\right)\right) \end{bmatrix}, \tag{7}$$

where $\overset{-}{\boldsymbol{H}}_4, \overset{+}{\boldsymbol{H}}_4$, and vec_4 are operators defined for quaternions, similarly to the ones defined for dual quaternions [2]. Since $dr_r^{\{\alpha\}}/dt = \alpha\dot{r}_r r_r^{\{\alpha-1\}}$ [2, Proposition 2.7.3], we use the Hamilton operators to obtain

$$\text{vec}_4\, \mathcal{P}\left(\frac{d\underline{\boldsymbol{x}}_r^{\{\alpha\}}}{dt}\right) = \overbrace{\alpha\overset{-}{\boldsymbol{H}}_4\left(r_r^{\{\alpha-1\}}\right)\boldsymbol{J}_{\mathcal{P}(\underline{\boldsymbol{x}}_r)}}^{\boldsymbol{J}_{\mathcal{P}\left(\underline{\boldsymbol{x}}_r^{\{\alpha\}}\right)}}\dot{\boldsymbol{q}}, \tag{8}$$

$$\text{vec}_4\, \mathcal{D}\left(\frac{d\underline{\boldsymbol{x}}_r^{\{\alpha\}}}{dt}\right) = \underbrace{\frac{\alpha}{2}\left(\overset{-}{\boldsymbol{H}}_4\left(r_r^{\{\alpha\}}\right)\boldsymbol{J}_p + \overset{+}{\boldsymbol{H}}_4\left(p_r\right)\boldsymbol{J}_{\mathcal{P}\left(\underline{\boldsymbol{x}}_r^{\{\alpha\}}\right)}\right)}_{\boldsymbol{J}_{\mathcal{D}\left(\underline{\boldsymbol{x}}_r^{\{\alpha\}}\right)}}\dot{\boldsymbol{q}}, \tag{9}$$

such that $\boldsymbol{J}_{\underline{\boldsymbol{x}}_r^{\{\alpha\}}} = \left[\boldsymbol{J}_{\mathcal{P}\left(\underline{\boldsymbol{x}}_r^{\{\alpha\}}\right)}^T\ \ \boldsymbol{J}_{\mathcal{D}\left(\underline{\boldsymbol{x}}_r^{\{\alpha\}}\right)}^T\right]^T$. Hence, using (5) and (6) with (8) and (9), the absolute Jacobian $\boldsymbol{J}_{\underline{\boldsymbol{x}}_a}$ is given by

$$\boldsymbol{J}_{\underline{\boldsymbol{x}}_a} = \begin{cases} \boldsymbol{J}_{\underline{\boldsymbol{x}}_{2\text{ext}}} & \text{if } \beta = 0, \\ \overset{-}{\boldsymbol{H}}_8\left(\underline{\boldsymbol{x}}_r^{\{\alpha\}}\right)\boldsymbol{J}_{\underline{\boldsymbol{x}}_{2\text{ext}}} + \overset{+}{\boldsymbol{H}}_8(\underline{\boldsymbol{x}}_2)\boldsymbol{J}_{\underline{\boldsymbol{x}}_r^{\{\alpha\}}} & \text{otherwise.} \end{cases}$$

4   General Task Descriptions Framework

Classifications of cooperative manipulation tasks are often based on aspects such as contact and motion trajectory [11], or hand roles and task symmetry [8]. Instead, we propose a classification from a control perspective that categorises manipulation tasks according to the most relevant manipulation mode from the ECDTS, the main desired control primitive, and the additional constraints necessary for the task execution. For example, the task of opening a jar can be defined as a serial manipulation task due to the asymmetrical coordination between the two hands. The primary desired control primitive for this task can be defined as a relative rotation between the end-effectors, whereas the task constraints due to the coupled nature of the task can be described as ensuring the relative position of the end-effectors is maintained and the jar is not tilted excessively during the task execution. The next section outlines some general primitives for the control objective and constraints using VFIs [9].

4.1   Classification Categories

Control Primitives. Depending on the primary goal of a cooperative manipulation task (e.g., a rotation to open a jar), a specific control primitive can be chosen for the desired task variable (x_d) of the constrained kinematic controller (1). Additionally, by relaxing the task objective from a full pose control, DoFs are released to potentially execute additional tasks. Table 1 summarizes the control primitives.

Table 1. Summary of control primitives

	Rel./Abs. distance	Rel./Abs. position	Rel./Abs. orientation	Rel./Abs. pose
Control primitive x_d	d	$vec_3\, p$	$vec_4\, r$	$vec_8\, \underline{\boldsymbol{x}}$
DoF	1	3	3	6

Task Constraints. Task constraints can be split into two main categories: constraints relative safety, such adherence of joint-limits and obstacle avoidance; and task-specific constraints, such as maintaining a specific relative pose to hold an object. These constraints are defined using the VFI framework [9], which requires distance functions between two collidable entities and the corresponding Jacobian matrices. Table 2 summarizes a few relevant constraints, and we refer the reader to [9] for more details on the VFI implementation.

4.2   Task Classification

Different manipulation tasks can be classified according to the categorization presented in Table 3. The categorization is not unique, and the same manipulation task might be defined using different modes and primitives.

Table 2. Summary of task (T1–T4) and safety (C1–C2) constraints [9,12,14].

(T1) Pose or Position	Ensure a firmly-grasped object is kept in a pose or position with the maximum deviation defined by $d_{\max}$.
(T2) Circular target region	Ensure the convergence of a point to a circular target region on a plane, with a radius denoted by $r_{p,c}$.
(T3) Tilt limit	Ensure that a line attached to the end-effector is inclined with respect to a reference line by at most an angle ϕ_{safe}.
(T4) Entry sphere constraint	Maintain a line within a predefined sphere of radius r_s.
(C1) Obstacle Avoidance	The safe distance to the object is denoted by $d_{\text{safe}} = 0$.
(C2) Joint and joint velocity limits	Enforce joint and joint velocity limits to comply with physical limitations or safety requirements.

Table 3. Non-exhaustive list of cooperative manipulation definitions. Manipulation modes are classified as orthogonal (O), serial (S), and parallel (P).

Example	Manipulation mode	Control primitive x_d	Task constraint
Picking up of an object by a single manipulator	O1	$\text{vec}_8\, \underline{\boldsymbol{x}}_r$ or $\text{vec}_8\, \underline{\boldsymbol{x}}_a$	None
Placing of object by a single manipulator on a flat horizontal surface	O2	d_r or d_a	T2 (absolute or relative target region), T3 (absolute or relative inclination)
Unscrewing a jar	S1	$\text{vec}_4\, r_r$	T1 (relative position), T3 (absolute inclination)
Peg-in-hole assembly, threading a needle	S2	$\begin{bmatrix} \text{vec}_3\, p_r \\ \text{vec}_8\, \underline{\boldsymbol{x}}_a \end{bmatrix}$	T3, T4
Hammering a nail	S3	$\begin{bmatrix} \text{vec}_3\, p_r \\ \text{vec}_8\, \underline{\boldsymbol{x}}_a \end{bmatrix}$	T1 (absolute pose)
Transporting a large/heavy object	P1	d_a	T1 (relative pose)
Carrying and keep an object upright	P2	d_a	T1 (relative pose), T3 (absolute inclination)

5   Simulation and Discussions

To illustrate the proposed framework, manipulation tasks for each mode were chosen from Table 3; namely, S3 (serial), P2 (parallel), and O1 and O2 (orthogonal). The tasks were simulated using CoppeliaSim[1] for visualisation and DQ Robotics [4] to model and control the manipulators and define the geometric primitives used in the constraints for the controller 1. The set-up for each task

---

[1] https://www.coppeliarobotics.com/coppeliaSim.

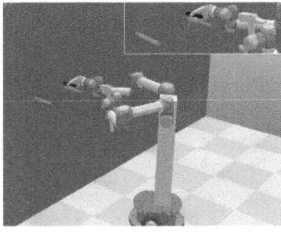

(a) Serial Manipulation Task.   (b) Parallel Manipulation Task.   (c) Orthogonal Manipulation Task

Fig. 3. In *(a)*, the two green cylinders define the desired configuration of the 'nail' represented by the black cylinder. In *(b)*, the green sphere defines the desired region for the centre of the object. In *(c)*, the two blue and green cuboids define the desired objects for the manipulators to pick up. The corresponding coloured disks define the desired regions for the cuboids to then be placed once picked-up, where the pink cones surrounding constrain how the end-effectors approach the target region. Additionally, for both *(b)* and *(c)*, the black vertical line must stay within the green cone that describes the maximum allowable tilting angle. In *(a)* and *(b)*, the red plane describes a wall/table-top for obstacle avoidance. (Color figure online)

is illustrated in Fig. 3 where a robot consisting of two UR5 manipulators with six DOFs attached to a pedestal on a holonomic base was used.

In the serial mode task shown in Fig. 3(a), initially a desired absolute $\underline{\boldsymbol{x}}_a = \underline{\boldsymbol{x}}_2$ and relative pose $\underline{\boldsymbol{x}}_r$ were defined, where $\underline{\boldsymbol{x}}_a$ described the pose of the end-effector holding the nail and $\underline{\boldsymbol{x}}_r$ described the relative pose to ensure the nail and hammer had the correct starting configuration. Then an absolute pose error constraint (T1) was specified to ensure the end-effector holding the nail stayed within a specific region. Finally, by varying a relative position control primitive, the 'hammering action' is generated. Throughout the task execution, an obstacle constraint was specified to ensure the robot did not collide with the wall. This task was then repeated for a different desired absolute pose.

For the parallel manipulation task, given the initial starting configuration shown in Fig. 3, an absolute distance control to a desired point is performed, while a relative pose constraint (T1) and a tilt constraint (T3) for the absolute frame are defined to ensure the tray is kept level and firmly grasped.

For the orthogonal manipulation task, three different poses were specified in the picking phase to ensure the correct grasping configuration and approach. More specifically, the first pose was defined as directly above the cuboids to ensure that during the approach the end-effector would not hit the object, the next being the actual grasping pose, and the third being lifting the cuboid back up from the table. Then, for the placing stage a distance control was defined in accordance with a constraint enforcing the convergence to a target region (T2). The constraints for each task and the parameters used are specified in Table 4.

The results in Fig. 4, 5, and 6 show that each task was successfully executed whilst all constraints were upheld.

Table 4. Parameters for the simulations.

Constraint	VFI gains	Parameter values
Serial manipulation task		
Absolute pose (T1)	$n_c = 1$	$d_{\max} = 0.01$
Collision avoidance with the wall (C1)	$n_c = 1$	$d_{\text{safe}} = 0$ m
Parallel manipulation task		
Absolute pose tilt limits (T3)	$n_c = 1$	$\phi_{\text{safe}} = 0.2$ rad
Relative pose (T1)	$n_c = 1$	$d_{\max} = 0.03$
Orthogonal manipulation task		
Circular target region (T2)	$n_c = 1$	$r_{p,c} = 0.04$ m
End-effector tilt limits (T3)	$n_c = 5$	$\phi_{\text{safe}} = 0.1$ rad
Collision avoidance with the table (C1)	$n_c = 5$	$d_{\text{safe}} = 0$ m

*For all simulations, the parameters $\eta = 10$ and $\lambda = 0.01$ in the controller (1) were chosen empirically to obtain a fast and smooth error convergence. The parameter η_c is used in the VFIs [9] used to generate the constraints in the controller (1).

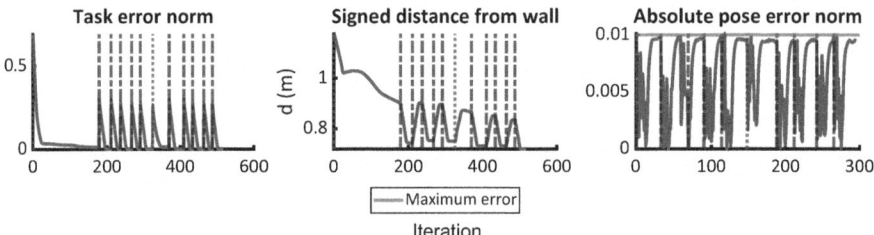

Fig. 4. Serial manipulation task. The vertical lines represent a change in the control objective, where the vertical dotted lines describe a relative and absolute *pose* control and the vertical dashed lines define a relative *position* control. The absolute pose error constraint (T1), was only imposed during the 'hammering' action. The control objective was changed once the task error had stabilised.

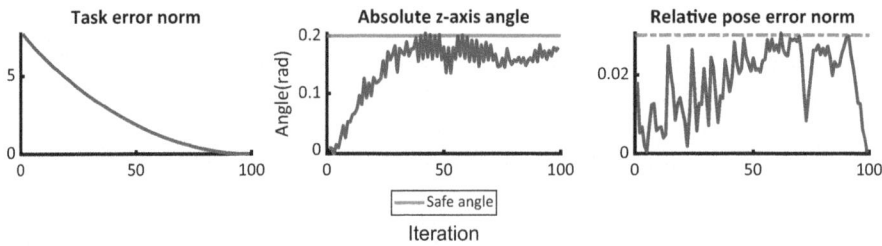

Fig. 5. Parallel manipulation task. The red dashed horizontal line represents the maximum relative pose error. (Color figure online)

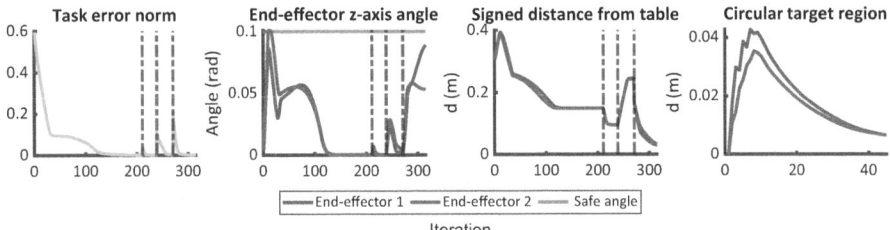

Fig. 6. Orthogonal manipulation task. Vertical dashed lines represent a change in the control objective, where the first two represent the stages of the picking phase, when end-effector poses are controlled. The last vertical dashed line represents the placing phase, where the point cone constraint is implemented and the end-effector distances to the table surface are controlled.

6 Conclusion

This work combined the extension proposed by [10] of the CTS with another formulation based on dual quaternion algebra, CDTS [3], to present the ECDTS. The ECDTS was shown to be more general than the CDTS framework, able to describe serial and parallel coordination in addition to uncoordinated motions within a single representation. By having a unified framework, it provides an abstraction that encompasses a greater number of tasks. We presented several examples of how VFIs coupled with the ECDTS can describe cooperative manipulation tasks that are subject to a wide range of constraints. Future works will focus on methods to automate the selection of the coordination mode dependent on the task and performance variables, such manipulability. Also, the representation of internal and external wrenches in a bimanual manipulation task, which is already formulated in CDTS [1], will be extended to the ECDTS.

References

1. Adorno, B.V.: Two-arm Manipulation: from manipulators to enhanced human-robot collaboration. Theses, Université Montpellier II - Sciences et Techniques du Languedoc, October 2011. https://theses.hal.science/tel-00641678
2. Adorno, B.V.: Robot Kinematic Modeling and Control Based on Dual Quaternion Algebra—Part I: Fundamentals (2017). https://hal.science/hal-01478225v1
3. Adorno, B.V., Fraisse, P., Druon, S.: Dual position control strategies using the cooperative dual task-space framework. In: IEEE/RSJ 2010 International Conference on Intelligent Robots and Systems, IROS 2010 - Conference Proceedings, pp. 3955–3960 (2010)
4. Adorno, B.V., Marinho, M.M.: DQ robotics: a library for robot modeling and control. IEEE Rob. Autom. Mag. **28**(3), 102–116 (2021)
5. Caccavale, F.: Cooperative manipulators. In: Baillieul, J., Samad, T. (eds.) Encyclopedia of Systems and Control, pp. 450–455. Springer, Cham (2021). https://doi.org/10.1007/978-3-030-44184-5_175

6. Chiacchio, P., Chiaverini, S., Siciliano, B.: Direct and inverse kinematics for coordinated motion tasks of a two-manipulator system. J. Dyn. Syst. Measur. Control **118**(4), 691–697 (1996). https://doi.org/10.1115/1.2802344

7. de Farias, C.M., Rocha, Y.G., Figueredo, L.F.C., Bernardes, M.C.: Design of singularity-robust and task-priority primitive controllers for cooperative manipulation using dual quaternion representation. In: 2017 IEEE Conference on Control Technology and Applications (CCTA), pp. 740–745 (2017)

8. Krebs, F., Asfour, T.: A bimanual manipulation taxonomy. IEEE Rob. Autom. Lett. **7**(4), 11031–11038 (2022)

9. Marinho, M.M., Adorno, B.V., Harada, K., Mitsuishi, M.: Dynamic active constraints for surgical robots using vector-field inequalities. IEEE Trans. Rob. **35**, 1166–1185 (2019)

10. Park, H.A., Lee, C.S.: Extended cooperative task space for manipulation tasks of humanoid robots. In: Proceedings - IEEE International Conference on Robotics and Automation, 2015-June, pp. 6088–6093 (2015)

11. Paulius, D., Huang, Y., Meloncon, J., Sun, Y.: Manipulation motion taxonomy and coding for robots. In: 2019 IEEE/RSJ International Conference on Intelligent Robots and Systems (IROS). IEEE, November 2019

12. Pereira, M.S., Adorno, B.V.: Manipulation task planning and motion control using task relaxations. J. Control Autom. Electric. Syst. **33**, 1103–1115 (2022)

13. Quiroz-Omaña, J.J., Adorno, B.V.: Bimanual mobile manipulation using the cooperative dual task-space framework and vector fields inequalities. In: Workshop on Applications of Dual Quaternion Algebra to Robotics, pp. 1–2 (2019). https://doi.org/10.5281/ZENODO.3566806

14. Quiroz-Omaña, J.J., Adorno, B.V.: Whole-body control with (self) collision avoidance using vector field inequalities. IEEE Rob. Autom. Lett. **4**(4), 4048–4053 (2019)

15. Smith, C., et al.: Dual arm manipulation-a survey. Robot. Auton. Syst. **60**(10), 1340–1353 (2012)

Mechanism Design

Design of a Robotic Trainer for Upper Body Physical Therapy

Yael Sznaidman[1,3(✉)], Maya Krakovski[1,3], Shirley Handelzalts[2,3], and Yael Edan[1,3]

[1] Department of Industrial Engineering and Management, Ben-Gurion University of the Negev, 8410501 Beer-Sheva, Israel
yaelszn@post.bgu.ac.il

[2] Department of Physical Therapy, Ben-Gurion University of the Negev, 8410501 Beer-Sheva, Israel

[3] Agriculture, Biological, Cognitive Robotics Initiative, Ben-Gurion University of the Negev, 8410501 Beer-Sheva, Israel

Abstract. This study introduces the design of a robotic system for upper-body training of individuals with lower-body orthopedic injuries undergoing rehabilitation. The system is an adaptation of 'Gymmy', a robotic system initially developed for physical and cognitive training of older adults. The paper outlines the prototype system designed based on a focus group and presents the results of an online video assessment by physical therapists, which further helped tailor 'Gymmy' to better suit physical therapy training. It discusses enhancements of the prototype system including performance feedback provision, methods to enhance training engagement, personalized training regimens, and data display features. These improvements aim to provide personalized rehabilitation experiences to improve engagement and performance.

Keywords: robotic trainer · rehabilitation · upper-body training · HRI

1 Introduction

Rehabilitation interventions incorporating upper-body resistance and flexibility training have shown to effectively reduce pain intensity, and enhance strength, muscle mass, and upper-limb functionality among wheelchair users [1]. Moreover, rehabilitation interventions targeting the upper body not only enhance the physical function of wheelchair users but also increase social participation and overall quality of life [2].

Technology has become increasingly involved in the realms of rehabilitation, physical training, and body strengthening, with a variety of applications such as video games, exergames, virtual reality, and mobile applications [3–5]. Among these technologies, robotic trainers have emerged as promising tools in rehabilitation and training, leading to enhanced patient performance and motivation [6]. In previous work, we developed 'Gymmy', a robotic trainer for upper extremity physical training, designated for the older adult population [7]. In this paper, we present the adaptation of 'Gymmy' to serve as a physical trainer for upper-body exercises in individuals with lower-body orthopedic

© The Author(s), under exclusive license to Springer Nature Switzerland AG 2025
M. N. Huda et al. (Eds.): TAROS 2024, LNAI 15052, pp. 157–163, 2025.
https://doi.org/10.1007/978-3-031-72062-8_14

injuries undergoing rehabilitation. These adaptations were based on feedback received from physical therapists. First, a prototype was designed based on a focus group. Then, the feedback and review from a questionnaire distributed among physical therapists who viewed a video of a prototype was used to improve the system.

2 The Prototype System

The prototype system was based on a previous system designed for physical and cognitive training for older adults [7]. The system includes a Poppy Torso robot to demonstrate exercises, a Realsense D435 camera to allow monitoring of users' performance by tracking the body joint positions, and a touch screen and speakers to provide feedback. A detailed technical description of the system's operation can be found in [7]. The training exercises program was modified to accommodate the requirements of the new system designated for physical therapy purposes based on a focus group of six senior physical therapists working in a rehabilitation center who experienced 'Gymmy'.

Out of 20 exercises recorded by physical therapists, 16 exercises were included in the system, based on their suitability for implementation by the robot and the feasibility of tracking proper movements using the camera (Fig. 1). Several exercises require strength training equipment such as a ball, weights, a stick, and a resistance band. However, while the system instructs the patient on which tool to use, the robot does not demonstrate its usage, as it is unable to lift weights or handle such resistance.

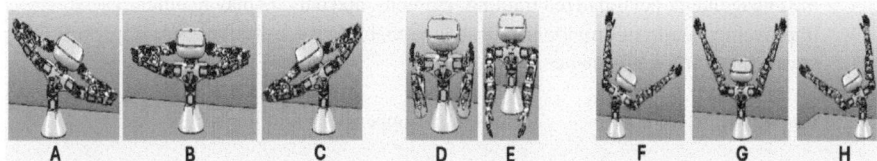

Fig. 1. Several examples of the exercises chosen for inclusion in the system: (A, B, C) - bending elbows behind the head and alternately leaning to the left and right; (D, E) - bending elbows with a ball between hands; (F, G, H) - stretching a resistance band between hands and alternately leaning to the left and right.

Upon entry, the robot greets the patient and presents training options, each comprising 2–4 exercises with specific equipment. The patient selects a session, gathers the required equipment, and signals when ready to begin. Each exercise consists of ten repetitions: one with verbal instructions followed by nine silent demonstrations. Patients mimic the robot while their performance is monitored by the system's camera, with successful repetitions audibly counted. Joint data and repetition counts are saved in an Excel file for research purposes. After completing the training, patients rate their perceived exertion using the Borg [8] scale and the workout menu reappears for further selection.

3   Methods

To evaluate the prototype system, we created an online video simulating a training session with the robot (https://youtu.be/w7OUOK_LFes). This video was attached to an online questionnaire that was distributed among physical therapists to receive their professional evaluation of the system along with specific recommendations for upper-body training for patients with lower-body orthopedic injuries. Questionnaires were distributed among physiotherapists from Ben Gurion University and Loewenstein Rehabilitation Hospital, as the research is being conducted in collaboration between these institutions. The diversity of patients at Loewenstein Hospital, in both age and type of injuries, allowed physical therapists to provide varied perspectives.

At the beginning of the questionnaire, after reviewing the information form, responders were requested to provide some background information, including their age, gender, completed degree, primary workplace, main field of practice, and years of professional experience. Subsequently, responders completed the pre-watching questionnaires. Then, they watched a 10-min video simulating a training session with the robot. After viewing the video, responders completed the questionnaire and addressed 2 additional questions. The study was approved by the department's ethical committee.

3.1   Measures

The pre-watching questionnaires assessed physical therapists' attitudes toward technology and robots using the Technology Adoption Propensity (TAP) [9] questionnaire and the Negative Attitudes Toward Robots Scale (NARS) [10], both on a 5-point Likert scale from "Strongly disagree" to "Strongly agree".

The post-watching questionnaires assessed respondents' perspectives on 'Gymmy' regarding *user experience, trust,* and *technophobia.* These indicators were adapted from a previous study [11], with several adjustments to fit the research methodology. Expected *user experience* was measured with the User Experience Questionnaire (UEQ), tailored to assess physical therapists' *quality evaluation* (QE) of 'Gymmy''s anticipated use, rated on a scale from -3 to $+3$. Respondents' *trust* was assessed using the Trust Perception Scale-HRI in a range between 0% and 100% in increments of 10%. The level of *technophobia* was assessed with a modified version of the Technophobia scale using a 5-point Likert scale. After completing the questionnaires, respondents rated their likelihood of recommending 'Gymmy' to patients and provided verbal feedback, including positive and negative aspects, along with suggestions for improvement.

3.2   Participants

Thirty-three physical therapists completed the questionnaire; however, seven were excluded due to incorrect responses to the distraction question. The final analysis included 26 respondents (17 females, 9 males), aged 28 to 62 (mean = 41.96, SD = 8.84), with 1 to 32 years of experience (mean = 13.65, SD = 8.6). Educational levels comprised 12 with Bachelor's degrees, 11 with Master's degrees, and 3 with Ph.Ds. In the multiple-choice questions, 17 responders reported working primarily in rehabilitation hospitals, with the main fields of practice being neurology (14) and orthopedics (11).

4 Results

User Experience Questionnaire (UEQ). Half of the respondents expressed neutrality regarding the robot's *attractiveness,* while 30.77% perceived it as attractive (mean = 0.186, SD = 1.338). In terms of *pragmatic quality evaluation,* which includes *dependability, efficiency,* and *perspicuity,* 65.38% were neutral, while 30.77% believed it would exhibit high pragmatic quality (mean = 0.788, SD = 0.816). Respondents rated the *hedonic quality evaluation,* represented by *stimulation* and *novelty,* as relatively low (26.92% low, 57.69% medium; mean = −0.187, SD = 1.277).

Trust Perception Scale-HRI. Respondents demonstrated high *trust in 'Gymmy''s performance* with 73.08% of respondents rated it as high (mean = 0.737, SD = 0.182). However, opinions about *trust in 'Gymmy''s social aspects* were diverse, with 26.92% expressing low trust, 38.46% medium, and 34.61% high (mean = 0.533, SD = 0.272).

Technophobia Scale. Respondents indicated low *personal failure,* reflecting *anticipated problems, frustrations, and failures* associated with the use of 'Gymmy' (mean = 1.993, SD = 0.619). Regarding the *ambiguity between human and machine interactions and the fear of 'Gymmy' dominating these interactions,* the mean rating remained neutral (mean = 3.346, SD = 0.794). Similarly, *convenience and enjoyment* yielded inconclusive results (mean = 3.179, SD = 0.818).

Level of Recommendation. The level of recommendation among physical therapists for training with 'Gymmy' for their patients was neutral (mean = 3.461, SD = 0.816).

Verbal Feedback. Some comments praised 'Gymmy''s good planes of motion and friendly demeanor, while a significant portion focused on areas for improvement. Many highlighted the lack of feedback of the robot, while others raised concerns about training monotony, a lack of motivation, and predictability. Criticisms included the generic nature of training, which lacked specificity for individual trainees, doubts about the advantages of using a robot over videos, and the absence of emotional and therapeutic support. Suggestions included adding instructional videos for complex exercises and changing the robot's voice to be less monotonous.

5 New System Design

Based on questionnaire responses, we identified four key areas for improvement and designed the new system to enhance the prototype accordingly. These improvements will add new features to the system and enhance the feedback algorithm and user interface, allowing for a more personalized and motivational robotic trainer.

Feedback Insufficiency. Numerous comments noted the robot's lack of feedback on the patient's performance in the prototype system. The prototype's only feedback is counting correct exercise repetitions.

In the updated system, the feedback algorithm will be enhanced. The counting feature will remain, supplemented by additional feedback during and after exercises. Upon exercise completion, users will receive motivational feedback, such as "Well done! You have successfully completed the exercise." During the exercise, if the patient falls behind by a certain number of repetitions (chosen randomly between 2 and 4 repetitions), verbal prompts like "You can do it," "One more," or "Faster!" will be triggered.

Non-specific Training for the Patient. In the prototype, patients select their training, but all sessions include identical exercises and equipment, regardless of individual abilities. This approach may not align with individual needs, making it necessary to adapt the system to meet physical therapy objectives and cater to each patient's abilities.

The new system will feature a pre-training interface to allow physical therapists to personalize each patient's regimen. They, along with the patient, will be able to select suitable exercises from all available options, including various equipment types. A performance tracking feature will display graphs from previous sessions, showing movement paths, training difficulty rating, and successful repetition counts, assisting in decision-making. Chosen exercises will be saved in the system per patient. To start a new training, patients will only need to prepare equipment, enter their ID, and begin training. The chosen exercises will then be presented categorically by equipment type.

Additionally, strength training equipment such as resistance bands, weights, and balls of varying strengths and weights will be provided to facilitate greater adaptation to each patient's needs, considering that individuals may have different strength levels.

Boring, Unmotivating, Monotonous, Predictable. In the current prototype system, each training program includes fixed exercises in a predetermined order, with specific equipment, and lacks motivating elements.

The adapted system will allow for the use of multiple types of equipment within a single training session, increasing training variability and engagement. To increase learning effects, exercises within each session will be introduced in a randomized order, both in terms of equipment type order and sequence of the exercises with the specific equipment type. Feedback provided at the end of exercises will be integrated along each exercise consistent with motor learning principles. Additionally, motivational statements will be integrated throughout the workout to boost user motivation. Furthermore, gamification elements will be implemented by introducing levels based on the number of successful repetitions achieved by users, with penalties for incomplete exercises. This gamified approach encourages users to progress through levels and potentially compete with others, incentivizing consistent training. To further enhance user engagement, patients will receive an email after each workout containing their current level status, and comparing successful repetitions with previous sessions.

Incomplete Motions and Inaccuracy in Palm and Finger Placement. Due to limitations in the robot's movement capabilities, achieving complete and accurate movements for all exercises is not feasible. As a solution, in the new system, in cases where 'Gymmy' will not be able to perform an exercise accurately, a video demonstration will be used instead. The remaining aspects of the system, including feedback provision, repetition tracking, and gamification features will remain unchanged.

6 Conclusion and Future Work

Ensuring the meticulous implementation of robotic trainer designs is critical for upper body physical rehabilitation to maintain sustainable system usage and optimize patient benefits. Through a focus group and an online video survey of a prototype upper-body robotic trainer for patients with lower-body injuries, four key improvement areas were identified by physical therapists: personalization, motivation, feedback provision, and motion accuracy. Design solutions were proposed for each aspect, to enable 'Gymmy' to provide motivation and real-time feedback, and to furnish performance data of the patients. Ongoing work is underway to fully incorporate gamification elements and email-sending features, as detailed in Sect. 5. IRB approval has been received, and experiments evaluating the new system design with patients and physical therapists will begin once development is completed. Additionally, in future work we aim to develop an adaptive system that learns from each patient, automatically tailoring the training to individual needs, including custom movement ranges for each patient.

Acknowledgments. Partial support was provided by Ben-Gurion University of the Negev through the Agricultural, Biological, and Cognitive Robotics Initiative, the Marcus Endowment Fund, and the W. Gunther Plaut Chair in Manufacturing Engineering. We acknowledge the work of Levav Shabtay, Inbar Duvdevani, and Coral Elmalach who contributed to the prototype design.

References

1. Serra-Añó, P., Pellicer-Chenoll, M., García-Massó, X., Morales, J., Giner-Pascual, M., González, L.M.: Effects of resistance training on strength, pain and shoulder functionality in paraplegics. Spinal Cord **50**(11), 827–831 (2012)
2. Kemp, B.J., Bateham, A.L., Mulroy, S.J., Thompson, L., Adkins, R.H., Kahan, J.S.: Effects of reduction in shoulder pain on quality of life and community activities among people living long-term with SCI paraplegia: a randomized control trial. J. Spinal Cord Med. **34**(3), 278–284 (2011)
3. Bantayan, B.: Enhancing muscular strength with resistance and high-intensity interval training using a fitness mobile app guide. JPAIR Multidiscip. Res. **51**(1), 1–15 (2023)
4. Montalbán, M.A., Arrogante, O., Montalbán, M.A., Arrogante, O.: Rehabilitation through virtual reality therapy after a stroke: a literature review. Rev. Cient. Enferm. Neurol. (English ed.) **52**, 19–27 (2020)
5. Pacheco, T.B.F., De Medeiros, C.S.P., De Oliveira, V.H.B., Vieira, E.R., De Cavalcanti, F.A.C.: Effectiveness of exergames for improving mobility and balance in older adults: a systematic review and meta-analysis. Syst. Rev. **9**, 1–14 (2020)
6. Salas-Monedero, M., et al.: Smoothness and efficiency metrics behavior after an upper extremity training with robic humanoid robot in paediatric spinal cord injured patients. Appl. Sci. **13**(8), 4979 (2023)
7. Krakovski, M., et al.: "Gymmy": designing and testing a robot for physical and cognitive training of older adults. Appl. Sci. **11**(14), 6431 (2021)
8. Borg, G.A.V.: Psychophysical bases of perceived exertion. Med. Sci. Sports Exerc. **14**(5), 377–381 (1982)
9. Ratchford, M., Barnhart, M.: Development and validation of the technology adoption propensity (TAP) index. J. Bus. Res. **65**(8), 1209–1215 (2012)

10. Nomura, T., Kanda, T., Suzuki, T., Kato, K.: Prediction of human behavior in human - robot interaction using psychological scales for anxiety and negative attitudes toward robots. IEEE T-RO **24**(2), 442–451 (2008)
11. Zafrani, O., Nimrod, G., Edan, Y.: Between fear and trust: older adults' evaluation of socially assistive robots. IJHCS **171**, 102981 (2023)

A Bio-inspired Jumping Robot: Design, Modelling and Experimental Tests

Harrison Elliott, Xiangyu An, and Mingfeng Wang$^{(\boxtimes)}$ (iD)

Department of Mechanical and Aerospace Engineering, Brunel University London,
London UB8 3PH, UK
Mingfeng.Wang@brunel.ac.uk

Abstract. This project investigates the development and optimization of a fully functional bioinspired jumping robot. It explores advancements within robotic locomotion that mimic the amazing jumping abilities of animals such as fleas, which can jump 200 times their body height! The significance of this study lies in its potential applications, for instance, search and rescue, military reconnaissance, and environmental monitoring. This project uses a combination of computer-aided design tools like Onshape, finite element analysis, and Unity simulations. These software's allows the project to effectively model and test the capabilities of the jumping mechanisms. With the benefit of real-time design adjustments and structural integrity evaluations, significantly boosting the prototyping process. The research focused on constructing a robot capable of overcoming physical obstacles, relative to its size. It was important to utilize the correct materials that optimize strength, weight, and energy efficiency to achieve this, therefore integrating carbon fibre was game-changing. The final design produced a functional robot demonstrating its capabilities to jump. Key findings include the robot's ability to jump approximately 190 mm, at an initial velocity of 3.7 m/s with a total efficiency of 52.1%. Therefore, future enhancements will aim to improve efficiency, landing stability and enable consecutive jumps.

Keywords: Jumping Robot · Bio-inspired Robotics · Mechanical Design · Simulation · Autonomous Systems

1 Introduction

Jumping as a form of locomotion provides robots with many unique benefits. To begin with, jumping allows robots to explore a larger pool of terrain that would be otherwise impossible for other forms of locomotion like wheeled or tracked robots. For example, exploring jungles and forests where there would be many challenging obstacles. Another benefit is that jumping does not have continual contact with the ground. This is beneficial as the probability of being stuck in surfaces such as sand and mud is heavily decreased. Jumping also provides a more speedy and agile form of locomotion when compared to walking or running. For example, fleas have an incredible capacity to jump, which allows them to reach heights that are 80 to 200 times greater than their body length [1].

© The Author(s), under exclusive license to Springer Nature Switzerland AG 2025
M. N. Huda et al. (Eds.): TAROS 2024, LNAI 15052, pp. 164–170, 2025.
https://doi.org/10.1007/978-3-031-72062-8_15

Due to the increased agility jumping provides, the robot can make rapid transitions from one location to another and can even sore over barriers that would otherwise have stopped traditional wheeled locomotion [2]. The relatively small size of jumping robots, further increases their overall agility, making them particularly helpful in situations that require navigation in confined locations. Jumping robots have a wide range of applications, including search and rescue operations. In the case of search and rescue jumping robots are excellent for navigating through debris and overcoming barriers like fallen trees within a disaster zone [3, 4]. As many jumping robots are small in nature, they are able to hop around and reach areas that otherwise would be difficult to access, making them exceptionally efficient for search and rescue missions [5].

In this work, a jumping robot is created inspired by nature, specifically a flea. In Sect. 2, the design is proposed with individual features. Section 3 shows the FEA analysis, and how the structure is affected by the spring. Section 4 shows the results, with the experimental and predicted jump heights of the robot. Conclusion with ideas for future work are presented in Sect. 5.

2   Design

In Fig. 1(a), a flea-inspired jumping robot is proposed where the mechanism incorporates a central compressive spring. This spring is positioned in the middle of the robot's body/legs. When preparing to jump the spring is stretched, through a pulley system and accumulating potential energy, causing the spring to rapidly contract on release. As the spring contracts, it pulls the robot's legs inwards with a significant force this in turn propels the robot into the air. With a single motor this robot is able to achieve its jumps. Furthermore, the geared arms of the robot allow for predictable movement/jumping. Six carbon fibre rods connect the robot together, with the addition of other supports, connecting the robot securely to prevent it from buckling.

The firing mechanism uses three gears one main gear which is directly connected to the motor, a planetary gear and winding gear. The planetary gear orbits around the motor gear, contacting or disengaging from the winding pulley gear. This is depending on the motor's rotational direction. As shown in Fig. 1(b), when the motor turns clockwise, the planet gear connects with the winding pulley gear. This then initiates the winding process, and when the motor rotates the other way, counterclockwise, the planetary gear separates from the winding pulley gear. This separation occurs as on the motor gear, there are teeth that only allow rotation in one direction, allowing the planetary gear to connect and disconnect from the winding gear. When rotated anti-clockwise these teeth lock onto the arm connecting the planetary gear, causing it to disengage from the winding gear. This in turn allows for the winding gear to freely spin, releasing the tension in the wire, sending the robot into its jumping motion. This active mechanism allows the robot to jump at different heights as the planetary gear can disconnect from the winding pulley gear at any given moment. Additionally, allowing the robot to have full control over its jumping motion with only one motor.

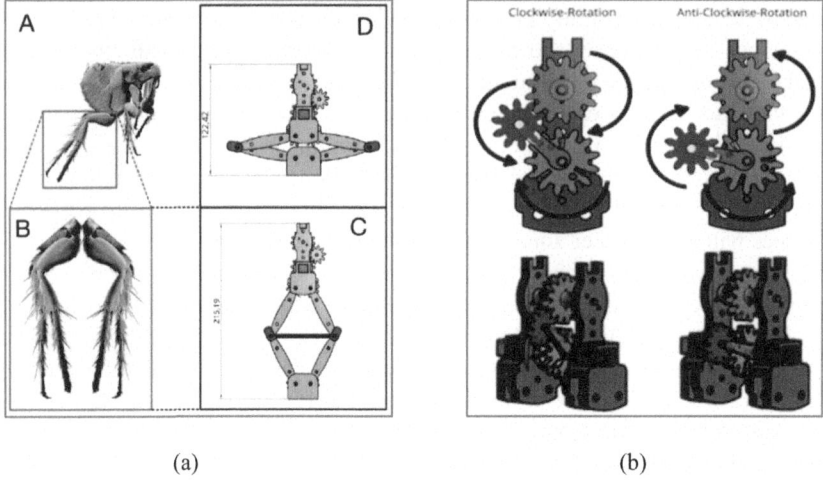

(a)   (b)

Fig. 1. Flea-inspired jumping robot: (a) overview and (b) firing mechanism with winding gear (Purple), planetary Gear (Blue) and motor Gear (Red) (Color figure online)

3 Modelling and Simulation

3.1 Finite Element Analysis (FEA)

Finite Element Analysis is an amazing tool, which can allow engineers to complete a virtual analysis of their designs before creating their prototype. Through simulations, designs can be tested with various forces allowing FEA to show where the potential weaknesses can occur. This can allow for product improvement early on in the design process. As a result, improving the overall final products. Also, reducing the cost of production as prototypes can be tested virtually. In this section, a 3D strucual analysis is performed in Onshape. These results are than compaired to the tested robot to validating these results.

3.2 Analysis Setup

The first step is to create a 3D CAD model of the chosen part which can then be directly inputted and analysed within On-Shape. In this case, a full assemble was used to simulate how stress and strain effect the robot's chassis. Unlike ANSYS, On shape does not require any meshing, as it automatically sets the mesh to your part, saving a great amount of time that would otherwise be spent on generating meshes. In this project the main materials used were carbon fibre and PLA, although other materials such as plywood were also used to test. After material section, setting all the applied loads and moments, to simulate the operating conditions. Once the above is set, the simulation can be run. When the simulation is complete, the results can be easily visualized and interpreted, viewing stress, displacements, deformation contours, and safety factors to assess the performance of the design.

3.3 Analysis of Results

Stress: the distribution of stress within the robot is illustrated in Fig. 2. The areas experiencing the highest stress are primarily located in the robot's gears and two carbon fiber rods holding the spring. This is attributed to the gear restraining the spring from releasing its energy, and the rods supporting the spring's. The greatest amount of stress within this robot was 39503 kPa and seen in these areas.

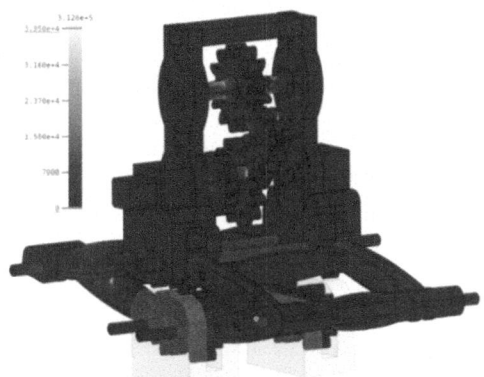

Fig. 2. FEA simulation results: Von Mises stress

Total Deformation: Remarkably, the arms show almost no deformation with PLA and zero with plywood, while the gears experience the most significant deformation, measured at 0.125 mm. This deformation is due to the gears functioning as a trigger mechanism, that prevents the robot from jumping and thereby sustaining the highest levels of stress. Nonetheless, this amount of deformation is considered acceptable, as it is minimal.

Safety Factor: The robot has a safety factor of 3.5, which is advantageous as it indicates that the robot can endure loads up to 3.5 times greater. This safety margin allows for the potential use of an even larger spring to achieve greater jumps than the current 20 cm. Additionally, a safety factor of 3.5 enhances the robot's reliability, ensuring its performance remains consistent under the increased stress of a larger spring. However, there would be a need for another motor or high gear ratio as more torque would be required.

4 Experimental Results

4.1 Experimental Setup

Figure 3 shows the experimental layout of the jumping robot prototype. At the heart of this robot it is controlled by a Arduino® nano, using an advance virtual RISC microcontroller, ATmega328. Connected to the robot is a high torque, 9 V DC motor, with a gear ratio of 1:1000, this is controlled by the motor module L298N. Additionally, an IR sensor is connected to the board, allowing the robot to be controlled by a remote. This is all powered through a 12 v battery.

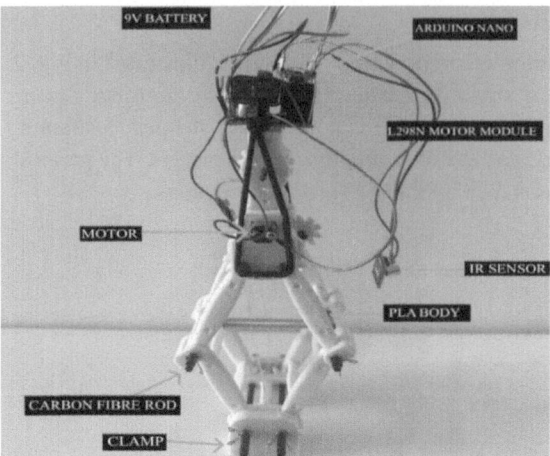

Fig. 3. Experimental layout

4.2 Jumping Experiment

To measure the jumping height of the robot accurately, a plane background with marking for each 10 mm jump interval, with a black line, with every 50 mm marked with text. A level floor was used to obtain accurate results.

The FEA results correlated with the finial build of the robot, as there was no visible deformation within the arms, however there was with the gears. Therefore, implementing carbon fibre rods, to replace the PLA rods, proved to solve this issue, removing any deformation within the gear box. Two materials were mainly tested, plywood and PLA. The plywood proved to be stronger overall however had increased resistance within the gears compared to the PLA gears. Furthermore, the plywood arms weighed considerably more than the PLA arms. As a result, PLA was chosen again as it reduced weight and decreased the friction between the gears. The final robot can approximately jump 190 mm high, Unlike the predicted over 1 m which unity estimated. This is likely due to the fact energy loss is not considered within the software. Due to the constraints of the spring, a simple solution to enhance the robot's jump height is to introduce stiffer springs and possibly another motor, which would store more potential energy and result in higher jumps. Additionally, the top-heavy nature of this robot affects its ability to land stably on its feet after a jump. As this robot is in the air it eventually wants to flip headfirst due to its high centre of mass. The placement of critical components like the battery, motor, and control module at the top of the robot contributes to this high centre of gravity (Fig. 4).

4.3 Loading Force Analysis

This section shows the hand calculations vs experimental seen in Table 1. Where generally the hand calculations showed a much larger predicted number than the actual results, this is due to not consider any energy loss, as seen the energy efficiency is just above 52% suggesting much energy is lost to friction.

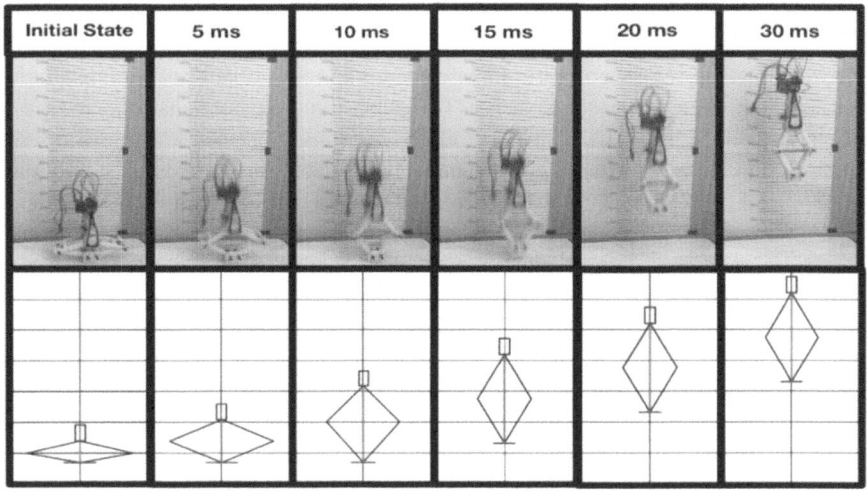

Fig. 4. Robot Jumping Motion

Table 1. Hand Calculations vs Experimental Results

	Calculated Jump height (m)	Acceleration (m/s^2)	Take-off Velocity (m/s)	Landing Impact Force (N)	Energy Efficiency (%)	Initial Stored Energy (J)
Predicted	0.365	51.2	3.8	414.4	52.1	1.03
Experiment	0.19	51.2	3.7	414.4	52.1	1.03

4.4 Comparison to Current Robots

Table 2 shows how the proposed robot performs the current robots such as the MiniWheg and the Scout. The proposed robot performed similar to that seen in the MiniWheg. The proposed Robot is also much taller and much heavier.

Table 2. Proposed Robot vs Current Robots

Robots	Mass (g)	Size (m)	Jumping height (m)
Scout [6]	200	0.115	0.57
MiniWheg [7]	191.4	0.104	0.18
Proposed Robot	287	0.215	0.19

5 Conclusion

This project has development a bio-inspired jumping robot, that draws its inspiration from fleas. Utilizing design tools such as OnShape, finite element analysis, and Unity simulations, this allowed for quick and effective prototyping. As a result, the robot achieved a maximum jump height of 190 mm, at an initial velocity of 3.7 m/s, with total energy efficiency of 52.1%. Despite not reaching the theoretical maximum of 365 mm, it can almost clear its own body height, showcasing the practical application of bio-inspired design in overcoming physical obstacles. The unique firing mechanism seen in Fig. 2, allowed for consistent jumps and at different heights. Carbon fiber was also crucial in optimizing the robot's structural integrity. Future enhancements will focus on improving landing stability, enabling consecutive jumps and most importantly achieving a more energy efficient robot.

Acknowledgement. This work was supported by Brunel Research Initiative & Enterprise Fund (BRIEF) and EPSRC Impact Acceleration Account (EP/X525510/1).

References

1. Mendoza, E., Azizi, E., Moen, D.S.: What explains vast differences in jumping power within a clade? Diversity, ecology and evolution of anuran jumping power. Funct. Ecol. **34**(5), 1053–1063 (2020)
2. Noh, M., Kim, S.W., An, S., Koh, J.S., Cho, K.J.: Flea-inspired catapult mechanism for miniature jumping robots. IEEE Trans. Robot. **28**(5), 1007–1018 (2012)
3. Yim, J.K., Fearing, R.S.: Precision jumping limits from flight-phase control in Salto-1P. In: 2018 IEEE/RSJ International Conference on Intelligent Robots and Systems (IROS), pp. 2229–2236 (2018)
4. Jung, G.P., Casarez, C.S., Jung, S.P., Fearing, R.S., Cho, K.J.: An integrated jumping-crawling robot using height-adjustable jumping module. In: 2016 IEEE International Conference on Robotics and Automation (ICRA), pp. 4680–4685 (2016)
5. Koh, J.S., Jung, S.P., Noh, M., Kim, S.W., Cho, K.J.: Flea inspired catapult mechanism with active energy storage and release for small scale jumping robot. In: 2013 IEEE International Conference on Robotics and Automation (ICRA), pp. 26–31 (2013)
6. Stoeter, S.A., Papanikolopoulos, N.: Kinematic motion model for jumping scout robots. IEEE Trans. Robot. **22**(2), 397–402 (2006)
7. Lambrecht, B.G., Horchler, A.D., Quinn, R.D.: A small, insect-inspired robot that runs and jumps. In: Proceedings of the 2005 IEEE International Conference on Robotics and Automation (ICRA), pp. 1240–1245 (2005)

BlimpleBee: The Helium Assisted Indoor Inspection Drone

Henry Hickson[1,2(✉)], Andrew T. Conn[1], and Hemma Philamore[1]

[1] University of Bristol, Bristol, UK
henry.hickson@bristol.ac.uk
[2] Bristol Robotics Laboratory, University of West England, Bristol, UK

Abstract. Exploring complex 3-dimensional spaces such as the ever-changing vegetated environment of a vertical farm, is a challenging task to automate. Whilst aerial robots are excellent at observing these types of spaces, they are limited in battery life and their rotor blades can be cause for safety concern. This paper introduces an alternative solution in the form of a lightweight, helium assisted drone. The BlimpleBee is thought to be the smallest of its kind at 0.55 m diameter and a total weight of 31.02 g, is able to navigate complex, restricted environments with efficiency. Its slow speed dynamics and hovering ability makes it ideal for inspection type tasks. Through a series of experiments, the performance of the BlimpleBee was characterised and shown to be a viable and promising solution for indoor inspection tasks provided some of the challenges around airflow disturbance can be overcome.

Keywords: Blimp · Drone · Helium-assisted · Indoor inspection

1 Introduction

Vertical farming is one potential solution to many of the food and population related environmental issues facing the human race [1]. Small-scale vertical farming enterprises in particular are a promising solution to sustainable food production. Food can be produced in climate-controlled environments at almost any location worldwide including urban areas, local to the consumers, which in turn reduces both reliance on arable land and lengthy transport practices [2]. This is the driving use case for the research discussed here.

Vertical farming is not without its challenges however, high startup and energy costs place a renewed importance on automation to improve efficiency and maximise yield for a given cost [1]. The monitoring of plant health and their environment is one of the manual, time-consuming processes that could be automated by a robotic system. This involves visually checking the plants for their growth progress and the surrounding growing areas for signs of mold or infection, a task that can be completed with a camera and machine vision pipeline.

© The Author(s), under exclusive license to Springer Nature Switzerland AG 2025
M. N. Huda et al. (Eds.): TAROS 2024, LNAI 15052, pp. 171–183, 2025.
https://doi.org/10.1007/978-3-031-72062-8_16

A key challenge in this task is to navigate the complex, ever-changing, 3-dimensional indoor environment whilst reducing the risk of damage to plants and human staff. Navigation is made ever more challenging by the vertical farm's compact layout with shelf height and distance between shelves minimised such that growing surface area is maximised for a given floor surface area. This is a driving factor in the design of any new small-scale vertical farm, it affects the amount of produce and critically the profitability of the enterprise which in urban areas, with high rent costs and small profit margins, is crucial to success. As such, any robot operating in these spaces must also be minimised in size and cost with little change to the existing infrastructure of the farm. Whilst vertical farming was the focus use case for this work, the same challenge extends to several other use cases for example, when minimising wasted space between shelves in warehouses, when inspecting confined spaces such as pipes and hard-to-reach structures or when used for security patrolling in human-focused spaces such as busy office corridors [3,4].

Blimps, lighter than air-supported, motorised airships [5], have been used for decades to reduce lifting forces in airborne vehicles to increase flight time or load-carrying capacity, all with a non-rigid hull and small propeller blades. Existing work however has not explored the challenges of minimising the size of these craft for small-spaced inspection tasks. The contributions of this work then are as follows:

1. An analysis of the design challenges around minimizing the volumetric size of autonomous blimps using off-the-shelf components.
2. An analysis of the control challenges for low-volume autonomous blimps using a low-cost, webcam-based controller.

To do so we present the design, build and test of BlimpleBee: a helium assisted aerial robot for the inspection of complex, indoor environments (Fig. 2). The blimp features a metallised polyester film envelope contained enough helium to lift a small electronics package, responsible for propulsion and control of the robot. The robot has excellent hovering capabilities and can perform slow speed inspections with higher electrical efficiency than typical aerial robots. A video of the BlimpleBee is available at tinyurl.com/blimplebee.

2   Related Work

Robot blimps have been considered for several applications including surveillance, inspection, crowd control, mapping and advertising [6]. There are examples of several different designs from quadcopters with above-mounted helium balloons [7] to custom designs such as the spherical balloon with motors mounted to each quarter as shown in [8] or a triple motor arrangement where two forward-facing motors control direction and a single downward-facing motor controls altitude [9].

In [5] a comparison was made between typical aerial drones and blimps for surveillance purposes. Blimps were shown to overcome many of the battery-related issues experienced by existing drones, increasing operating time from an average of 20–30 min to up to 8 h. The downside to this marked improvement however was that the blimps tested were not optimised for size, resulting in hull sizes significantly larger than typical quadcopters and significantly more challenging to control.

A comparison of existing size-driven autonomous blimp designs is given in Table 1. These designs vary from the 5 m long tethered blimp with an 8-h battery life designed for wildlife monitoring [10] down to sub-1-metre sizes designed for education [9], indoor surveillance [11] and human-robot interaction [12].

The GT-MAB [12–14] is perhaps the most established design, the smallest yet with impressive flight times. Despite this, the 0.7 m diameter hull and 0.45 m height are still restrictive for navigating small-scale vertical farms with typical shelf spacing of 1 m and shelf heights of 0.5 m [15]. Once plants are placed on the shelves, the available height to navigate the space will be reduced to less than the 0.45 m height of the GT-MAB. Similar constraints can be found, in pipe monitoring, confined structural inspections, warehousing etc. [3,4]. Whilst the hull size could be reduced with battery size, the exponential loss of lift would mean that the existing GT-MAB gondola design can not be lifted by a hull of comparable size to the BlimpleBee.

Table 1. A comparison of size driven blimp designs from existing literature.

Reference	Length (m)	Height (m)	Flight Time
[10]	5	1.8	8 h
[7]	2	2	60 min
[9]	0.91	∼0.6	-
[8]	∼0.8	∼0.5	12 min
[11]	0.75	0.55	48 min
[12–14] (GT-MAB)	0.7	∼0.45	120 min

The control methodology of helium-assisted drones has also been explored in past studies. In [16], a deep learning controller demonstrated excellent path-following characteristics. Similarly, a differentiator-based control methodology was used in [17] for altitude control, an area that is typically much more challenging than quad-rotor counterparts. Differentiator control was in this case chosen for its applicability to low-mass micro-controllers. [18] followed up with a detailed review of a disturbance compensation algorithm, essential for smaller blimps which are more susceptible to changes in airflow in their near vicinity. Similarly, fuzzy logic has also been shown to be applicable [19] as it can to some extent account for drifting motions again due to airflow disturbances. Simpler methods were used in various other designs including PID controllers which once tuned can provide good results [9] with low computational power.

While many of the above studies have successfully demonstrated the autonomous blimp concept, none have met the small size requirements of a vertical farm environment where typical shelf spacing can be as low as 0.5 m. Reducing the size has a payload penalty but with the increasing availability of low-mass electrical components, it is plausible that autonomous blimps can be reduced to a size suitable for more intricate inspection tasks.

3   Materials and Methods

3.1   BlimpleBee Robot

The BlimpleBee and a detailed view of the lightweight electronics package is given in Fig. 2. The primary objective of this design was to minimise size and mass. This section details the design decisions to minimise mass whilst maintaining a maneuverable, flight system (Fig. 1).

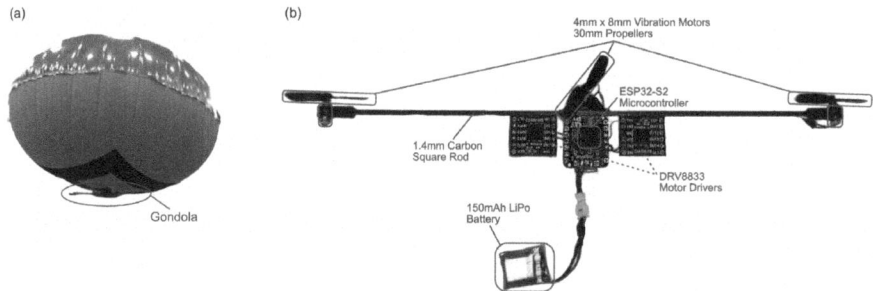

Fig. 1. (a) BlimpleBee robot with hull and gondola below (b) BlimpleBee gondola including the ESP32-S2 QT-PY breakout board, 2x DRV8833 breakout boards, 3×4 mm $\times 8$ mm vibration motors with 30 mm diameter propellers and carbon fibre structural beam.

The drone features a 0.55 m diameter, 0.24 m height circular hull with a material mass of 20.5 g and a net buoyancy of 11 g at 22.5 °C. The hull is made from metallised PET (polyethylene terephthalate), identified in [9] as an excellent retainer of small helium molecules due to the close-knit structure of the particles in the metallic coating. The two-piece circular construction minimises the number of seams to further reduce gas leakage. Other more volume-efficient shapes were considered but the complexity in manufacturing these shapes meant that the likelihood of helium leakage would be far higher. Despite these measures, there is still an inevitable leakage rate recorded in the region of 0.25 g per day (subject to environmental conditions).

Control is handled by an ESP-32 QT-PY, a 2 g breakout board mounted microcontroller with inbuilt memory and a single core, 240 MHz chip. On either side, two DRV8833 motor driver breakout boards are mounted, each used to drive the respective right and left motors and one central motor used to control altitude.

To simplify the base gondola design, no sensors were included in this iteration of the design. The addition of a camera or similar sensor would be required to complete the inspection tasks, existing lightweight modules from the micro-air vehicle industry range from 0.2 g to 2 g [20]. It is possible that this could be offset with a custom circuit board design.

The electronics are attached to a single structural element using a lightweight adhesive, a 1.4 mm × 1.4 mm hollow square carbon-fibre strut, 200 mm long. The low mass beam provided enough separation between the two motors to induce turning moments large enough for proximity manoeuvres. Aside from the primary power bus, all cables are enamel-coated signal wires. Two 4 mm × 8 mm vibration motors are used for their low mass, mounted to each was a 30 mm diameter plastic propeller. The motors are attached with 3D-printed PLA mounts. The gondola is placed at the longitudinal centre of mass of the hull to minimise the total height of the hull when in the flying position. The mass of each component is given in Table 1.

Table 2. A breakdown of component masses.

Item	Mass (grams)	Quantity
ESP32-S2 QT-PY	2.24	1
DRV8833 breakout board	1.30	2
4 m × 8 mm vibration motor	0.65	3
30 mm diameter propeller	0.09	3
3D printed fwd/bwd motor mount	0.18	2
3D printed downward motor mount	0.06	1
1.4 mm × 1.4 mm carbon fibre beam	0.63	1
150 mAh battery and connector	2.07	1
Cables and solder	0.33	1
Total mass	**10.51**	**grams**

The battery was sized to match the buoyancy of the hull minus the weight of the gondola, just over 2 g, enough for a 150 mAh LiPo battery. Battery use was dependent on the buoyancy provided by the hull which in turn is proportional to the volume of helium gas within that hull (Table 2).

Perfect neutral buoyancy allows the robot to operate with far less input from the downward-facing motor as altitude can be maintained with very little thrust. However, due to fluctuations in helium volume, it is rarely the case where perfect neutral buoyancy is achieved. Aside from gas leaks; temperature and atmospheric pressure can cause fluctuations in the volume of gas in the hull. As a result, there is a level of variability in buoyancy throughout the operational time of the blimp. Following a change in temperature, pressure or simply gas leaked since full inflation, the propeller can be rotated to counteract any negative buoyancy. At the recorded temperature of 22.5 °C, there is a 0.5 g-positive buoyancy allowance for adding sensors such as a lightweight camera.

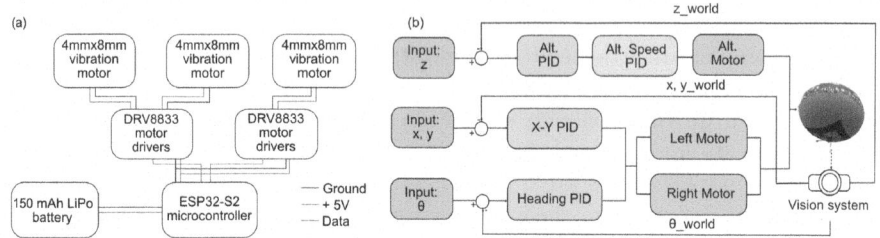

Fig. 2. (a) System diagram for the BlimpleBee gondola design. (b) PID controller diagram for the altitude and x-y positioning control.

3.2   Testbed

The testbed was built to mimic a small size space ready for inspection. A two-camera vision tracking system was used to localise the blimp in 3D space. This system is low-cost and requires minimal infrastructural changes. The off-board processing for localisation also further reduces the mass of the gondola as no localisation sensing is required onboard the robot. Both cameras were connected via USB cable to the testbed server PC which processed incoming video feeds and returns the Blimps x, y, z coordinates. The coordinates were relayed back to the blimp via a dedicated 5 GHz WiFi network. The blimp could request a positional update at any moment through an HTTP request to the testbed's custom built API, hosted on the server. The API returned the latest known positional data as an HTTP response message. The messaging protocols align with the FAST API web framework.

The positions of the cameras are paramount to the success of this method. Camera 1 was mounted on a frame above the testbed, for tracking x-y position. Camera 2 was mounted 1.5 m from the ground, at the centre of one edge of the testbed, for tracking altitude. Each camera had a 120-degree field of view which gave a 4.5 m × 3.5 m × 3.5 m visible testing region. Localisation was completed through the use of fiducial markers. For this study, just two markers were used, one mounted to the underside of the hull, just in front of the gondola, and one mounted to the front of the hull. These were processed using the open-source ArUco tag library in OpenCV.

3.3   Control

To assess the challenges in controlling a blimp of this size, a PID control loop was developed. By using a simple and widely recognised control methodology at this stage in the development of the system, we were able to capture a set of baseline results to characterise the controllability of the blimp under known and repeatable conditions. The controller was made up of three PID loops, one for altitude, one for forward/backward speed, and one for heading.

The controller was implemented on the ESP-32 microcontroller. Routing could be programmed in two ways, the first is via an on-board controller with a series of hardcoded objectives, the second via the testbed API whereby a centralised controller was implemented on the testbed server and the blimp requested waypoints using the HTTP protocol. The latter was used for all testing to better control the experiments from a single location without needing to reprogram the robot for every iteration. Positional data requests from the vision system provided feedback against the demand signal for all three PID controllers. The heading and speed control loops used standard single-loop control, with positional feedback provided by the above-mounted camera.

The altitude control loop was a nested distance-speed controller where the demand was two-fold, first a target altitude, in metres above ground level, and second an altitude velocity of $0\,\text{m/s}$. Altitude velocity was taken as the rate of change of altitude and was computed at each time step. This nested approach is needed to account for changes in buoyancy that may occur throughout the operation of the blimp, for example, due to gas leakages or changes in temperature. By tuning for both speed and altitude, the system can correct for these by slowing the rate of change of altitude when approaching the target position.

4   Experimental Setup

To characterise the performance of the BlimpleBee drone, a series of experiments were carried out. Each experiment was designed with the vertical farm inspection use case in mind, the primary capability for such a task being the ability to locomote around a 3-dimensional space for a given time period. As such four experiments were designed to test the speed, battery life, vertical locomotion and horizontal planar position holding. All experiments were completed in controlled test conditions in the testbed. Disturbances such as circulating air were minimised and with a constant temperature of $22.5\,^{\circ}\text{C}$ was maintained throughout each test.

Test 1: Velocity. Forward velocity was assessed by programming the robot to follow a straight line with various pulse-width modulation (PWM) input signals increasing from 25 to 250, just shy of the maximum PWM of 255. Only the X-Y PID loop and forward-facing motors were used. The experiment was repeated ten times for each speed and in each direction to reduce the effect of any circulating air disturbances.

The experiment was conducted whilst the blimp was at neutral buoyancy with an altitude of 1.5 m above ground.

Test 2: Battery Life. The ideal operating condition for minimising power consumption is one where the drone flies with perfect neutral buoyancy in which case little to no thrust is required to maintain a constant altitude. However, to reduce the time between helium refills, the BlimpleBee is designed to be able to operate with some negative buoyancy. Therefore, battery life was tested in two conditions: neutral buoyancy and with a negative buoyancy of 1 g.

The blimp was programmed to fly continuously between two waypoints, the path following performance was irrelevant for this test. Altitude was constant at 1.5 m above ground level which with perfect neutral buoyancy required no input from the downward facing motor but with -1 g buoyancy required a near constant downward thrust. A range of speeds were tested by simply setting the PWM input signals to increase from 25 to 250. Each interval was repeated 10 times.

Test 3: Altitude Control. To test the BlimpleBee's ability to vary altitude in response to a controlled input, a tethered altitude experiment was carried out. To remove any complex position holding control, a low friction vertical tether was attached to the rear of the BlimpleBee to ensure it remained in the same X-Y location. The tether, a low friction wire, ran upright from ceiling to floor. A 50 mm diameter, circular loop of the same low friction wire was attached to the back of the blimp and looped around the tether. The result was a low-friction sliding mechanism that kept the blimp in the same x-y position, removing any influence from airflow. A square wave input was sent to the controller specifying a varying altitude input between two points with a 20 s pause at each altitude. Two conditions were assessed, neutral buoyancy and -1 g.

Test 4: X-Y Position Holding. To test speed and heading control, a position-holding test was devised. The aim was for the blimp to hold its position over a 20-s period. The blimp was held at the zero position and let go when the timer started, the blimp was expected to naturally drift from the starting point, with the speed and heading controllers used to manoeuvre the blimp back to the start position and to hold that position. This is both a test of the controller's accuracy and reaction time to changes due to drift, a common challenge in blimp control. The experiment was carried out with neutral buoyancy at 1.5 m above ground.

5   Results

Test 1: Velocity. The results for the velocity experiments are given in Fig. 3 (a) The maximum straight line speed recorded was 0.446/s. In general inspection tasks are likely to take place at slow speed, particularly in delicate environments like vertical farms. Hence, speeds are considered to be sufficient, in fact, it would likely operate at speeds far below the maximum.

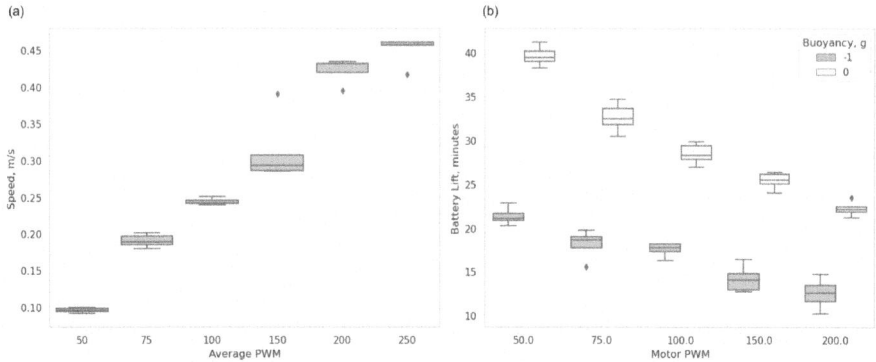

Fig. 3. (a) Results from the speed experiments demonstrating the range of achievable speeds between a PWM of 0 and the maximum PWM of 255. (b) Results from the battery life experiments at two different buoyancy levels, 0g and −1 g. At −1 g, battery life is significantly reduced by the constant use of the downward-facing motor.

Test 2: Battery Life. The results of the battery life experiment are shown in Fig. 3 (b). For any battery dependent robot, operational time is limited by both the battery size and the downtime spent charging it. In this case, the operational time at low speeds is comparable to similarly sized aerial robots with a maximum continuous operation of 39.5 min. When taking into account the far quicker charging time for a battery of this size, the advantage in total operational time becomes far more prominent.

In contrast, as buoyancy reduces, the thrust needed to maintain altitude quickly reduces the battery life. At its maximum output, thrust from the altitude propeller measured as 2.28 g, meaning that theoretically, the BlimpleBee can operate with a buoyancy between 0 and −2 g. However, operating with a buoyancy below −1 g results in rapid battery depletion due to the constant thrust requirements to maintain altitude.

There are options to additionally improve battery life. No propeller flow optimisation was carried out, nor have the motor inputs been optimised for any single forward speed. With a specially designed propulsion arrangement, battery life for each given speed may increase. A reduction in mass through custom circuitry or a low mass envelope material (for example a thinner grade metallised PET) could also allow for a larger battery to be placed on board.

As alluded to previously, increasing the volume of the helium envelope and the battery would significantly increase operating time. The penalty in that case would be in accessibility which for use in vertical farming applications for example would be severely limiting. Other use cases may be more applicable to larger sized or different shaped envelopes.

Test 3: Altitude Control. The results for the altitude control experiments are given in Fig. 4. The nested altitude-velocity controller demonstrated reasonable

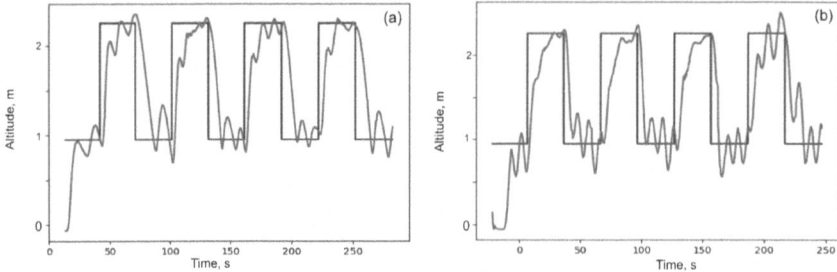

Fig. 4. Response curve for the altitude controller at a buoyancy neutral buoyancy (a) and with a buoyancy of −1 g (b). Despite the slower rise time in the −1 g case, the nested controller adapts well to changes in buoyancy.

results at the two different buoyancy levels, showing its resilience to changing buoyancy throughout operation as was the aim for the nested implementation.

There is an intrinsic challenge in controlling altitude smoothly in an uncontrolled environment and whilst air flow was minimised, the ultra-light weight blimp was sensitive to small changes in environmental conditions. As such there is a variance in altitude around the target altitudes. Further tuning on the controller gains through some level of optimisation techniques may help to reduce this.

Test 4: Position Holding. The result of the position holding test is given in Fig. 5. Similarly to the altitude control, small airflow disturbances made for challenging position holding. As soon as the blimp was released from its starting position, it began to drift. At just 31.02 g total mass and a net weight varying between 0 and 2 g, the BlimpleBee is highly susceptible to disturbances in the air (ventilation etc.). Due to the forward-back propulsion system, a particular issue is lateral disturbance to which the BlimpleBee is slow to recover as it needs to rotate up to 90° to retain its previous position.

Our initial hypothesis that the smaller size would make the blimp less susceptible to airflow disturbances then has proven not true. Whilst there is less surface area to catch moving air, the extreme weight limitations of such for such a small hull mean that the blimp is reactive to even the smallest disturbances. Whilst challenging to correct, the low weight blimp is also easily manoeuvrable under minimal propeller thrust. By investigating two areas we would expect to see significant improvement in reacting to those disturbances. The first is the control methodology which needs to react faster and more accurately to position changes, the second is the propeller layout which could be modified to allow the blimp to crab sideways without needing to keep rotating. The latter may incur a weight penalty but this could be offset by the reduced thrust loading from not having to rotate for every direction change.

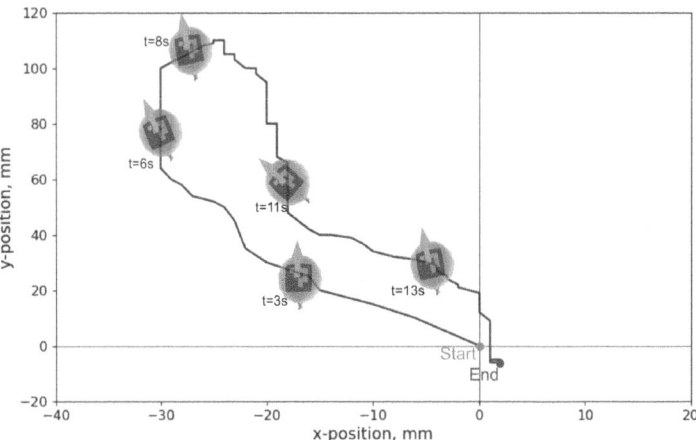

Fig. 5. A tracked position holding plot for the BlimpleBee attempting to hold a position at x = 0, y = 0 for 20 s. The not to scale overlayed images of the robot demonstrate its heading at discrete timesteps throughout the experiment.

6   Conclusion

This study has presented the design, build, and characterisation of the Blimple-Bee robot, a helium-assisted, indoor inspection drone. The lightweight gondola and 0.55 m diameter, 0.24 m height envelope is thought to be one of the smallest of its kind. Performance testing showed promise that with even a simple set of controllers, the BlimpleBee was able to navigate a 3D space with reasonable accuracy. Battery life testing gave good results with operating times greater than average for small-sized aerial robots.

The addition of a camera would allow for a more in-depth investigation into the applicability to inspection type tasks, with experiments in this area a plausible next development. Experiments should also be carried out in vertical farm environments. Future developments will therefore aim to fit a camera module, complete a series of vertical farm specific inspection-based experiments, and test custom circuitry to further reduce mass reductions.

Overall, this study demonstrates the applicability of helium-assisted drones to low-speed, indoor locomotion and shows promise for inspection tasks.

Acknowledgements. This work was funded and delivered under the support of the EPSRC Centre for Doctoral Training in Future Autonomous, and Robotic Systems (FARSCOPE) - Ref: EP/L015293/1.

References

1. Benke, K., Tomkins, B.: Future food-production systems: vertical farming and controlled-environment agriculture. Sustain. Sci. Pract. Policy **13**(1), 13–26 (2017). https://doi.org/10.1080/15487733.2017.1394054

2. Kalantari, F., Tahir, O.M., Joni, R.A., Fatemi, E.: Opportunities and challenges in sustainability of vertical farming: a review. J. Landscape Ecol. **11**(1), 35–60 (2018). https://doi.org/10.1515/jlecol-2017-0016

3. Jeong, K., Kwon, J., Do, S.L., Lee, D., Kim, S.: A synthetic review of UAS-based facility condition monitoring. Drones **6**(12), 420 (2022)

4. Zhang, B., Tsuchiya, S., Lim, H.: Development of a lightweight octocopter drone for monitoring complex indoor environment. In: 2021 6th Asia-Pacific Conference on Intelligent Robot Systems (ACIRS), pp. 1–5, July 2021

5. Adorni, E., Rozhok, A., Revetria, R., Ivanov, M.: Literature review on drones used in the surveillance field. In: Proceedings of the International Multiconference of Engineers and Computer Scientists, October 2021

6. Mashoedah, M., Dhewa, O.A., Roviyan, Z.S., Leo, D., Masyitoh, S.L.: Crowd detection system using blimp drones as an effort to mitigate the spread of Covid-19 based on internet of things. Elinvo (Electron. Inf. Vocat. Educ.) **6**(2), 157–165 (2021). https://doi.org/10.21831/elinvo.v6i2.43873

7. Papa, U., Ponte, S., Del Core, G.: Conceptual design of a small hybrid unmanned aircraft system. J. Adv. Transp. **2017**, 1–10 (2017). https://doi.org/10.1155/2017/9834247

8. Sadasivan, N.: Design and realization of an unmanned aerial rotorcraft vehicle using pressurized inflatable structure. Int. J. Aviat. Aeronaut. Aerosp. **6**(4), 3 (2019). https://doi.org/10.15394/ijaaa.2019.1352

9. Gorjup, G., Liarokapis, M.: A low-cost, open-source, robotic airship for education and research. IEEE Access **8**, 70713–70721 (2020). https://doi.org/10.1109/ACCESS.2020.2986772

10. Adams, K., Broad, A., Ruiz-García, D., Davis, A.R.: Continuous wildlife monitoring using blimps as an aerial platform: a case study observing marine megafauna. Aust. Zool. **40**(3), 407–415 (2020). https://doi.org/10.7882/AZ.2020.004

11. Macias, G., Lee, K.: Design and analysis of a helium-assisted hybrid drone for flight time enhancement (2021). https://doi.org/10.20944/preprints202109.0483.v1

12. Yao, N., et al.: Autonomous flying blimp interaction with human in an indoor space. Front. Inf. Technol. Electron. Eng. **20**(1), 45–59 (2019). https://doi.org/10.1631/FITEE.1800587

13. Seguin, L., Zheng, J., Li, A., Tao, Q., Zhang, F.: A deep learning approach to localization for navigation on a miniature autonomous blimp. In: IEEE 16th International Conference on Control and Automation (ICCA), pp. 1130–1136 (2020). https://doi.org/10.1109/ICCA51439.2020.9264514

14. Tao, Q., Wang, J., Xu, Z., Lin, T.X., Yuan, Y., Zhang, F.: Swing-reducing flight control system for an underactuated indoor miniature autonomous blimp. IEEE/ASME Trans. Mechatron. **26**(4), 1895–1904 (2021). https://doi.org/10.1109/TMECH.2021.3073966

15. Chaichana, C., Man, A., Wicharuck, S., Mona, Y., Rinchumphu, D.: Modelling of annual sunlight availability on vertical shelves: a case study in Thailand. Energy Rep. **8**, 1136–1143 (2022). https://doi.org/10.1016/j.egyr.2022.07.077

16. Liu, Y.T., Price, E., Goldschmid, P., Black, M.J., Ahmad, A.: Autonomous blimp control using deep reinforcement learning (2021). arXiv preprint arXiv:2109.10719. https://doi.org/10.48550/arXiv.2109.10719

17. Wang, Y., Zheng, G., Efimov, D., Perruquetti, W.: Differentiator application in altitude control for an indoor blimp robot. Int. J. Control **91**(9), 2121–2130 (2018). https://doi.org/10.1080/00207179.2018.1441549
18. Wang, Y., Zheng, G., Efimov, D., Perruquetti, W.: Disturbance compensation based controller for an indoor blimp robot. Robot. Auton. Syst. **124**, 103402 (2020). https://doi.org/10.1016/j.robot.2019.103402
19. Al-Jarrah, R., Roth, H.: Developed blimp robot based on ultrasonic sensors using possibilities distribution and fuzzy logic. J. Autom. Control Eng. **1**(2), 1–7 (2013). https://doi.org/10.12720/joace.1.2.119-125
20. BitCraze AB: Datasheet AI-deck color camera module - Rev 1 (2021)

Effects of Passive Ankle on Bipedal Robot Walking Locomotion

Xiangyu An, Mayo Adetoro⬭, and Mingfeng Wang$^{(\boxtimes)}$⬭

Department of Mechanical and Aerospace Engineering, Brunel University London, London UB8 3PH, UK
Mingfeng.Wang@brunel.ac.uk

Abstract. This study explores the effect of robot locomotion with equipping passive ankle joints. The designed bipedal robot consists of two five-bar linkage legs with torsional springs equipped on both ankle joints, which is constructed and elaborated in SolidWorks® environment. Inverse kinematic analysis for legs are computed in MATLAB® to generate the input values of hip and knee motors for robot locomotion. The simulations are conducted in SolidWorks® motion analysis environment with the aims of evaluating and comparing the dynamic walking performances between the robot model with passive ankle joints and the one with active ankle joints. The proposed robot model is expected to perform continuous walking on flat surface in the simulation and the simulation results are analyzed in three categories including contact force between feet and the ground, motor torques, and motor energy consumptions during the locomotion. The results show the robot with passive configuration has more stable walking performance with lower contact force during locomotion. In the meantime, the average motor torques and average motor energy consumptions of passive configuration during simulation are lower than the ones with active configuration.

Keywords: Bipedal robot · Passive ankle design · dynamic walking

1 Introduction

Recent achievements of technology in bipedal robotics over the past two decades have positioned these robots as pivotal solutions to global challenges such as aging populations and rising labor costs [1]. As bipedal robots are designed to mimic human movement, their applications are expanding across various fields such as industrial operations, healthcare, space exploration, and hazardous material handling [2, 3]. Within the design of bipedal robots, compliant mechanisms have played an important role to enhance their efficiency by using elastic elements to store and release energy during the locomotion cycle, thus reducing active power needs, improving terrain adaptability, and mirroring human walking dynamics for natural gait and increased stability [4].

Compliant elements applied in bipedal robots can be mainly grouped into three categories. The first utilizes compressed air, such as pneumatic cylinders and pneumatic artificial muscles (PAM), which offer high compliance and driving force. Mowgli is a

M. N. Huda et al. (Eds.): TAROS 2024, LNAI 15052, pp. 184–194, 2025.
https://doi.org/10.1007/978-3-031-72062-8_17

lightweight robot that equipped with PAMs as musculoskeletal structures, capable of jumping heights exceeding half its body height [5]. The second category includes coil springs, which simulate leg muscles to efficiently store and release energy. This method is popular for its simplicity and effectiveness in energy management, as demonstrated by KenKen, a one-legged robot inspired by a dog's hindlimb with a design that allows springs to convert kinetic energy efficiently [6]. Same method is also applied to these works [7–10]. The final category involves using compliant structures and novel materials to manufacture lightweight elements with high energy capacity and elasticity, such as the University of Pennsylvania's RHex Robot Platforms, which feature adjustable compliance in their composite C-shaped legs controlled by a self-locking actuation system and monitored via a rotary sensor for precise stiffness adjustments [11–13].

Research on passive mechanisms in bipedal robots predominantly centers on the legs, torso, and spine, with few studies exploring the integration of passive and compliant mechanisms in ankle joints [14]. The ankle is important in walking and jumping, providing compliance, shock absorption, and energy storage. It plays a key role in human gait, stabilizing the body, adapting to various terrains, and conserving energy through its elastic properties. Especially in uneven terrain locomotion, the ankle's ability to adjust to surface angles significantly reduces the computational demand on control systems, enhancing stability and responsiveness. This adaptability is essential in robots designed for complex settings like search and rescue [15]. Moreover, the passive ankle mechanism boosts energy efficiency by storing and releasing elastic energy, thereby lowering power demands from actuators and extending operational capabilities [16]. However, integrating passive ankles faces challenges in balancing compliance with control and ensuring stability across diverse terrains and activities.

This paper proposes a bipedal robot with passive ankle joints by using coil springs and evaluates the effectiveness during its locomotion. In Sect. 2, the mechanical design of the robot and the leg inverse kinematics are discussed. Section 3 presents a comparison study on the performance of the proposed bipedal robot with proposed passive ankle joints and active ones via dynamic simulation. Section 4 presents the discussion and conclusion.

2   Methodology

2.1   Mechanical Design of the Bipedal Robot with Passive Ankles

The bipedal robot is designed and constructed in SolidWorks® as shown in Fig. 1, consisting of two five-bar linkage legs and one robot body equipped with four servo motors. The hip joints are directly driven by the motors located in the bottom of the robot body and the knee joints are tendon-driven by the upper motors in the body. In addition, to gain more energy, a compression spring is equipped on the middle segment of the leg as detailed in Fig. 1(b). The ankle joints of the robot are passive joints, and a pair of torsional springs is equipped at each ankle joint to store and release energy as shown in Fig. 1(c). To ensure the robot can have stable locomotion performance, flat feet are considered and selected in the paper.

In Table 1, the dimensions of each part are presented. L_1, L_2, L_3 represent the length of each segment of the leg, and H_1 represents the normal distance between the hip joint

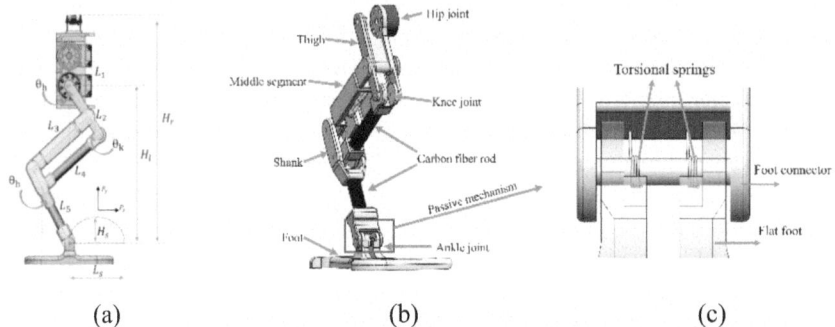

(a)   (b)   (c)

Fig. 1. The design of the proposed robot: (a) overview, (b) leg mechanism, and (c) foot with passive element

Table 1. Dimension parameters in Fig. 1 (in mm)

L_1	L_2	L_3	L_s	H_r	H_l	H_s	Foot
80	110	110	240	205	50	60	140×90

and ankle joint and H_r represents the total height of the robot. H_s and L_s represent the height and length of one footstep, respectively. The semi-curved flat foot is designed to provide sufficient support during the single leg standing phase and offer lighter weight than the full-size flat foot. Torsional springs are applied as the passive energy storage and release mechanism and installed at the ankle joint. The ankle torsional springs play a significant role in robot locomotion. They will store the energy during the stance phase of the leg and release the energy at the moment of the swing phase of the leg. They also provide support forces for the ankle joints during the stance phase. Therefore, the stiffness of the torsional spring must be strong and accurate. The required spring stiffness can be calculated by using the method of Morasso and Schieppati to approximate the minimum value of ankle stiffness [14].

$$c_a = mgl \tag{1}$$

where C_a is the required ankle stiffness and m is the mass of the robot, and the g represents the gravity of earth. The required minimum value of the ankle stiffness is 41 Nmm/deg for the design of the robot.

2.2 Inverse Kinematic Analysis of Leg Mechanism

To evaluate the mobility of the proposed design, the inverse kinematic analysis is conducted. Figure 2a shows the configuration of the five-bar linkage leg and the schematic diagram is shown in Fig. 2b. The Degrees of Freedom (DoF) of each leg can be calculated as

$$DOF = M \cdot \left(N_l - 1 - N_j\right) + \sum_{K=1}^{N_j} f_k = 3 \times (5 - 1 - 5) + 5 = 2 \tag{2}$$

where M is DoF of a link in 2D space, N_l, N_j, f_k are the number of links, joints and DoF of joint k, respectively.

The DoF of each leg is two and it leads to the bipedal locomotion is restricted to the sagittal plane due to the limitation of the mechanical design. The kinematic analysis of the leg system can be calculated in cartesian coordinates. Two Cartesian coordinate systems $\{H\}$: O_H-$X_H Y_H Z_H$ and $\{C\}$: O_C-$X_C Y_C Z_C$ are fixed on the centres (i.e. points H and C) of the hip joint and the ankle joint, respectively. The joint points are denoted by H, A, B, C and the length of links HA, AB, BC are denoted by L_1, L_2, L_3. Link HA is connected with the hip joint and driven by the motor directly. Link AB is connected with link HA at another end. Link BC is connected with link AB and the foot with torsional spring are attached at the end of the link BC. The angle α is the initial angle of the thigh relative to the body. The angle β is the initial angle of link AB and link HA. The angle of γ is the initial angle between the projection line of point B and linkage BC. Since the link HA is parallel to link BC due to mechanical design, we can obtain α = γ.

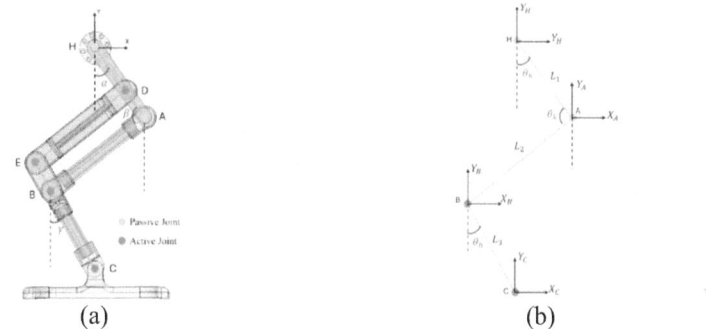

Fig. 2. Configuration of single leg: a) CAD model and b) schematic diagram

In Fig. 2b, the θ_h and the θ_k represent the input values for the hip and knee motors during locomotion. To obtain the expressions the θ_h and the θ_k, the inverse kinematic analysis has been conducted. The trajectories of the point C in Cartesian coordinate have been predefined and discussed in the next section. The expressions of the θ_h and θ_k can be calculated by employing the following equations.

The position of point C in cartesian coordinate with respect to the origin H need be represented:

$$\theta_h = arcsin \frac{(L_1+L_3)^2-L_2^2+X_C^2+Y_C^2}{2\times(L_1+L_3)\times\sqrt{X_C^2+Y_C^2}}$$
$$\theta_h = arcsin \frac{(L_1+L_3)^2-L_2^2-X_C^2-Y_C^2}{2\times(L_1+L_3)\times\sqrt{X_C^2+Y_C^2}} - \theta_h \tag{3}$$

Based on the mathematical model and analyses performed in this section, the input angle values of hip and knee joints can be obtained and applied to the trajectory generation of the ankle of bipedal robot.

3 Walking Simulation

The walking simulation is carried out in SolidWorks® motion analysis environment to compute the proper contact properties between the ground and the robot feet. The gait of the robot in the simulation is predefined and discussed in the trajectory plan simulation by referring to our previous works [17, 18].

3.1 Trajectory Plan

The value of the angle θ_h and θ_k serve as inputs governing the of the hip and knee motors, for the bipedal robot's locomotion. As detailed in the preceding section, the equations to obtain these data have been discussed. The computations are executed using MATLAB®, with subsequent validation conducted through SolidWorks® motion analysis.

Figure 3a shows the overview of predefined trajectory of both ankles, starting with a half step in right foot and ending with another half in left foot. Specifically, in the starting half step, the right foot moves along A-B-C as shown in Fig. 3b while the left foot moves from A to B as shown in Fig. 3c. It follows with cycle full steps in both feet, i.e., right foot with C-D-E-C in Fig. 3b and left foot with B-C-D-B in Fig. 3c. The ending half step is reverse of starting half with the right foot moving from C to A in Fig. 3b while the left one moving along B-E-A. The shape of the trajectory in the air is a semi-elliptical curve, where L_s and H_l represent the length and height of a full step set as $L_s = 60$ mm and $H_l = 50$ mm, respectively.

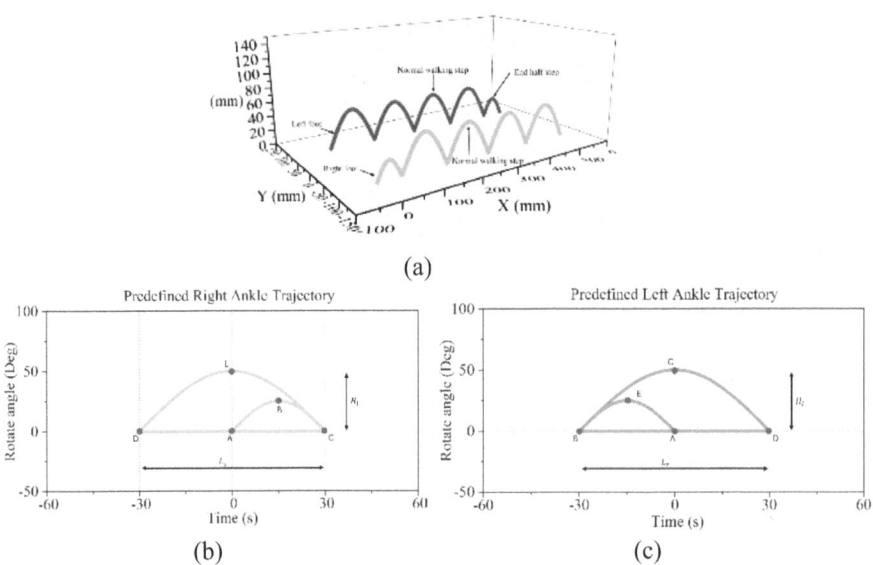

Fig. 3. The walking trajectories (a) Overview of both ankle trajectories (b) right ankle trajectory (c) left ankle trajectory

3.2 Walking Simulation with Passive Ankle Joints

As mentioned in previous section, the limited DoFs of the robot results that the front plane of the robot will be constrained in the simulation as shown in Fig. 4a. Therefore, the robot locomotion is contrained in sagittal plane. The passive ankle consists of a set of torsional springs where the joint stiffness is set as 100 Nmm/deg to support the robot locomotion as Fig. 1c demonstrated.

The robot with passive ankle joints is expected to walk five steps on a flat steel surface with the step cycle of 3 s/step in the simulation. The duration of the simulation is 15 s in total. Figure 4b illustrates the input angular displacements for the hip and knee motors for both right and left side and Fig. 4c shows the computed walking trajectory.

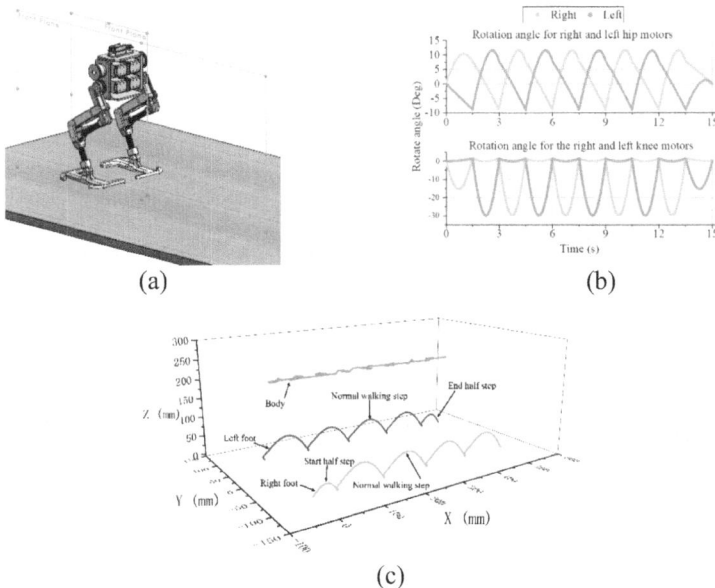

(a)   (b)

(c)

Fig. 4. The simulation environment input and output (a) Robot constraint (b) Input angular displacements for hip and motors (c) The computed walking trajectories for robot with passive ankle joints

3.3 Walking with Active Ankle Joints

The second walking simulation is carried out to test and compare the locomotion performances of the robot equipped with ankle motors with the robot equipped with passive ankle joints. Because of the mechanical design, the thigh is parallel with the shank of the robot. Therefore, the input angular displacement of the ankle motors is the same as the input of the hip motors. The simulation setup is the same as the first simulation and the requirement of the active walking simulation is the same as the passive walking simulation.

Figure 5 shows the comparisons between predefined trajectories and computed trajectories for two ankle configurations. It shows that there are offsets between predefined and computed trajectories and the one with passive ankle joints performs much less offsets as compared to the one with active joints, details can be found in Table 2.

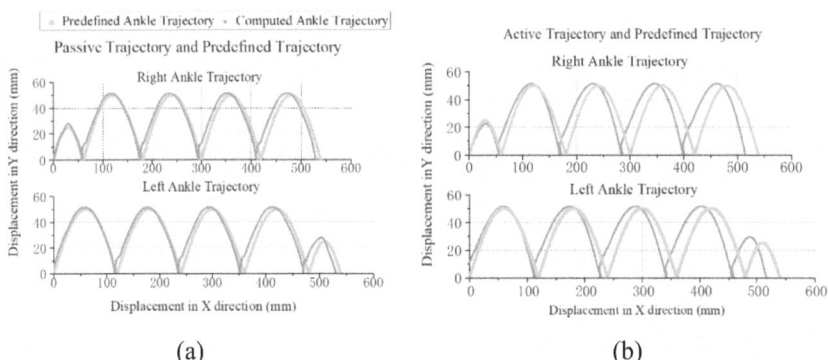

Fig. 5. Walking distance and offset for predefined trajectory and computed trajectories in X direction: (a) Comparison between predefined trajectory and computed passive trajectory (b) Comparison between predefined trajectory and computed active trajectory

Table 2. The walking offset between predefined trajectory and computed trajectories at each step in X direction (mm)

Step	Offset between predefined trajectory and computed passive trajectory		Offset between predefined trajectory and computed active trajectory	
	Right ankle	Left ankle	Right ankle	Left ankle
1	2.36	4.56	3.21	7.24
2	4.11	5.18	8.83	12.9
3	5.76	8.19	14.18	17.85
4	8.18	10.16	19.64	21.21
5	9.72	9.27	24.15	23.74

3.4 Simulation Results and Discussion

Simulation results are presented in terms of the walking accuracy, contact forces, motor torques and motor energy consumption during locomotion. These results can be considered as performance characteristics for evaluating and comparing for the robot equipped with passive ankle joints.

As shown in Fig. 5 and Table 2, the walking distance of each step for both passive configuration robot and active configuration robot are recorded and compared with values from predefined trajectory. It is found that with the increase of the walking distance, the distance offsets of both configurations are increased, and the passive mode has less offsets as compared to the active mode. The right ankle offset at the first step for passive configuration is 2.36 mm and the offset for the active configuration is 3.21 mm. The offset at the last step increases to 9.72 mm for passive configuration and 24.15 mm for active configuration. The left ankle offsets show the same result. Moreover, the offset for passive ankle in each step is smaller than the offsets for the active configuration. It can be concluded that the walking accuracy of the robot with passive ankle joints is higher than the one with active ankle joints.

Figure 6 illustrates the contact force between the right foot and the ground. Due to the similarity between the right and left feet, the analysis focuses on the right foot's contact force. Both passive and active ankle joints exhibit cyclical and similar patterns, with the contact force ranging from 15 N to 30 N in both configurations. The detailed figure presents the contact force for one full step from 6 s to 9 s. A full step is defined from the moment the contact force reaches zero to when it begins to increase from zero. Between 7.5 s and 9 s, the contact force for both configurations' ranges from 20 N to 30 N, indicating ground contact and balance maintenance.

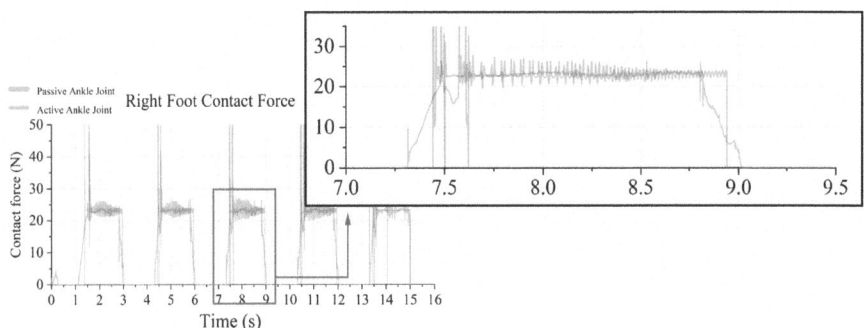

Fig. 6. The contact force of right foot with zoomed view for one full step

The blue line (passive ankle joint) is more convergent than the orange line (active ankle joint), with lower contact force values for the passive configuration during this period. This suggests that the robot with a passive ankle joint maintains balance more easily and exhibits more stable performance with lower contact force. The movements of the active ankle joint are predefined, ensuring that the foot remains parallel to the ground, allowing for direct up-and-down motion. Consequently, the contact force for the active ankle joint transits directly from 0 N to 20 N. In contrast, the passive ankle joint's design limits direct control, causing the foot's edge to contact the ground first during landing and to leave last during departure. This results in the observed slopes in the passive ankle joint's contact force graph, representing gradual increases and decreases in contact force. This sloped pattern facilitates smoother landings for the robot.

Figure 7 presents the motor torques for the right hip and right knee. Because of the similarity between right leg and left leg, we only discuss the motors on the robot's right side. The detailed figure shows the motor torques from 6 s to 9 s which shows the motor torques of 1 full step. It can be observed that the blue line (passive joint configuration) is more convergent than the orange line (active joint configuration) and the average motor torques of hip and knee motor for the robot with passive ankle joints is lower than the ones with active ankle joints.

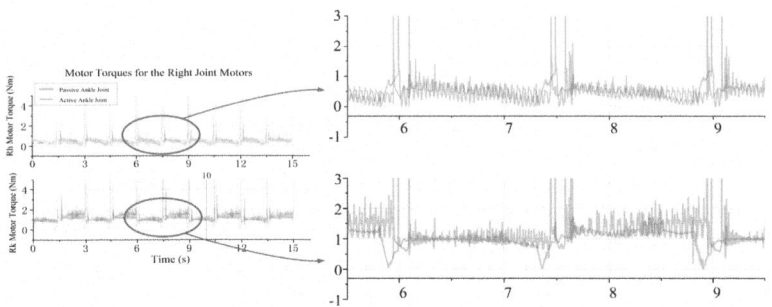

Fig. 7. The motor torque for the Right Hip (RH) motor and Right Knee (RK) motor

Figure 8 shows the motor energy consumptions for both configurations at each step. The results show that most of the energy consumptions of passive configuration in each step are lower than the energy consumptions of active configuration. But there are three values for the passive configuration slightly higher than the values for the active configuration. Therefore, the average energy consumption for each motor is calculated and compared to evaluate the performance of two configuration joints. Table 3 shows the average energy consumption in each motor. And it can be concluded that the energy consumptions in passive configuration are lower than the energy consumptions in active configuration.

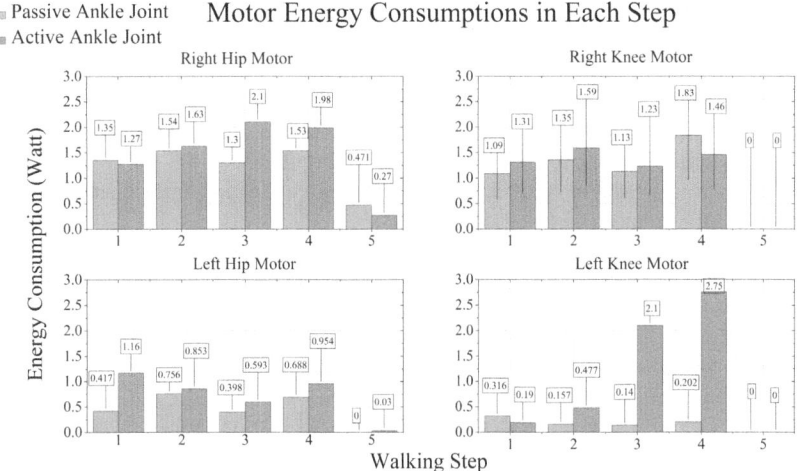

Fig. 8. Comparison of motor energy consumption for passive ankle configuration and active ankle configuration

Table 3. The Average Energy Consumption for Each Motor

	RH Motor	RK Motor	LH Motor	LK Motor
Passive Ankle Joint	1.238	1.08	0.452	0.163
Active Ankle Joint	1.45	1.118	0.718	1.103

4 Conclusion

In this work, a comparison study has been carried out to investigate the effect of passive and active ankle joints on the bipedal robot during walking. To evaluate the performance of the passive ankle configuration locomotion, a robot with active ankle joints is tested as a reference.

The walking simulations have shown that the robot with passive ankle configuration exhibited much less offset during locomotion, which indicates higher walking accuracy as compared to the one with active configuration. The robot with active ankle joints occurs with higher contact force values during walking, in other words, the active ankle configuration needs larger force to maintain the robot's balance. Moreover, the motor torques and energy consumptions for the robot with passive ankle joint is significantly lower than the ones with active ankle joints. In the future work, more locomotion types such as jumping and running will be conducted via simulation and experimental tests.

References

1. Tong, Y., Liu, H., Zhang, Z.: Advancements in humanoid robots: a comprehensive review and future prospects. IEEE/CAA J. Automatica Sinica. **11**, 301–328 (2024). https://doi.org/10.1109/JAS.2023.124140

2. Wahde, M., Pettersson, J.: A brief review of bipedal robotics research. In: Proceedings of the 8th UK Mechatronics Forum International Conference. pp. 480–488 (2002)
3. Brooks, R.A.: Prospects for human level intelligence for humanoid robots. In: Proceedings of the First International Symposium on Humanoid Robots (HURO-96), pp. 17–24 (1996)
4. Zhou, X., Bi, S.: A survey of bio-inspired compliant legged robot designs. Bioinspir Biomim. 7 (2012). https://doi.org/10.1088/1748-3182/7/4/041001
5. Niiyama, R., Nagakubo, A., Kuniyoshi, Y.: Mowgli: a bipedal jumping and landing robot with an artificial musculoskeletal system (2007)
6. Hyon, S.H.: Development of a biologically inspired hopping robot-"Kenken". In: IEEE International Conference on Robotics and Automation (Cat. No. 02CH37292), vol. 4, pp. 3984–3991. IEEE (2002)
7. Zhang, Z.Q., Chen, D.S., Chen, K.W.: Analysis and comparison of three leg models for bionic locust robot based on landing buffering performance. Sci. China Technol. Sci. **59**, 1413–1427 (2016). https://doi.org/10.1007/s11431-016-6100-8
8. Nguyen, Q.V., Park, H.C.: Design and demonstration of a locust-like jumping mechanism for small-scale robots. J. Bionic Eng. **9**, 271–281 (2012). https://doi.org/10.1016/S1672-652 9(11)60121-2
9. Burdick, J., Fiorini, P.: Minimalist jumping robots for celestial exploration (2003)
10. Vanderborght, B., Tsagarakis, N.G., Van Ham, R., Thorson, I., Caldwell, D.G.: MACCEPA 2.0: compliant actuator used for energy efficient hopping robot Chobino1D. Auton. Robots **31**, 55–65 (2011). https://doi.org/10.1007/s10514-011-9230-7
11. Brown, B., Garth, Z.: The bow leg hopping robot. In: IEEE International Conference on Robotics and Automation, pp. 781–786 (1999)
12. Galloway, K.C., Clark, J.E., Koditschek, D.E.: Design of a tunable stiffness composite leg for dynamic locomotion. In: International Design Engineering Technical Conferences and Computers and Information in Engineering Conference, pp. 215–222 (2009)
13. Galloway, K.C.: Passive variable compliance for dynamic legged robots (2010). http://reposi tory.upenn.edu/edissertations/246
14. Manoonpong, P., Kulvicius, T., Wörgötter, F., Kunze, L., Renjewski, D., Seyfarth, A.: Compliant ankles and flat feet for improved self-stabilization and passive dynamics of the biped robot "RunBot." In: 11th IEEE-RAS International Conference on Humanoid Robot,. pp. 276–281. IEEE (2011)
15. Rashty, A.M.N., Sharbafi, M.A., Mohseni, O., Seyfarth, A.: Role of compliant mechanics and motor control in hopping - from human to robot. Sci. Rep. **14**, 6820 (2024). https://doi.org/10.1038/s41598-024-57149-0
16. Liu, Z., Gao, J., Rao, X., Ding, S., Liu, D.: Complex dynamics of the passive biped robot with flat feet: Gait bifurcation, intermittency and crisis. Mech. Mach. Theory. **191** (2024). https://doi.org/10.1016/j.mechmachtheory.2023.105500
17. Wang, M., Ceccarelli, M., Carbone, G.: A feasibility study on the design and walking operation of a biped locomotor via dynamic simulation. Front. Mech. Eng. **11**, 144–158 (2016). https://doi.org/10.1007/s11465-016-0391-0
18. Wang, M., Ceccarelli, M., Carbone, G.: Design and development of the cassino biped locomotor. J. Mech. Robot.Mech Robot. **12**, 031001 (2020). https://doi.org/10.1115/1.404 5181

Development of Underactuated Geometric Compliant (UGC) Module with Variable Radial for Robotic Applications

Mark Krysov⦿ and Seyed Amir Tafrishi$^{(\boxtimes)}$⦿

School of Engineering, Cardiff University, Cardiff CF24 3AA, UK
{krysovm,tafrishisa}@cardiff.ac.uk

Abstract. This paper introduces a novel underactuated geometric compliant (UGC) robot and investigates the behaviors of underactuated compliant modules with variable radial stiffness, aiming to enhance the versatility and functionality of UGC robots. We initiate the study by designing and fabricating various compliant semi-rigid geometric joints, each tailored to a specific design objective. These joints undergo physical testing to validate their stiffness characteristics and returnable angles as durability factors. Subsequently, we develop a mathematical model based on Gaussian process regression to incorporate the different geometric joint characteristics, including thickness, facilitating the development of fully functional prototypes with easy-to-3D print models. After analyzing individual joints, we present various configurational combinations to construct the overall UGC module for robotics applications. Our final prototype UGC can dynamically alter its radius, reducing to 80–85% of its original value while maintaining structural integrity and operational efficiency. This study discusses potential abilities, challenges, and limitations associated with employing UGC modules, offering valuable insights for future research and developments in UGC robotics.

Keywords: Geometric compliant joints · Underacuated modular robots · soft robotics · Compliant mechanisms design

1 Introduction

As the field of robotics increasingly embraces soft and compliant mechanisms, there is a growing recognition of their significance within research studies [6]. These attributes, characterized by flexibility and compliance, not only enhance the safety of robotic platforms but also afford them greater degrees of freedom in body deployment. However, the attainment of such compliance and flexibility often necessitates intricate and voluminous actuation mechanisms, thereby presenting formidable hurdles for practical implementation in robotic systems. There are challenges in how can robot bodies achieve *modularity* and *compliance*

© The Author(s), under exclusive license to Springer Nature Switzerland AG 2025
M. N. Huda et al. (Eds.): TAROS 2024, LNAI 15052, pp. 195–207, 2025.
https://doi.org/10.1007/978-3-031-72062-8_18

while operating within constraints imposed by a limited number of actuators, known for *underactuation* [5].

Compliance and softness are crucial for creating safer and more flexible mechanisms [12] and robotic systems [10], offering advantages across various scenarios. One such scenario is the inspection and maintenance of power plants, where traditional intrusive methods are both difficult and expensive to implement [7,15]. Utilizing a robot capable of adapting to the environment's shape, like a soft and compliant robot, while also withstanding extreme conditions like a rigid robot, would enable tasks to be performed directly within the system [4]. This would primarily involve the robot travelling through piping to an area of interest and using its ability to change shape to lock its position so that any needed inspection or repairs can be carried out in a stable environment. With the completion of the task, the robot may then re-form itself and continue to the next task/objective location. Another potential application is the exploration of cramped and hostile environments, where sharp objects pose a danger to soft robots and challenge traditional wheeled or legged robots' ability to traverse. This would include cave systems for search and rescue or research where using personnel would be unreasonable. However, the challenge lies in both design perspective and material choice [8], as selecting the appropriate semi-rigid material, such as Polylactic acid (PLA) with easy-to-3D-print models, for compliant robot surface structures remains a significant challenge. Silicon-based materials, although commonly used, are fragile and struggle to endure high force, limiting their suitability for directly controllable surfaces. However, a benefit of the use of compliance is the ability to function at different scales, where drastic redesigns are not required when changing the size of the design, even into microscopic levels.

Another aspect is underactuated robotics that, despite their greater intricacy compared to fully actuated counterparts, offer a multitude of advantages encompassing energy efficiency, material conservation, and space optimization [2,5]. The underactuated robots exhibit heightened efficiency and flexibility in select scenarios, particularly when coupled with precise control mechanisms. This efficacy is notably evident in intricate tasks like locomotion [16] or shape transformation [1] in constrained environments. Underactuated robots play a pivotal role in conservation efforts, as their diminished actuation requirements translate to reduced mass, volume, and energy consumption [2,13,14]. This principle mirrors biological structures in nature; for instance, snakes or birds utilize their multi-skeletal bodies with minimal actuation to execute complex, flexible, and multi-purpose morphological transformations for traversing or moving in space. The primary objective in underactuated systems is to incorporate the least actuator numbers capable of not only adjusting stiffness but also facilitating changes in the body's size or form to navigate challenging environments, akin to the locomotion strategies observed in soft-bodied animals like snakes. However, in the robotics field, this capability is still an open problem of how geometric compliance can be integrated with underactuation to reshape (change size) or transform with a smaller number of actuators.

In this paper, our motivation stems from creating geometrically simple-to-3D print bodies that not only possess compliance and deformability but also utilize

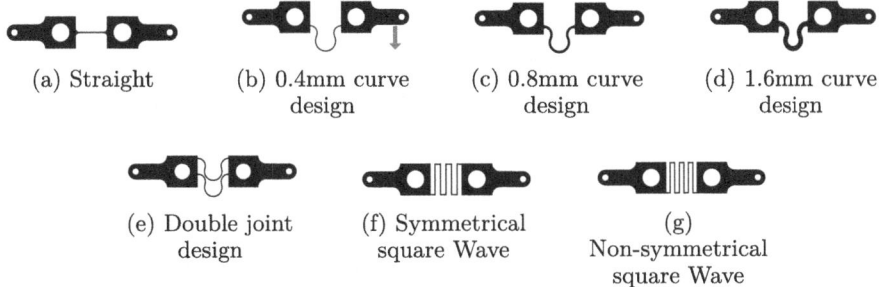

(a) Straight

(b) 0.4mm curve design

(c) 0.8mm curve design

(d) 1.6mm curve design

(e) Double joint design

(f) Symmetrical square Wave

(g) Non-symmetrical square Wave

Fig. 1. Series of designed geometric compliant joints.

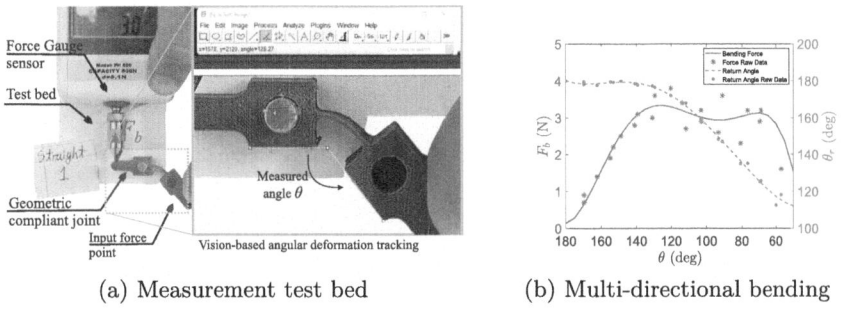

(a) Measurement test bed

(b) Multi-directional bending

Fig. 2. (a) The measurement of each joint for testing stiffness and return angle,(b) The force (blue line) and return angle (red line) versus the deformation angle for the straight joint. (Color figure online)

underactuated systems to reshape or resize. Therefore, our contributions are as follows:

- Conducting an in-depth study on compliant geometric plastic (semi-rigid PLA) joint designs and their ease of 3D printing in Sect. 2. Also, analyzing the behavior of joints concerning stiffness and recovery angle.
- Developing a model using the Gaussian Regression Process to identify variable stiffness and return angle with varying thickness in Sect. 3.
- Proposing the final compact printable design, including various module design variations, of an underactuated geometric compliant (UGC) module with a motor actuator in Sect. 4.

2   Geometric Compliant Joint Design and Tests

This section delves into fundamental studies on compliance and flexibility across various geometric joint designs. By examining how geometry impacts joint stiffness and recovery, we lay the groundwork for understanding the optimal combinations necessary to achieve a variable radius module design. Also, we explore how it can be easy to 3D print the whole module.

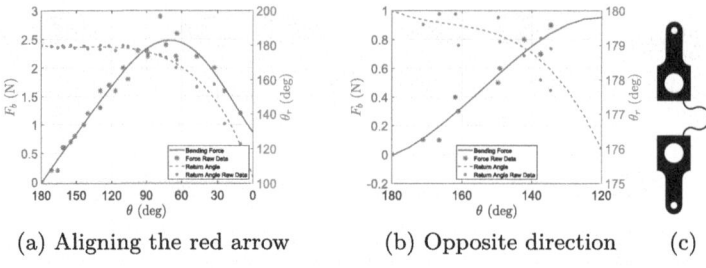

(a) Aligning the red arrow   (b) Opposite direction   (c)

Fig. 3. The force (blue line) and return angle (red line) versus the deformation angle for $T = 0.4$ mm curve joint. (Color figure online)

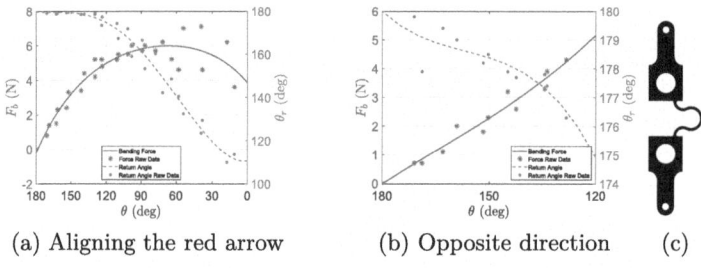

(a) Aligning the red arrow   (b) Opposite direction   (c)

Fig. 4. The force (blue line) and return angle (red line) versus the deformation angle for $T = 0.8$ mm curve joint. (Color figure online)

We have considered a variety of geometric joint forms, as shown in Fig. 1, fabricated using 3D printed PLA materials. Each joint is measured and analyzed based on the configuration depicted in Fig. 2-(a). During this process, the force exerted on each designed geometric compliant joint is measured using a Force gauss sensor placed at one end of the joint. Additionally, we measure the angular bending θ using computer vision techniques, specifically ImageJ Fiji. The depicted first straight linear connection, shown as a joint in Fig. 1-(a), displayed a material stress-strain relationship [11], resulting in a predictable yield point. Figure 2-(b) illustrates the stiffness change concerning force F_b and return angle θ, along with the recovery/return angle θ_r after the impacted force. The measured data are averaged and fitted with high-order polynomial curves to illustrate the overall behaviour of stiffness and return angle. Beyond an angle of $\theta = 135°$, the joint exhibited plastic deformation, impeding its return to the full recovery angle of $\theta_r = 180°$. This premature onset of plastic deformation restricts joint motion and heightens the risk of fatigue damage, potentially leading to premature failure. Furthermore, the test revealed uneven deformation and the formation of a crease off-center, shifting the joint's rotation center during testing. While the initial stages of the test demonstrated consistent force requirements to reach specific angles, post-yield point data displayed significant variability, signifying unpredictability and rendering it unsuitable for the design phase.

The second geometric joint, the decentralized curved (wave) with a $T = 0.4$ mm thickness design (Fig. 1–(b)), employs a strategy of increasing material sep-

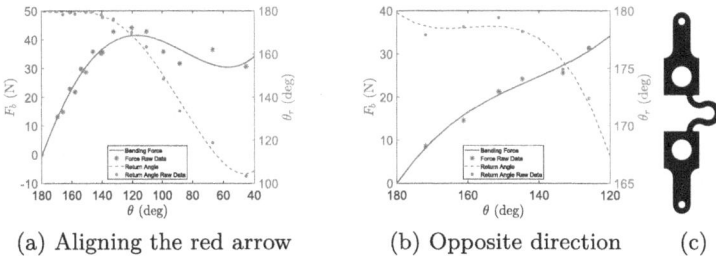

(a) Aligning the red arrow   (b) Opposite direction   (c)

Fig. 5. The force (blue line) and return angle (red line) versus the deformation angle for $T = 1.6$ mm curve joint. (Color figure online)

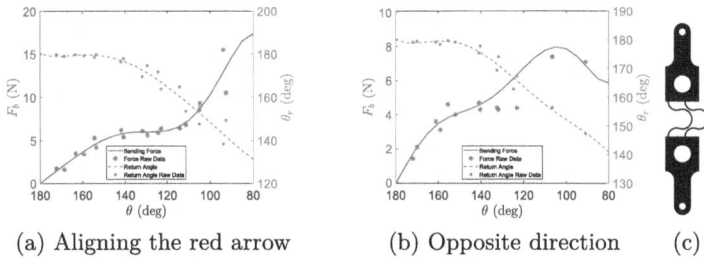

(a) Aligning the red arrow   (b) Opposite direction   (c)

Fig. 6. The force (blue line) and return angle (red line) versus the deformation angle for the double curve joint. (Color figure online)

aration between plates in one direction while eliminating sharp corners to reduce stress concentrations and potential failure. This approach effectively extends the straight joint, distributing force over a larger material area, thereby reducing stress on individual joint elements. Figure 3 illustrates the behaviour of this joint in both directions, demonstrating prolonged angle preservation up to approximately $\theta = 90°$ and a linear stiffness pattern, enabling a wide range of motion. Additionally, it requires low force, with a maximum of $F_b = 2.9$ N. Reversing the bending direction yields a similar response, with tests concluding due to template self-contact. Consistency between test runs, even beyond the yield point, confirms stable repeatability.

The next two joints (Fig. 1(c)–(d)) belong to the same family as $T = 0.4$ mm curve joint, with increasing thicknesses. Testing aimed to understand how joint behaviour changes with thickness variation, allowing for fine-tuning of compliant modular robots' responses via localized resistances. Thicker joint designs (Figs. 4-5) exhibited higher force requirements, reaching a maximum of $F_b = 7.1$ N, and yielded at around $\theta = 140°$, earlier than expected. The response was similar in the reverse direction, but beyond the yield point, results showed deviation, indicating some inconsistency between tests. Delamination of the joint connection, consisting of two wall layers due to a 0.4 mm print nozzle, could be a contributing factor. The return angle starts to present smaller changes with linear form as the layers get thicker in joint design.

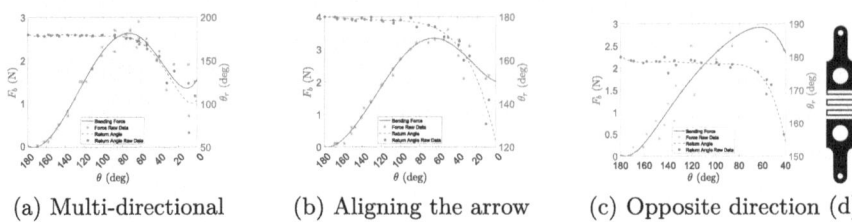

(a) Multi-directional   (b) Aligning the arrow   (c) Opposite direction (d)

Fig. 7. (a) The force (blue line) and return angle (red line) versus the deformation angle for the symmetrical square wave joint, (b)-(c) The force (blue line) and return angle (red line) versus the deformation angle for the non-symmetrical wave. (Color figure online)

The double curve joint with a parallel design (Fig. 1(e)) reached its yield point at around $\theta = 150°$ as shown in Fig. 6. Its peak force is challenging to quantify as it typically operates at $F_b = 6.8\,\text{N}$. However, due to self-contacting at $\theta = 110°$, the force dramatically increases to $F_b = 15.5\,\text{N}$. This joint also experienced early damage, with the central joint being pulled apart unevenly. In one test, the proximity of the two curves caused them to fuse during printing, effectively creating a straight joint, and leading to additional damage.

Inspired by studies on the behaviour of square-based waveforms [3] (Fig. 1(f)-(g)), we integrated this new joint type into our analysis. Here, "non-symmetrical joint" refers to directional bending not occurring at the same connection point upon switching. This joint style exhibited exceptional performance, showing a gradual return angle decrease only starting at $\theta = 70°$ (Fig. 7(a)), allowing it to reach challenging angles without fatigue. It requires low actuation force, facilitating motor operation, and demonstrates near-linearity (prior to the yield point), simplifying prediction. Consistency across tests and behaviour suggests its reliability, potentially making it the most optimal joint design. The joint functions normally without interference despite self-contact starting at $\theta = 150°$ as shown in Fig. 7(b)-(c). Compared to symmetrical and non-symmetrical joints, the non-linearity of the latter presents connections from different ends, with a larger return angle at around $\theta = 40°$.

3   UGC Mechanism Modeling

3.1   Stiffness and Return Angle Modeling of Joints

In this section, we model the stiffness and return angle of the targeted joints using the Gaussian Process Regression (GPR) method. This GPR allows us to have two models the multi-input nonlinear model: a) inputs as the thickness and angle and output as the force. b) input was the thickness and given angle and output was potential recovery/return angle. Note that polynomial curve fitting limitations necessitate the shift to use Gaussian regression since it offers greater accuracy in value prediction. This enables precise calculation of force requirements across various designs, ensuring optimal functionality.

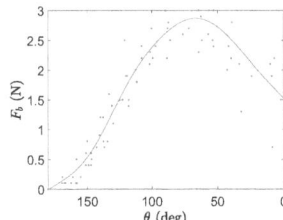 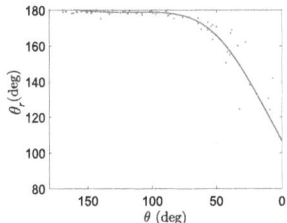

Fig. 8. The trained GP model for angle input θ and output of body force F_b and return angle θ_r of square wave joint.

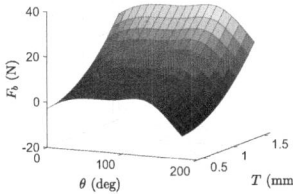 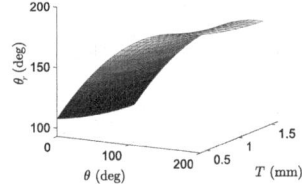

Fig. 9. The trained GP model for angle θ and joint thickness T inputs and output of body force F_b and return angle θ_r of curve wave joint.

Consider a training set $\{(x_i, y_i); i = 1, 2, ..., n\}$, where $x_i \in \mathbb{R}^{d \times d}$ and $y_i \in \mathbb{R}$, drawn from an unknown distribution for the training data [9]. The GPR model of the form is considered: first case as $x = [T, \theta]$, $y_1 = F_b$, the second case with an output of $y_2 = \theta_r$ in the following form

$$y_i = x^T \beta + \varepsilon, \tag{1}$$

where $\varepsilon \sim \mathcal{N}(0, \sigma^2)$. The error variance σ^2 and the coefficients β are estimated from the data. The GPR has a set of random variables such that any finite number of them have a joint Gaussian distribution. GP function is defined by its mean function $m(x)$ and covariance function $k(x, x')$ which is the Kernel-based covariance matrix in our study and is computed as squared exponentials. For the GPR model in (1), from Matlab toolbox gave us the $\beta = [-2.4933, 0.1164, 0, -0.0007, 8.4377]$ and $\varepsilon = 1.9272$ model parameters with covariance matrix below 10^{-3} values for the wave joints. And, we obtained the following $\beta = [1.6940, 0.0225, -0.0002]$ and $\varepsilon = 0.2916$ values for the square joints. Note that the Gaussian regression model for the curve design is more complicated as it has an extra dimension of thickness that was measured from $T \in [0.4, 1.6]$ mm in 0.4 mm steps.

By using Gaussian regression, a much better fit was calculated that allowed for more accurate force prediction when it was required in the design stage. As seen in Fig. 8, the fit follows a much more accurate form to the data, without creating oscillation as seen in the polynomial equations's fitting model. It is also easier to reach the point of self-contact for the wave design at 150° where there

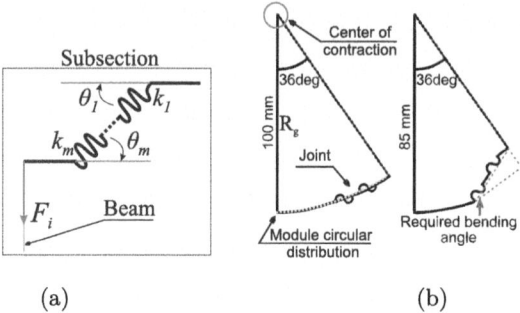

Fig. 10. (a) The modelled series of joint connections (b) Diagram showing angle required to be bent to achieve a reduction in diameter for expected UGC module.

is a slight force increase compared to the angle. Figure 8 also clearly shows that the joint can be rotated freely until a full 90° bend has been achieved at which point the material will begin to yield.

The Gaussian regression model as shown in Fig. 9 can effectively encapsulate the recorded data and serve as a model for calculating the expected force from angle deformation and joint thickness. However, the regression model becomes obsolete for angle values above 150° or below 30° due to a lack of recorded data in these regions, particularly the extremely high force values observed in the $T = 1.2$ mm and $T = 1.6$ mm tests.

3.2   Modeling the Circular Modules with Joint Connections

To obtain a generalised calculation for connected distributed geometric compliant joints, the required torque from motor τ_m for n sections of the connected beam/cables, we can define it as a summation of left and right subsections force, $F_i = F_l + F_r$ (see example of Fig. 12), which results in

$$\tau_m = \frac{F_m}{r_m} = \frac{1}{r_m} \left[\sum_{i=1}^{n} F_i \right], \tag{2}$$

where r_m is the radius of the centralized rotational actuator. Please note that spring deformation, both axial and angular, complicates the stiffness behaviour and radius changes. Thus, the force required will differ based on the radius of the approximated circle of module R_g. Next, to determine each section's force F_i (having symmetric subsections) using Fig. 10(a), with a series of i number of springs k_i, resulting in the following equations

$$F_i = F_l + F_r \approx 2 \left[\sum_{i=1}^{m} \left(\frac{k_i \Delta \theta}{R(t)} \right) \right], \tag{3}$$

where $\Delta \theta = \theta_i - \theta_{i-1}$ and $R(t) \in [r_m, R_g]$ are the relative angular change between the first, last spring, and changing outer radius of the module. Also, note that each spring stiffness is $k_i = \tau_i / \Delta \theta = (R(t) F_i)/\Delta \theta$.

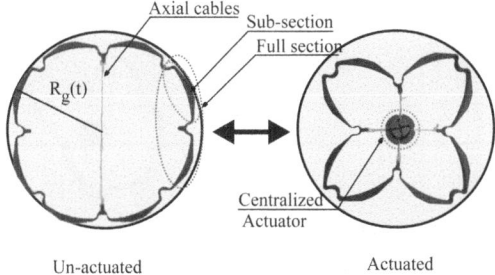

Fig. 11. First prototype using curved joints in singular distribution form.

4 Underactuated Geometric Compliant Module

In this section, we present our finalized single 3D printed UGC modules and compare their results. Lastly, we discuss the successful development of our UGC module results and analyze their behaviour.

4.1 Passive Design and Tests

Our first design is shown in Fig. 11, used curved joints with two cable connections in the centre. This design required a large amount of force to actuate and did not decrease in radius efficiently. This is due to the shape change that caused part of the geometry to bloom outwards while the rest moved in. Our targeted aim is to reduce the R_g outer radius of the module based on a contraction of beams/cables.

The next design used a series of square wave joints in linear series to create collapsible areas on the ring as shown in Fig. 12. In this first prototype, by pulling in on the cables, these sections would be pulled into the centre, essentially removing that area of the circumference (compressing the module) and pulling the outer rings closer together. This way, there should be a minimal deformation in the main geometry. This design would also require less force in order to achieve the same reduction in radius as the previous design. However, as seen in Fig. 12, during actuation, due to the lower force requirements, the geometry failed to maintain structural rigidity which caused extreme 3D arching. This would be missing multiple layers of these rings together to constrain any movement in the vertical axis and create non-symmetric compression of the module which is another important observation from the designed UGC module.

Another passive compression of UGC design is in Fig. 13, which was briefly tested using a pulley system in order to drag a section of the out layer under a previous layer. The cable would contact the outer shell and be redirected perpendicular to a connection piece located on a neighbouring section. This design did function but required constant manual rotational from the centralised actuator location to maintain its shape. Such a design is not naturally stable in a circular position when actuated, due to the inconstancy actuation method and

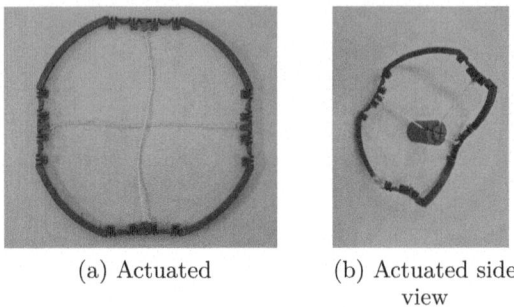

(a) Actuated

(b) Actuated side
view

Fig. 12. Passive prototype of passive actuation with series of square wave model.

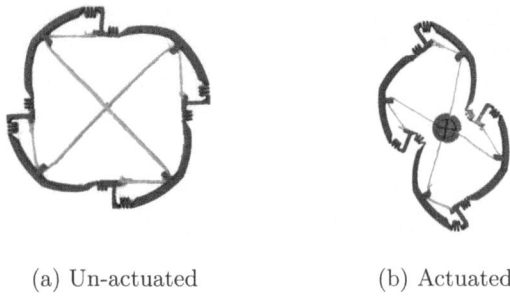

(a) Un-actuated

(b) Actuated

Fig. 13. Prototype pulley design with perpendicular square wave models.

high dependency on order of contraction from cables. The redirection also added friction in the system which greatly increased the force required for it to function.

4.2 Active Design and Evaluation

For the final successful prototype featuring an actuator as the UGC module, the objective was to achieve approximately an 80–85% reduction in diameter to demonstrate the effectiveness of UGC models. Based on hierarchical test designs and experiments conducted with passive modules, it was determined that employing a 5-section ring (as depicted in Fig. 14) would enable more consistent deformation throughout the ring, drawing from our experience with the perpendicular square wave mode (illustrated in Fig. 13). The initial diameter was set to be 200 mm mostly due to the limitation of the print bed of 3D printer. In order to prevent the bending that appeared in previous prototypes, two layers of these rings were printed and connected using vertical connectors that would slot together and keep the centralized actuator properly balanced in centre of the modular total body. This would also allow for internal space inside the structure to better integrate the motor mounting components.

To calculate the force required to actuate the ring design, the number of joints and the angle required for each joint, derived from Eq. (2) at the motor actuator

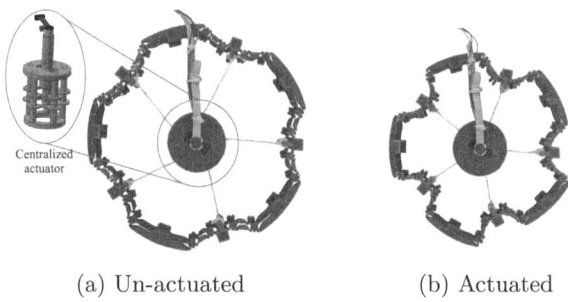

(a) Un-actuated   (b) Actuated

Fig. 14. Showing both actuate and not actuated states showing 80–85% radius change.

end, are essential. This total force aids in determining the spindle diameter to prevent motor overload. The ring, depicted in Fig. 10, consists of 5 mirrored repeating modules, allowing analysis of each 36° section individually as Fig. 10-b. In its uncompressed state, each section has a radius of 100 mm and an arc distance of 62.8 mm. To reduce the radius to around 85%, the arc distance must decrease to 53.4 mm, resulting in a reduction in circumference per section. The change in shape can be simplified into a triangle, where the hypotenuse represents half the original arc length, and the adjacent length is the arc distance minus the required change. With these lengths, the required bending angle can be determined, indicating that all joints need to bend accordingly for the module diameter to decrease. Although this triangle simplification is an estimate, it provides insight into the required angle. By knowing the angle requirements, the required motor torque can be determined by calculating (2)–(3). At this angle, the return angle is also provided, with negligible damage expected at these forces as expected from GPR model (1) as shown in Fig. 8.

In the detail of computation, across both rings, there are a total of 40 joints, each requiring 1.05 N of actuation force, resulting in a total force requirement of $F_m \approx 42N$ from (3). The motor utilized was a single 6V DC Maxon motor (making the model underactuated), with a torque of 0.08 Nm. Dividing the torque by the force yields a spindle radius of 2 mm. To prevent torsional damage to the spindle, the prototype spindle was printed with 100% infill and a radius of $r_m = 3$ mm. This approach was deemed acceptable due to several factors: slightly overshoot on voltage of the motor to enable shorter bursts of increased torque to compensate, and slightly modifying the wave joints with smoother transitions to enhance their performance, thereby facilitating smoother operation. This control problem in future will be carefully studied to have successful feedback control to bring the R_g to specific radiuses $R(t)$. The spindle housing is directly connected to the motor, serving as a cable redirect, ensuring that the cable emanates from as central a point as possible and preventing the housing from freely spinning instead of actuating the module. With the correct spindle radius, the motor successfully actuated the prototype to the required 80–85%.

5   Conclusion

This study first examines different compliant geometric joints. Subsequently, we introduce a novel underactuated geometric compliant (UGC) module capable of adjusting its radius. The final prototype successfully achieved a reduction to 80–85% of its initial value. Moreover, the research meticulously analyzes various geometries for compliant joints, optimizing the design to minimize actuation components while enhancing strength and actuation consistency. These achievements demonstrate the efficacy of these methods, highlighting their significance in integrating soft and hard robotics within hybrid systems. This integration particularly benefits snake-like robots, enhancing their adaptive capabilities to conform to the surrounding environment.

We plan to integrate the developed UGC into compact forms for the creation of shape and size-changing snake robots. We will also explore the scalability of these joint in different materials e.g., silicon and TPE, designs, both by decreasing and increasing their sizes, for various robotic platforms.

References

1. Firouzeh, A., Salehian, S.S.M., Billard, A., Paik, J.: An under actuated robotic arm with adjustable stiffness shape memory polymer joints. In: IEEE International Conference on Robotics and Automation (ICRA), pp. 2536–2543 (2015)
2. He, B., Wang, S., Liu, Y.: Underactuated robotics: a review. Int. J. Adv. Rob. Syst. **16**(4), 1729881419862164 (2019)
3. Howell, L.L., et al. (eds.): Handbook of Compliant Mechanisms. John Wiley & Sons, Hoboken (2013)
4. Liu, J., Tong, Y., Liu, J.: Review of snake robots in constrained environments. Robot. Auton. Syst. **141**, 103785 (2021)
5. Liu, P., Huda, M.N., Sun, L., Yu, H.: A survey on underactuated robotic systems: bio-inspiration, trajectory planning and control. Mechatronics **72**, 102443 (2020)
6. Manti, M., Cacucciolo, V., Cianchetti, M.: Stiffening in soft robotics: a review of the state of the art. IEEE Rob. Autom. Maga. **23**(3), 93–106 (2016)
7. Mavinkurve, U., Kanada, A., Tafrishi, S.A., Honda, K., Nakashima, Y., Yamamoto, M.: Geared rod-driven continuum robot with woodpecker-inspired extension mechanism and imu-based force sensing. IEEE Rob. Autom. Lett. **9**(1), 135–142 (2024)
8. Pagoli, A., Chapelle, F., Corrales-Ramon, J.A., Mezouar, Y., Lapusta, Y.: Review of soft fluidic actuators: classification and materials modeling analysis. Smart Mater. Struct. **31**(1), 013001 (2021)
9. Rasmussen, C.E., Williams, C.K.I.: Gaussian Processes for Machine Learning. MIT Press, Cambridge (2006)
10. Rus, D., Tolley, M.T.: Design, fabrication and control of soft robots. Nature **521**(7553), 467–475 (2015)
11. Schwab, R., Harter, A.C.: Extracting true stresses and strains from nominal stresses and strains in tensile testing. Strain **57** (2021)
12. Tafrishi, S.A., Hirata, Y.: Spiro: a compliant spiral spring-damper joint actuator with energy-based sliding-mode controller. IEEE/ASME Trans. Mechatron. **29**(2), 947–959 (2024)

13. Tafrishi, S.A., Svinin, M., Yamamoto, M.: Singularity-free inverse dynamics for underactuated systems with a rotating mass. In: IEEE International Conference on Robotics and Automation (ICRA), pp. 3981–3987 (2020)
14. Tafrishi, S.A., Svinin, M., Yamamoto, M.: Inverse dynamics of underactuated planar manipulators without inertial coupling singularities. Multibody Syst. Dyn. **52**(4), 407–429 (2021). https://doi.org/10.1007/s11044-021-09788-8
15. Wang, M., Yuan, J., Bao, S., Du, L., Ma, S.: Robots for pipeline inspection tasks-a survey of design philosophy and implementation technologies. In: 5th Asian Conference on Artificial Intelligence Technology (ACAIT), pp. 427–432 (2021)
16. Xu, R., Liu, C.: Tracked robot with underactuated tension-driven RRP transformable mechanism: ideas and design. Front. Mech. Eng. **19**(1), 4 (2024)

Design and Motion Analysis of a Reconfigurable Pendulum-Based Rolling Disk Robot with Magnetic Coupling

Ollie Wiltshire⬤ and Seyed Amir Tafrishi$^{(\boxtimes)}$⬤

School of Engineering, Cardiff University, Queen's Buildings, The Parade, Cardiff
CF24 3AA, UK
{wiltshireoj, tafrishisa}@cardiff.ac.uk

Abstract. Reconfigurable robots are at the forefront of robotics innovation due to their unmatched versatility and adaptability in addressing various tasks through collaborative operations. This paper explores the design and implementation of a novel pendulum-based magnetic coupling system within a reconfigurable disk robot. Diverging from traditional designs, this system emphasizes enhancing coupling strength while maintaining the compactness of the outer shell. We employ parametric optimization techniques, including magnetic array simulations, to improve coupling performance. Additionally, we conduct a comprehensive analysis of the rolling robot's motion to assess its operational effectiveness in the coupling mechanism. This examination reveals intriguing new motion patterns driven by frictional and sliding effects between the rolling disk modules and the ground. Furthermore, the new setup introduces a novel problem in the area of nonprehensile manipulation.

Keywords: Rolling robots · Magnetic coupling mechanism · Modular reconfigurable robots · Motion analysis

1 Introduction

Reconfigurable robots possess multiple units/modules that can interact and alter their shape or configuration through external connections. However, these systems often struggle with limitations in versatility and the ability to achieve movement independently. To address this challenge, there is significant interest in coupling rolling robots as modular components. Rolling robots represent a novel form of mobile robotics [1,15], enabling displacement of the entire system while being internally actuated. Nonetheless, the development of coupling mechanisms without external components remains an open challenge in the field of reconfigurable robots [11].

Modular Self-Recongifurable Robot (MSRR) [11] is a category of mobile robots. MSRRs are constructed as individual modules that house controllers,

© The Author(s), under exclusive license to Springer Nature Switzerland AG 2025
M. N. Huda et al. (Eds.): TAROS 2024, LNAI 15052, pp. 208–219, 2025.
https://doi.org/10.1007/978-3-031-72062-8_19

sensors, and actuation systems, the individual modules can reconfigure benefiting MSRR-type systems to a diverse set of tasks or traversing unpredictable environments [6,18]. The modules interconnect using docking systems, summarised in Table 1, including mechanical surfaces (such as locking pins or latches), magnetic interaction, and grippers, including hybrid docking combinations. Reconfiguration takes place autonomously or by manual intervention. MSRRs would prove themselves invaluable in search and rescue, environmental monitoring, and harmful or dangerous environments. Due to their modular nature and reconfigurability [6,11], allows the robots to adjust their morphology to maximize stability and/or manoeuvrability. A reconfigurable robot's ability to dynamically adapt makes for a very versatile robotic system. Therefore using a modular rolling robot at the centre of the reconfigurable system could prove to be beneficial and the rolling motion could be an effective option [4]. However, a proper coupling method between each of the rolling modules needs to be carefully considered and is still an open problem.

Rolling robots during motion don't remain fixed to the surface they're upon and are subject to slippage, therefore the state of the robot is lost during movement [9], therefore prediction of the dynamics is required for accurate control. This motion is a form of non-prehensile manipulation. Non-prehensile manipulation is a form of robot manipulation which involves no direct grasp of the object under control [5,9] which poses significant control challenges due to various factors. The object manipulated is subject to slippage and friction, leading to unpredictable dynamics during operation. While some research has focused on the manipulation of rolling and balancing disk on disk [10], other studies have sought to simplify the control of a disk on a beam [12]. These studies underscore the complexities inherent in non-prehensile manipulation control. Understanding these challenges is paramount for achieving accurate control of the disk modules. Unlike previous approaches that rely on external actuation for balance, the modular disk module introduced in this paper is internally actuated. However, existing literature does not adequately address the complexities arising from the independent rotation of two modular disk bodies, both in mechanism design and control. This paper aims to showcase new motion control challenges posed by modular mobile disk robots for future non-prehensile manipulation problems in more application-based scenarios. Thus, the contributions of this paper are as follows:

- Introduction of a novel internalised pendulum-based magnetic coupling mechanism for rolling reconfigurable disk robots.
- Investigation of magnetic coupling configurations for pendulum-actuated disk modules.
- Demonstration of the distinctive motion capabilities of the robot through the implementation of tuned scenario-based PD controllers.
- Presentation and discussion of a new reconfigurable robotic platform poised to address future challenges in non-prehensile manipulation problems.

This paper is organised as follows. Section 2 introduces the robot design and methods used to optimise the magnetic array, detailing the algorithm used to

Table 1. Advantages and disadvantages of common coupling systems

Coupling Types	Advantages	Disadvantages
Permanent Magnet - An array of permanent magnets [8]	– Self-alignment docking capabilities, – Quick attachment & good redundancy – Possible multiple docking orientations	– Weak connection – Fixed permanent coupling – Limited coupled modules and weight-dependent
Electromagnet - An electromagnet coil with/without permanent magnets e.g. [4,13].	– Stronger connection than permanent magnets – Controllable decoupling	– High power consumption and dependency – Overheating possibility – Larger internal space and weight
Gripper - A mechanical gripper e.g. [3].	– Strong fixed connection – Direct manipulation of orientation	– Single orientation Constraint on degrees of freedom – Power consumption
Mechanical Locking - Mechanisms such as latches, lips and cones e.g. [19].	– Strong and rigid connection – Impact resistant – Extendable module number	– Limited by orientation/type of coupling – Potential alignment/coupling errors
Manual Reconfiguration - Manually re-configures to different geometries [7]	– Strong and ridged connection – Fixable with bolts – More complex geometries achievable	– Not self-reconfigurable – Limits in obstacles traverse – No dynamic adaption

control the robots' motion and the verification method used to check the stability of the PD controller. Section 3 presents the potential motion behaviours of the reconfigurable disk robot, analysing the modules independently and together collaboratively. Finally Sect. 4 concludes the findings of the work so far, improvements to be made and intended future works.

2   Robot Design and Modelling

This section centres on the design of the rolling reconfigurable disk modules featuring a magnetic coupling mechanism. Initially, the section outlines the design of the modular bodies, it provides an in-depth examination of magnetic configuration design to ensure sufficient coupling strength between multiple disk modules. Additionally, employing a pendulum-actuated model that approximates the modules as a cart-pendulum system for stable module control with feedback control.

The main body design of the modular robot features a Dynamixel XM430 high-precision servo motor positioned at the centre, connected to the pendulum-based docking system, as illustrated in Fig. 1. To present the concept of basic

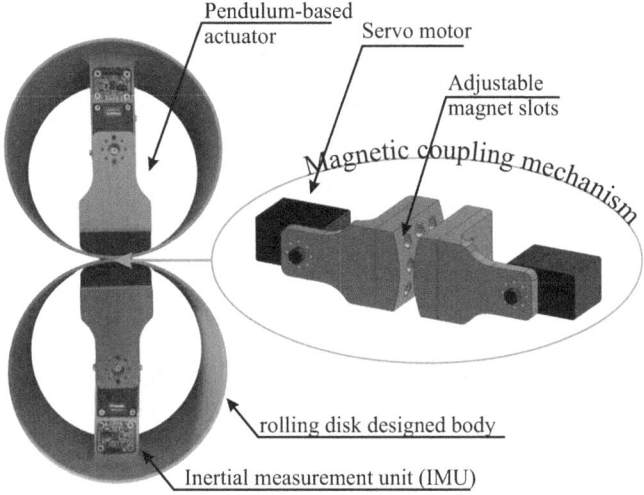

Fig. 1. Main body design of reconfigurable rolling disk robot

behaviours and analyze coupling, the model is simplified to a disk configuration. Here, we consider two modules operating independently of each other, each utilising one motor. Additional bodies can be incorporated by attaching passive magnets in different alignments or by adding extra motors to increase degrees of freedom (DoF). It's important to note that the focus of this study is primarily on the fundamental problem of motion behaviours on coupling and spinning and rolling due to friction, which introduces a new platform for non-prehensile manipulation problem [9]. The robot is controlled with an Arduino MEGA for interfacing with the IMU sensor using a Kalman filter. Meanwhile, the Dynamixel system employs serial bus communication, necessitating conversion via a U2D2 converter for communication with MATLAB which handles the program containing the control loop and data collection.

The outer disk thickness underwent optimisation based on simulation data of the magnetic array. Figure 2a illustrates the iterations of the design compared to the magnetic flux at the proximity of the disk as issues arose when insufficient flux was reaching beyond the body, therefore the coupling strength was insufficient causing spontaneous decoupling during high-velocity motions. Conversely, if the coupling force between two modules is excessively strong, frictional effects between the bodies dominate, inhibiting the disks' ability to actuate. In addition, the magnetic force due to the magnet array can be found as

$$F = \frac{\mu_0 H^2 A}{2}, \tag{1}$$

where $\mu_0 = 4\pi \times 10^{-7}$ m^{-3} kg^{-1} s^4 A^2 is the permittivity of free space, $H = 501.34 \times 10^3$ Am^{-1} is the magnetising field and $A = 1.37 \times 10^{-4}$ m^2 cross-sectional area of magnets. The magnetic force (1) is approximated as 7.24 N supporting

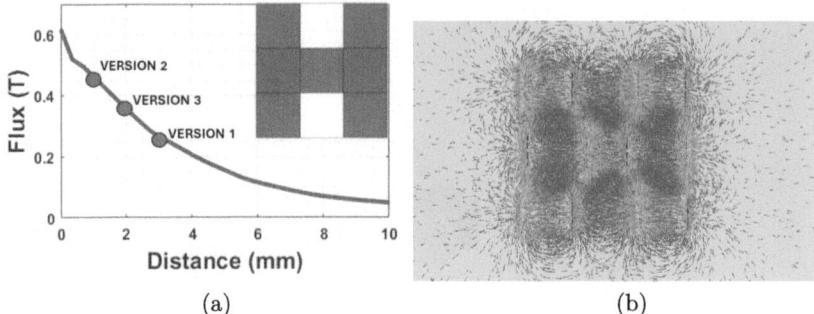

Fig. 2. a) Simulation results for magnetic flux versed different distances Blue = Positive/Red = Negative Pole) b) Magnetic flux formation. (Color figure online)

a load up to 0.731 kg. The mass of each module is 0.266 kg which is sufficient to support 2 modules.

To explore the potential effects of magnetic array configuration, we studied how the pendulum-based mechanism is independently affected by design variables. At the end of the pendulum lies a magnetic array, serving as the coupling mechanism to another module with opposing polarity. This setup ensures a robust coupling force between modules while preserving enough slip to enable rotational movement between the bodies. Extensive evaluation was conducted on the magnetic array to determine the optimal polarity arrangement, aiming to enhance the coupling force generated by the array, as depicted in Fig. 3. This involved simulating the magnetic array in the SolidWorks EMS package, where configurations such as "H" or "X" types were tested, with polarity adjustments made to assess the resulting magnetic flux (in Tesla) at a distance of 1–2 cm from the array's centre. It is clear the cross-form configuration has a more steady flux at a shorter distance. Thus, we selected the final array in the form of an "H", where the polarity of the outer two centre magnets was reversed. This adjustment resulted in an increased maximum magnetic flux due to the magnets' orientation at near distance, which also exhibited a similar fall-off in flux density relative to distance allowing easier decoupling.

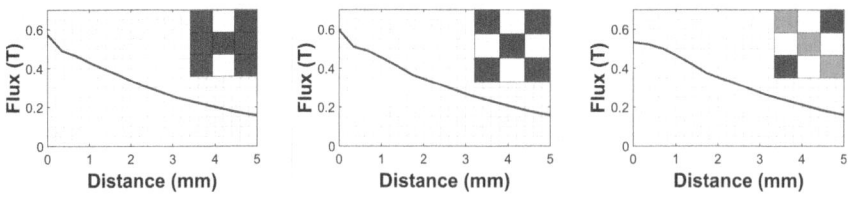

Fig. 3. Simulated polarity arrangement magnetic flux data. (Blue = Positive/Red = Negative Pole). (Color figure online)

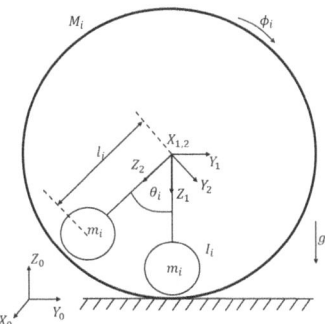

Fig. 4. Modules reference model

2.1   Rigid Body Model and Closed-Loop System

The model of the module is defined with reference to the geometry shown in the diagram in Fig. 4. The center of the pendulum is referenced to the world coordinate frame $X_0 Y_0 Z_0$, and the disk is referenced to $X_1 Y_1 Z_1$. The actuated i-th disk's motor states are defined by the angle and angular velocity of $\{\theta_i, \dot{\theta}_i\}$ with a pendulum mass of m_i between the reference frames $X_1 Y_1 Z_1$ and $X_2 Y_2 Z_2$ about which it rotates. The rotation of the i-th disk, $\{\phi_i, \dot{\phi}_i\}$, representing the angular orientation and angular velocity, is measured through the IMU sensor with respect to $X_0 Y_0 Z_0$.

Rolling pendulum-based (rotational mass) disk modules have inertial-coupling [16,17] models that do not allow a linearised model of the system and partial linearization is required [14]. Here to simplify our problem and allow us to find proper control gains for different motion patterns by using PID, we assume the pendulum always stays in a lower semi-circle. This assumption allows us to approximate the rolling motion with an axial one similar to the cart-base system [2]. Under this assumption and linearization around equilibrium $\theta_i = 0$, we can find the state functions depending on the availability of gravitational force and perpendicular input force on pendulum F_i as presented in the following

$$T_p(s) = \frac{\theta_i}{F_i} = \frac{(m_i l_i/q)s}{s^3 + \frac{b_i(I_i+m_i l_i^2)}{q}s^2 - \left(\frac{(M_i+m_i)m_i g l_i}{q}\right)s - \frac{b_i m_i g l_i}{q}} \qquad (2)$$

where $q = (M_i+m_i)(I_i+m_i l_i^2) - (m_i l_i)^2$. For our designed module the real values are $l_i = 60$ mm, $M_i = 0.172$ kg, $m_i = 0.094$ kg, $I_i = 3.384 \times 10^{-3}$ kgm^{-3}. Where M_i is the mass of the i-th module body and I_i is the mass moment of inertia for the pendulum the other properties are defined as per the module model as Fig. 4. Please note that when the disk body is located at the side aligned with the plane, there is no gravity factor, meaning $g = 0$. Additionally, the angular orientation of the pendulum and the entire body is considered the same, $\theta_i \approx \varphi_i$ during rolling action, due to slow changes in velocity and the large mass of the pendulum compared to the disk. Now, we can check the poles of the model to adjust the gain for stable and robust convergence. Because the aim focuses on controlling the pendulum especially when it is coupled, we can find the feedback T_f control of (2) as follows:

$$T_f(s) = \frac{T_p(s)}{1 + C(s)\, T_p(s)}$$

$$= \frac{\left(\frac{l_i\, m_i}{q}\right) s}{s^3 + \frac{\left[b_i(I_i + m_i\, l_i^2) + K_{d,i}\, l_i\, m_i\right]}{q} s^2 + \frac{\left[K_{p,i}\, l_i\, m_i - (M_i + m_i) m_i g l_i\right]}{q} s - \frac{b_i\, g\, l_i\, m_i}{q}} \tag{3}$$

where $C(s) = K_{p,i} + K_{d,i}\, s$. The Routh-Hurwitz stability criterion on roots of the closed-loop systems allows us to find acceptable values, always assuming pendulum weight way larger than spherical disk (cart) $m_i < M_i$ meaning $q > 0$ for $K_{p,i}$ and $K_{d,i}$ which can be in generalised form as follows: $b_i(I_i + m_i l_i^2) + K_{d,i} l_i m_i > 0$ and $K_{p,i} l_i m_i - (M_i + m_i) m_i g l_i > 0$.

3   Robot Motion Behaviour Study

This section analyzes the potential motion patterns of the reconfigurable rolling disk, exploring both independent motions and collaborative coupled motion behaviours between the two modules.

The robot employs closed-loop proportional-derivative (PD) control (3) to achieve a predetermined angle dictated by the planned trajectory for specific motions or tests. In this setup, the controller receives an angle input and converts it into a velocity output according to Eq. (4), aligning the actual angle θ_i with the desired angle as per the reference model (Fig. 4).

$$u_i = K_{p,i}\left[\theta_i(t) - \theta_{d,i}\right] + K_{d,i}\left[\dot{\theta}_i(t) - \dot{\theta}_{d,i}\right] \tag{4}$$

The PD controller for the i-th module was specifically designed to enable adjustments to the speed of the motors while retaining angular control of the pendulum. Operating the motors in velocity control mode, the PD controller generates a smooth output curve for motor speed, preventing spontaneous decoupling during movement. Moreover, during significant set point changes, the motor speed is moderated. The gains in Eq. (4) are set to $K_{p,i} = 2.5$, while the derivative gain is set to $K_{d,i} = 0.5$, ensuring a prompt response without overshoot.

In the first test, we look for the independent pendulum-based rotation as shown in Fig. 5. The design uses a pendulum-based actuator using the controller to rotate the pendulum by θ_i degrees causing the body to rotate in the opposing direction due to the current design once the pendulum reaches the physical stop of the servo bracket conservation of angular momentum takes over and the disk rolls along the table further. The results indicate that the pendulum controlled by a PD controller successfully follows the desired angular rotations, while the presence of mass imbalance permits a certain relative rotation of the entire body in the same values. However, it's crucial to highlight that the PD controller

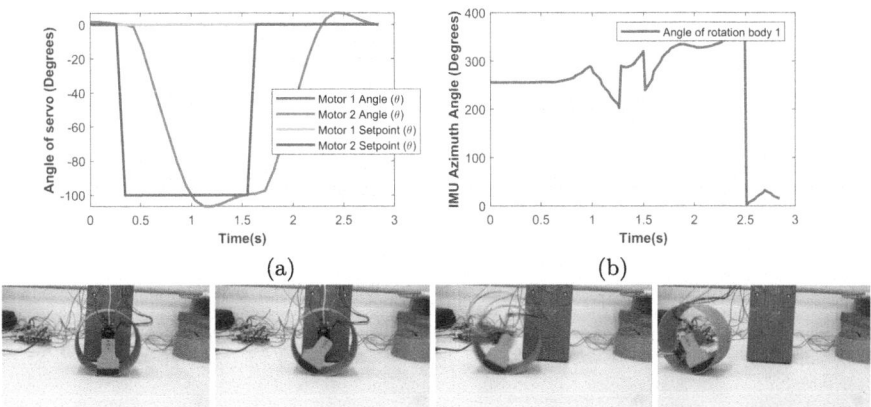

Fig. 5. Pendulum Mode: the top figure are a) the results of angular motion for two modules, b) IMU data for the moving module, and the bottom figure is the video snapshots of the robot in pendulum mode

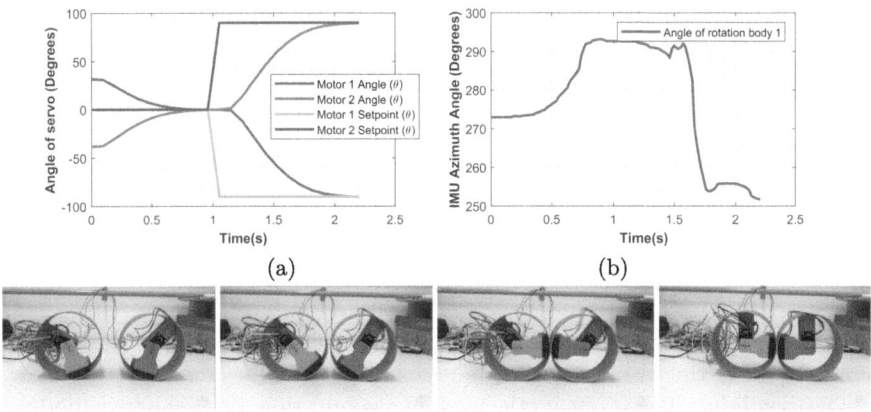

Fig. 6. Coupling Mode: the top figure are a) the results of angular motion for two modules, b) IMU data for moving module, and the bottom figure is the video snapshots of robot coupling

alone may not accurately position the disk shell, especially during high-velocity motions where the conservation of angular momentum introduces complexity.

Another crucial function of the disk modules is coupling, either when laid flat on a surface or while actuating using the mass imbalance of the pendulum, as depicted in Fig. 6. The mass imbalance of the two disk modules propels them together, enabling coupling outside the effective area of the magnets. This capability is vital in a reconfigurable system, as modules may become separated during real operation and need to be manoeuvred back together. Additionally, the significance of slippage in the interaction between disk modules is illustrated in

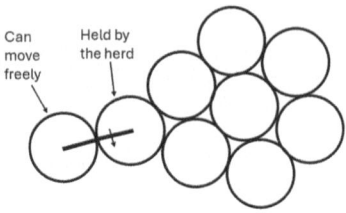

Fig. 7. Illustration of module held by herd of modules

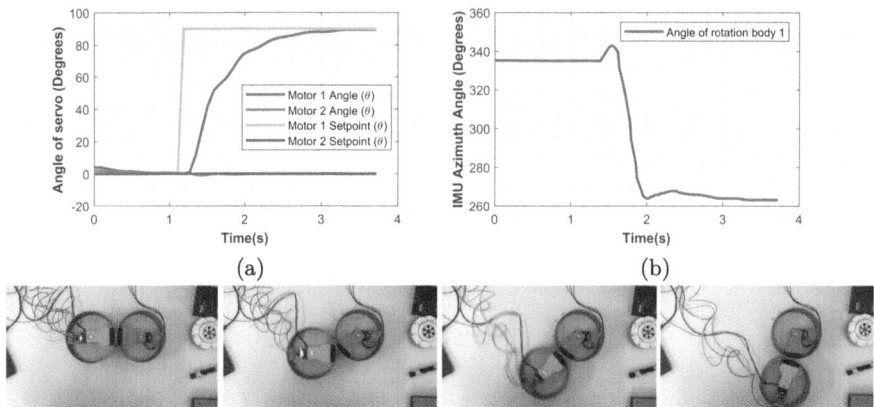

Fig. 8. Joint mode: the top figure are a) the results of angular motion for two modules, b) IMU data for moving module, the bottom figure is the video snapshots of the robot in joint mode

the last snapshot. They can rotate around themselves upon coupling, altering the configuration of the disks, but this doesn't induce any movement on the plane.

The pendulum-based magnetic coupling connection between two modules can function as a fixed link, allowing one body to rotate another about its outer sphere, provided it remains fixed relative to the other. For instance, in the scenario depicted in Fig. 7, only two modules were constructed. To simulate this scenario, the right-hand module, as shown in Fig. 8, was secured to the table while the other module remained free to move. The ability to rearrange and actuate around each other is essential for categorizing it as a reconfigurable system. With additional modules, it could alter its morphology to perform a variety of tasks. This motion also highlights the significance of mass during coupling; if the actively coupled modules lack sufficient mass coverage, their role as the base will be altered.

Another movement pattern emerges for the disk robot when both modules are free from the table. If both modules' servos rotate in opposing directions (Fig. 9a), the relative frictional effect and force of the coupling array between the modules generate a different type of motion. Conversely, if the motors rotate

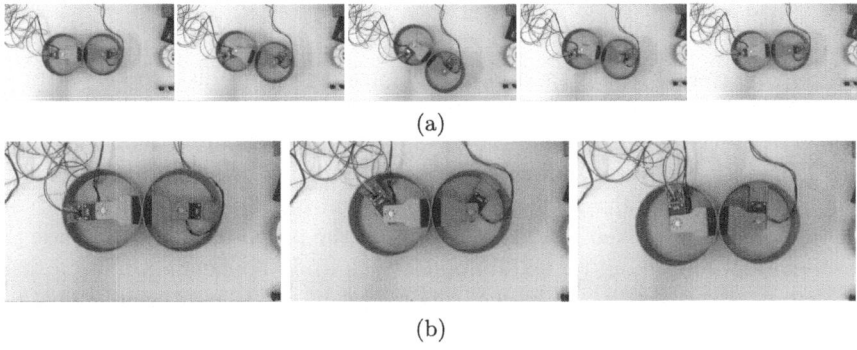

(a)

(b)

Fig. 9. a) Video snapshots of disk spinning relative to each other motors set to spin clockwise b) Video snapshots of disk spinning relative to each other motors set to spin clockwise-anticlockwise

in the same direction (Fig. 9b), an alternative movement pattern emerges. In particular, this motion pattern is intriguing because if it could be fully classified for control purposes, it could facilitate reorientation by spinning and reconfiguration of the modules. This is because it induces relative movement between each of the disks rather than the entire module moving, which would be necessary for reconfiguration.

Finally, an attempt was made to self-balance the two modules on top of each other, as shown in Fig. 10. Using the previous PD controller and IMU data to keep the two pendulums directly in line with the centre of gravity and adjusting to rotate the outer bodies to an upright position to remain balanced, this simplified approach proved insufficient in steady-state mode due to the complex dynamics from slippage $\theta_i \neq \varphi_i$ and magnetic force present in the design. Therefore, a more advanced controller is required for balancing motion in this non-prehensile manipulation problem [10] as future work.

Fig. 10. Disks balanced on top of one another

4   Conclusion

This paper presents a design of a reconfigurable rolling disk robot featuring a pendulum-based coupling system. The design enhances coupling strength through optimisation and simulation techniques while remaining internally actuated. The motion analysis showcases various interesting behaviours of the rolling robot modules, including independent pendulum-based rotation, module coupling and acting as a fixed link. The abilities of the proposed robot showcase the possible adaptability and potential applications from independent motion tasks to collaborative tasks as a modular robot swarm. Overall this paper lays the groundwork for future works towards systems capable of dynamically adapting to diverse tasks and environments with internalised rolling mechanisms.

In future works, additional research into refining the control algorithms to address control issues with high-velocity motions, using methods such as energy-based controllers to encompass the whole rolling disk in the model possibly leading to more advanced motions such as balancing, improving the control of the non-prehensile nature of the problem, once this achieved moving forward to more real-world surfaces and scenarios. Furthermore, improvements could be made to the coupling mechanism to improve robustness and increase DoF in the motion of the pendulum.

References

1. Armour, R.H., Vincent, J.F.V.: Rolling in nature and robotics: a review. J. Bionic Eng. **3**(4), 195–208 (2006). https://doi.org/10.1016/S1672-6529(07)60003-1
2. Bai, Y., Svinin, M., Yamamoto, M.: Motion planning for a hoop-pendulum type of underactuated systems. In: IEEE International Conference on Robotics and Automation (ICRA), pp. 2739–2744 (2016)
3. Gross, R., Bonani, M., Mondada, F., Dorigo, M.: Autonomous self-assembly in swarm-bots. IEEE Trans. Rob. **22**(6), 1115–1130 (2006)
4. Liang, G., Luo, H., Li, M., Qian, H., Lam, T.L.: Freebot: a freeform modular self-reconfigurable robot with arbitrary connection point-design and implementation. In: IEEE/RSJ International Conference on Intelligent Robots and Systems (IROS), pp. 6506–6513. IEEE (2020)
5. Mason, M.T.: Progress in nonprehensile manipulation. Int. J. Rob. Res. **18**(11), 1129–1141 (1999)
6. Moubarak, P., Ben-Tzvi, P.: Modular and reconfigurable mobile robotics. Robot. Auton. Syst. **60**(12), 1648–1663 (2012)
7. Möckel, R., Jaquier, C., Drapel, K., Dittrich, E., Upegui, A., Ijspeert, A.: Exploring adaptive locomotion with yamor, a novel autonomous modular robot with bluetooth interface. Ind. Rob. Int. J. **33** (2006)
8. Romanishin, J.W., Gilpin, K., Rus, D.: M-blocks: momentum-driven, magnetic modular robots. In: 2013 IEEE/RSJ International Conference on Intelligent Robots and Systems, pp. 4288–4295 (2013)
9. Ruggiero, F., Lippiello, V., Siciliano, B.: Nonprehensile dynamic manipulation: a survey. IEEE Rob. Autom. Lett. **3**(3), 1711 1718 (2018)
10. Ryu, J.C., Ruggiero, F., Lynch, K.M.: Control of nonprehensile rolling manipulation: balancing a disk on a disk. IEEE Trans. Rob. **29**(5), 1152–1161 (2013)

11. Seo, J., Paik, J., Yim, M.: Modular reconfigurable robotics. Ann. Rev. Control Rob. Auton. Syst. **2**, 63–88 (2019)
12. Serra, D., Ruggiero, F., Donaire, A., Buonocore, L.R., Lippiello, V., Siciliano, B.: Control of nonprehensile planar rolling manipulation: a passivity-based approach. IEEE Trans. Rob. **35**(2), 317–329 (2019)
13. Shiu, M.C., Lee, H.T., Lian, F.L., Fu, L.C.: Design of 2d modular robot based on magnetic force analysis, pp. 1 – 6 (2008)
14. Spong, M.: Partial feedback linearization of underactuated mechanical systems. In: Proceedings of IEEE/RSJ International Conference on Intelligent Robots and Systems (IROS'94), vol. 1, pp. 314–321 (1994)
15. Tafrishi, S.A., Svinin, M., Esmaeilzadeh, E., Yamamoto, M.: Design, modeling, and motion analysis of a novel fluid actuated spherical rolling robot. J. Mech. Rob. **11**(4), 041010 (2019)
16. Tafrishi, S.A., Svinin, M., Yamamoto, M.: Singularity-free inverse dynamics for underactuated systems with a rotating mass. In: 2020 IEEE International Conference on Robotics and Automation (ICRA), pp. 3981–3987 (2020)
17. Tafrishi, S.A., Svinin, M., Yamamoto, M.: Inverse dynamics of underactuated planar manipulators without inertial coupling singularities. Multibody Syst. Dyn. **52**(4), 407–429 (2021). https://doi.org/10.1007/s11044-021-09788-8
18. Yim, M., Zhang, Y., Duff, D.: Modular robots. IEEE Spectr. **39**(2), 30–34 (2002)
19. Zhang, H., Deng, Z., Wang, W., Zhang, J., Zong, G.: Locomotion capabilities of a novel reconfigurable robot with 3 dof active joints for rugged terrain. In: 2006 IEEE/RSJ International Conference on Intelligent Robots and Systems, pp. 5588–5593 (2006)

Soft Robotics

Contact-Rich Task Learning on an Articulated Soft Robot Arm Through Simulation

Laurenz Elstner[1]($\boxtimes$) (ID), Erik Kyrkjebø[1] (ID), and Martin F. Stoelen[1,2] (ID)

[1] Department of Computer Science, Electrical Engineering and Mathematical Sciences, Western Norway University of Applied Sciences, Førde, Norway
`laurenz.elstner@hvl.no`
[2] Fieldwork Robotics Ltd., Cambridge, UK

Abstract. Traditional rigid robots face limitations in exploration during learning processes in dynamic and unknown environments where contact and impacts are common. Soft robots that leverage compliant materials, for example as part of variable stiffness actuators (VSAs), offer promising avenues for enhanced adaptability and safety. Deep Reinforcement Learning (RL), while proven to be a powerful tool, requires a lot of training data that is time-consuming to gather. This paper introduces a novel approach to simulating robots driven by antagonistic VSAs, as well as full-task learning with a soft robot arm in three environments with different contact types: path following (no contact), door opening (kinematic constraint), and surface wiping (continuous contact). We show that the tendon-driven robot can learn all three tasks, even if learning in a very low-level actuation space. The stretchable tendons enables it to perform relatively well on the contact-rich tasks. Further, using the proposed variable stiffness controller the robot is able to develop a stable behaviour in terms of forces applied to the environment. The results are promising for extending the approach to the physical robot.

Keywords: Soft Robot Simulation · Contact-Rich Task Learning · Deep Reinforcement Learning · Variable Stiffness Actuators

1 Introduction

Humans actively employ both imitation and exploration when learning new skills. The user starts by imitating the movements of the expert teacher (in the case of infants, the parents) and then optimizes it (since everyone's morphology is a little bit different) through trial and error. The rigid structure of a classic robot is fundamentally flawed for doing exploration without causing damage to either itself or its surroundings. Soft materials are a vital part of the body composition in natural agents. Even animals with rigid structures (exo- or endoskeleton) consist mainly of soft materials [7]. However, the non-linearity and hysteresis effect of soft materials makes soft robots hard to control precisely,

M. N. Huda et al. (Eds.): TAROS 2024, LNAI 15052, pp. 223–235, 2025.
https://doi.org/10.1007/978-3-031-72062-8_20

Fig. 1. The GH360 is a 7 DOF articulated soft robotic arm with 6 antagonistic VSAs. It is an adaptation of the GH2 developed by Fieldwork Robotics (Cambridge, UK). (a) The real arm opening a door. (b) The elbow joint of the arm. Each VSA joint has active pulleys connected to the motors on each side that actuate through the tendons the passive pulleys located at the joint.

and even harder to use for learning full tasks. Also, it has been shown that introducing compliance can reduce the complexity of task execution when learning constrained or contact-rich tasks [11].

Collaborative robots (cobots) are equipped with force/torque sensors in the joints that can be used to create a virtual mass-spring-damper system through impedance control. Since the spring is simulated, these properties can be changed to give different stiffness profiles. However, such systems have a limited bandwidth when encountering fast impacts/shocks. This limited ability to respond in time can result in damage to either the robot or the object it collides with. Variable Stiffness Actuators (VSAs) are similarly able to change the spring properties of the system, but can also better absorb the force peaks from impacts/shocks by introducing soft materials [13]. While there are different designs for VSAs, they generally fall into two approaches, and typically require at least two motors per joint. The simplest setup uses one motor for adjusting the position and a second motor to adjust the stiffness. In this approach, the output power of the actuator is limited by the position motor, but allows for a less powerful, lighter second motor. An example of this approach is the MACCEPA 2.0 [16] which is used in legged locomotion. The second approach for VSAs is to use an agnostic/antagonistic setup where two motors are connected through spring-like tendons to each side of the joint. Movement is generated by moving both motors in the same direction, and stiffness is changed by moving the motors in opposite directions. This type of VSA design is used for example in the GummiArm [14], an open-source human-inspired robot arm. The output power of this type of actuator is limited to one of the motors depending on the direction of movement. Therefore both motors generally need to have the same size. The advantage of tendon-driven systems is that the motors do not have to be located in the joint, but can be placed closer to the base reducing the dynamic forces of the robot. In a bidirectional setup, each motor has tendons connected to either side of the joint, and the potential output power is doubled during low stiffness levels, as shown in the DLR Hand Arm System [3].

Deep Reinforcement Learning (RL) has emerged as a powerful tool for robots to learn complex tasks from high-dimensional sensory inputs. The goal of RL is to train a policy π that maximises the expected rewards given in a partially observable Markov decision process. The policy π receives observations O, that represent part of the state S, which it uses to choose actions A. The environment gives rewards R to the policy based on state S. In robotics S, O and A are generally continuous and high-dimensional. Therefore, deep neural networks are used for the policy. The actions given by the policy are a reference signal that often still has to be mapped to actuation commands on the robot. The choice of that interface, also known as the action space, can have big impacts on robustness, task performance and learning efficiency [11]. Previous results have looked at both low-level interfaces like joint torque [10], but also higher-level interfaces like joint space [4] or task space commands [9].

Compliance in joints allows for safe movements in unknown and dynamic environments, and can make the control easier for contact-rich tasks [11]. Classic position-based robot control strategies are not suitable for interaction tasks as they may lead to high-impact forces and generate stiff movements which contradicts the reason for using soft materials [12]. While there is a lot of work on different controller designs that allow for high-performance trajectory tracking while keeping the compliant properties of the robot [6], the area of learning complete tasks with this type of robot is relatively unexplored.

While deep RL has shown impressive results, it requires a lot of training data to be successful, which can be expensive to gather on a real robot. Instead, training in simulation runs orders of magnitude faster than what is possible with a physical robot and enables experiments to be reset without human intervention. General physics engines often operate under the assumption that the robot and the environment are rigid. Soft robots require simulation of deformable objects, as well as novel modes of actuation such as tendons, pneumatic devices or heat transfer. MuJoCo [15] is an open-source physics engine known for its contact stability and its ability to simulate soft materials. It has previously been used to learn to solve a Rubiks cube with a 24 DOF tendon-actuated robotics hand [1].

In this paper, we want to explore learning constrained and contact-rich tasks on a human-inspired soft robot arm in simulation. We evaluate a variable-stiffness actuated robot arm against an arm with the same kinematic structure, but with a single motor in each joint in the following environments: path following (no contacts), door opening (kinematic constraints) and surface wiping (continuous contact). The robotic platform used is the GH360 (see Fig. 1), which is an adaptation from the open-source GummiArm [14], and the GH2 arm designed by Fieldwork Robotics Ltd (Cambridge, UK). It has 7 joints, 6 of which use bi-directional antagonistic VSAs. The simulation is done in MuJoCo as it allows for muscle-tendon actuation. For the learning, we use the robosuite [17] framework which comes with several benchmark environments, robot implementations and features the OpenAI gym interface, which is a popular interface for

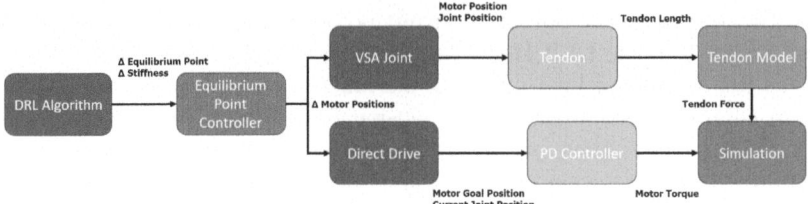

Fig. 2. Block-diagram of the simulated equilibrium and stiffness controller. The input to the controller is Δ Equilibrium Point and Stiffness which is transformed into motor positions using Eq. 1 and 2. The tendon length is then calculated using the motor and joint position with Eq. 3 and 4, and then further translated into tendon actuation force using tensile test data of the composite tendon [2].

RL. The tendon-driven version of the arm uses an equilibrium stiffness controller as action space which is discussed in more detail in the following section. The stiff version of the arm uses both a joint space and an operational space variable impedance controller. The main contributions of this paper are a novel approach to learning in simulation with agonist-antagonist VSAs, full task learning on a soft robot arm, and the addition of a tendon-driven robot arm to robosuite.

2   Simulation of Antagonistic Variable Stiffness Actuators in MuJoCo

The simulation software MuJoCo [15] allows the user to include tendons in the robot model. These tendons can wrap around objects but aren't able to be coiled up on a pulley, since at a certain point the tendon would jump to the other side of the object. Therefore, in the simulation model of the robot used here, the tendon is actuated directly, rather than being simulated as a motor that drives a pulley. To be able to later transfer learned behaviours from the simulation model to the real robot, a control architecture was developed that translates motor movements into approximate tendon forces. A visualization of the control system can be seen in Fig. 2.

The VSAs used for the joints of the GH360 arm use an agonistic/antagonistic setup with two soft tendons on each side connected to an active pulley that is moved by the motor and a passive pulley that moves the joint as shown in Fig. 1b. The tendons behave like non-linear springs during extension which allows for a change of stiffness by moving the motors in opposite directions. When properly setup, the actuator can change joint position without changing stiffness when both motors move together, and change stiffness without changing the joint position when the motors move against each other. The result is a simple controller for each joint which uses the change in equilibrium position and stiffness as inputs, and outputs motor goal positions using

$$m_1(t) = m_1(t-1) + \Delta ep(t) + \Delta s(t) \tag{1}$$

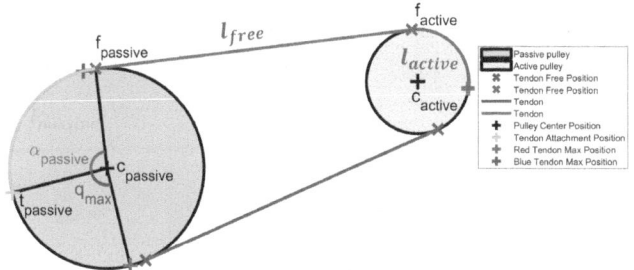

Fig. 3. 2D pulley setup for calculating tendon lengths from [2].

$$m_2(t) = m_2(t-1) + \Delta ep(t) - \Delta s(t) \tag{2}$$

where $m_i(t)$ is the position of motor i at time step t, $\Delta ep(t)$ is the change in equilibrium position and $\Delta s(t)$ is the change in stiffness.

In [2], tensile test data was gathered to characterise the resistance-behaviour of the tendons on the GH360 during extension cycles as well as the mathematics developed to calculate the length of tendons on the joints. In general, the tendon length is divided into 3 sections (see Fig. 3): l_{active} is the length of the tendon touching the active pulley, $l_{passive}$ is the length of the tendon in contact with the passive pulley and l_{free} is the free hanging tendon between the pulleys. l_{free} always stays the same no matter the rotational positions of the pulleys and is calculated as the euclidean distance between $f_{passive}$ and f_{active}. The passive pulley is mounted at the centre of rotation of the joint, and therefore the length of the tendon in contact with the passive pulley can be calculated using

$$l_{passive}(t) = r_{passive}(\alpha_{passive} + q(t)) \tag{3}$$

where $r_{passive}$ is the radius of the passive pulley. The angle $\alpha_{passive}$ is the angle between where the tendon starts touching the passive pulley and where the tendon is mounted on the pulley when the joint is in zero position, and $q(t)$ is the current joint angle which is received from the simulation environment during runtime. The active pulley is connected to the motor and therefore controls the movement of the joint. The length of the tendon touching the active pulley is calculated using

$$l_{active}(t) = r_{active}(\alpha_{active} + m(t)) \tag{4}$$

where r_{active} is the radius of the active pulley, α_{active} is the angle of the active pulley that is covered by the tendon when the motor is in zero position, and $m(t)$ is the current motor position.

The length of the tendon is translated into tendon actuation force using tensile test data of the composite tendons on the arm [2]. The GH360 also features one joint, namely the forearm roll, that is not tendon-actuated but rather directly driven by a motor. This joint therefore receives only one action from the deep RL network which is the change of motor position.

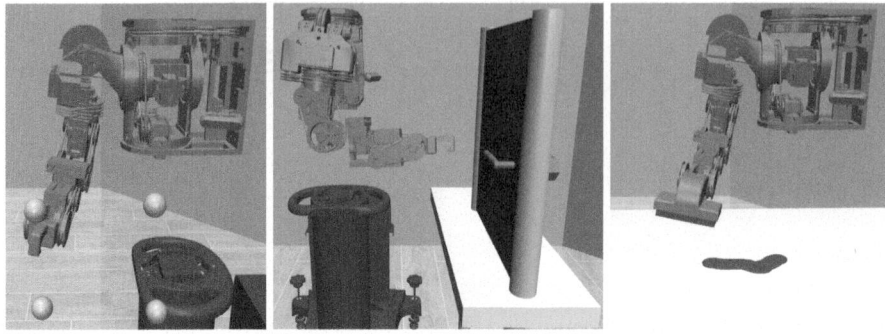

Fig. 4. The learning performance is evaluated on simulation environments with different contact types: (a) path following (no contact), (b) door opening (kinematic constraint), (c) surface wiping (continuous contact)

3   Evaluation Setup

The experiments were designed to evaluate different action spaces in three environments with different types of contact. The GH360 robot was tested with 3 different variants of the equilibrium and stiffness controller described in Sect. 2: variable stiffness, fixed low co-contraction, and fixed high co-contraction. If comparing the results with a classic, non-soft standard cobot, the difference in kinematic structure would influence the results in terms of more or less favourable base position and initial pose. Instead, we used the kinematic model of the GH360, and placed one motor at the centre of each joint. This rigid version of the robot was then tested with variable impedance control in both joint and operational space. In all experiments, the policy sent actions to the controllers at a rate of 20 Hz. The evaluation criteria used were as follows:

- **Sample efficiency and task completion** - We look at whether the task has been completed and use the reward score over time to evaluate sample efficiency.
- **Physical effort** - Since we are dealing with contact-rich tasks, we are also interested in the wrenches applied to the environment.

The learning in all experiments was done by a soft-actor critic algorithm with automatic temperature tuning [5]. Both the policy and the critic networks are multilayer perceptrons (MLPs) with two fully connected layers of 512 neurons each. The number of neurons per layer was chosen empirically since the default of 256 in [5] did not give the desired performance. The full list of hyperparameters is shown in Table 1. The size of the action spaces are as follows: 7 for equilibrium control without variable stiffness, 13 for equilibrium control with variable stiffness, 14 for variable impedance control in joint space and 12 for variable impedance control in operational space. In all environments, the policy gets the following robot-specific observations: joint positions, joint velocities, tool-center-point (TCP) position and orientation.

3.1   Path Following

In this environment, the robot is trained to reach a set of predefined positions in space. It doesn't require any interaction with any objects and therefore has no contacts. We use this environment to evaluate how controllers that work well in contact-rich environments perform in terms of accuracy. There are 4 via-points placed in predefined locations, as shown in Fig. 4a, with their position being randomized by ± 10 mm in x, y and z. The robot end-effector has to reach these positions in a fixed order. The via-point is counted as reached if the TCP is within 10 mm of the target location. The reward $r \in R$ is designed to give gradually more reward the closer to task completion the robot gets and is calculated using

$$r = 0.25 * p + 0.25 * (1 - tanh(10 * d_i)) \tag{5}$$

where p is the number of points already reached and d_i is the distance between TCP and the current target via-point. The environmental-specific observations given to the policy are as follows: the position of all via-points, the difference between all via-points positions and the TCP position, and the percentage of how many points have been reached so far.

3.2   Door Opening

The door-opening scenario consists of a short trajectory tracking phase to reach the door handle and a contact phase where the robot has to twist the handle and open the door during which the movement of the robot is constrained by the limited DOF of the handle and the door hinge. To solve this task, the robot is equipped with a hook. The door is placed on a table with a fixed position, as shown in Fig. 4b. The door position and orientation on the table are randomized by ± 10 mm in x and y, and ± 0.1 rad around z. The task is completed if the door hinge reaches an angle of more than 0.3 rad. In this case, the policy will get the full reward of 1. Otherwise, the reward is gradually shaped using

$$r = 0.2 * (1 - tanh(10 * d)) + 0.2 * g + 0.2 * |\frac{q_c}{q_g}| \tag{6}$$

where d is the distance between the TCP and the door handle, g is a boolean variable which is 1 if the inside of the hook is in contact with the door handle, q_c is the current handle angle, and q_g is the goal handle angle to unlatch the door. The observations given to the policy in the door-opening environment are as follows: the position of the door and the handle, the difference between the handle and TCP position, and the joint angles of the handle and hinge joint.

3.3   Surface Wiping

The goal of this environment is to remove any "dirt" on the surface of a table. The robot is equipped with a tool that resembles a whiteboard eraser, as shown in Fig. 4c. Each dirt particle is modelled as a cylinder with a radius of 20 mm and a height of 1 mm. During each episode 30 dirt particles are placed randomly in a way that creates one continuous line. The robot has to press the eraser tool into the table to remove particles. The reward system is structured in a way that encourages realistic force application while keeping the robot and the environment safe through the following criteria:

- For every particle wiped at this time-step r += 1.75.
- If the surface is not clean yet: r += $0.1755 * (1 - tanh(5 * d)$ where d is the distance between the TCP and the mean position of all remaining dirt particles.
- If the eraser tool is in contact with the table r += 0.00351.
- If the force f applied to the table is >60 N, then r -= $0.001755 * f$
- If the force f applied on the table is >0.5 N, r += $0.00351 + 0.00351 * f$
- If the task has been completed r += 0.9
- If the robot is within 0.1 rad of any joint limit, or in collision with any part of the body except the tool all other rewards will be ignored, and only $r = -0.351$ is given.

The numbers above are calculated using a normalization factor which ensures, that the average reward over an entire episode will always be < 1.0. The observations specific to the surface-wiping environment are as follows: the mean position and the radius of all remaining "dirt" particles, the percentage of wiped particles, and the difference between particle mean position and TCP position.

Table 1. SAC Hyperparameters.

Parameter	Value
optimizer	ADAM [8]
activation function	ReLU
policy and α learning rate	0.001
q-function learning rate	0.0003
discount factor	0.99
replay buffer size	10^6
number of samples per minibatch	128
target smoothing coefficient	0.005
target update interval	5

4   Results and Discussion

To evaluate the learning performance of each controller type in the different environments, we measured the average reward per epoch, which is the average over 40 episodes. Each episode is 500 policy steps long. The experiments were repeated three times and the lines and shading in Fig. 5 resemble the mean and variance. We also evaluated the trained policies on their average success rate in completing the task over 10 episodes after training, which are shown in Table 2.

Path Following: The learning rates for this environment, visualized in Fig. 5a, are fairly stable. Both the operational space (OSC) and the joint position controller outperform the tendon-driven controllers, which was expected. This is also reflected in the success rates of the trained policies during evaluation, shown in Table 2. It can be seen that the equilibrium controller with a high fixed stiffness reached a similar reward level to the joint position controller towards the end of the learning. It can also be seen that the high-stiffness controller outperforms the other two equilibrium controllers. This is to be expected, given that generally stiffer joints are better for accuracy. That the variable stiffness controller performs worse than the controller with low stiffness suggests that either the training time was not long enough, or that more neurons per layer in the policy network are required. It is also worth noting that the policy does not receive any observations on motor positions. This is done to keep the observation space the same for all control types. Knowing the motor positions might help with learning also the correct stiffness levels.

Door Opening: The learning rates of this environment in Fig. 5b suggest that again operation-space and joint position controller perform the best. The success rates after training, on the other hand, show that the performance of the equilibrium variable stiffness controller is on par with the operation space controller when it comes to the reliability of opening the door. The difference in the height of the reward curve is likely explained by the difference in maximum step size for the different controllers. Since the equilibrium point controller is limited to smaller movements per policy step, it reaches the door open state later in time, resulting in a smaller overall reward over the episode. This is true for all environments but is especially visible for the door opening.

Surface Wiping: The learning curves for all controllers seem unstable in this environment, as can be seen in Fig. 5c. The reason why the drops are so big is related to the reward structure, where when the robot gets close to a joint limit or has collisions with anything other than the wiping tool, all other rewards are overwritten by a negative reward. However, during the evaluation of the policies, we also noticed that on some runs the trained policies would have a 100% success rate, whereas on others they would have close to 0%. We, therefore tested both the policy with the best average reward over 5 episodes and the policy at the end of the learning, and chose whichever performed better. With this, all control types performed with a high success rate, with the variable stiffness controller being the lowest at 86.67%. The instability in the learning curve suggests that more training time could be needed, as well as that the robot is positioned in a way that solving the task requires getting close to joint limits or collisions. It

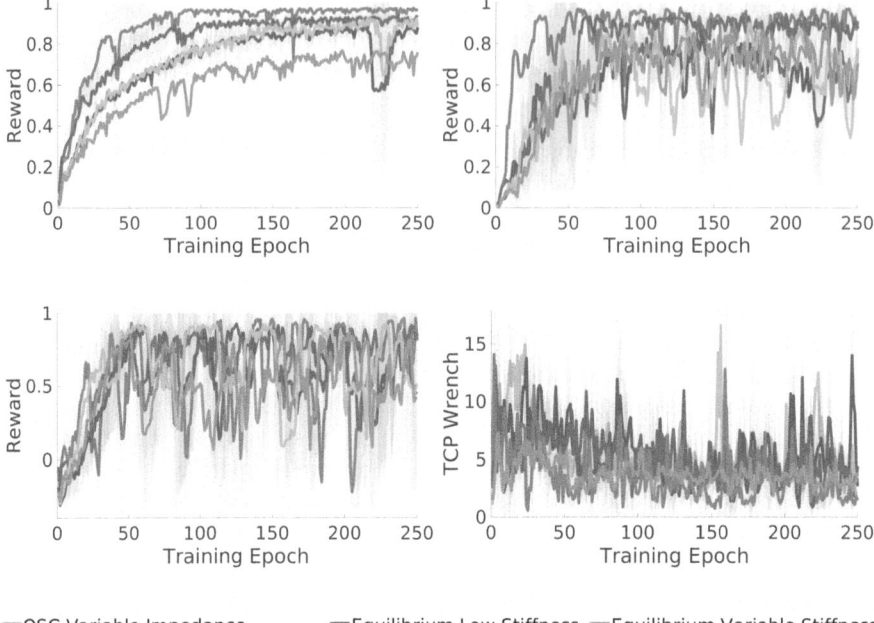

Fig. 5. Training curves of simulation environments with different contact types: (a) path following (no contact), (b) door opening (kinematic constraint), (c) surface wiping (continuous contact). (d) The total wrenches per episode while learning surface wiping. The values are averaged per epoch (40 episodes).

is interesting to notice that the joint position controller looks to have the most stable learning curve. It is believed by the authors that this is related to running into joint limits being punished so severely. Since this is framed in joint space, it is easier for a joint space controller to learn to stay within those limits.

When training contact-rich tasks in simulation, unsafe behaviours during the learning process can be neglected. However, it is still of interest to evaluate how safe the learning behaviours of the different control types would be to learn within the real world, since one-shot transfers are not always possible, or the robot might have to adapt to a variation of the task like wiping a surface that is not flat. Figure 5d shows the total amount of forces and torques applied by the end-effector over an episode in the wiping environment. The values are averaged per epoch. It can be seen that the equilibrium variable stiffness controller shows the most stable behaviour. This suggests that it would also be the safest controller to learn contact-rich tasks within the real world. From the overall results, we believe that the space in which the rewards are framed has a big effect on the learning performance of the different control types. Most goals are defined in operational space resulting in the overall best performance by that controller.

The equilibrium controller works in motor space which in this case is an even lower-level action space than joint space.

Table 2. Success rates of the different control types after training. The values represent the mean of 3 policies evaluated over 10 episodes. OSC and Joint Position Variable Impedance are the controllers used on the rigid arm. Equilibrium Low, High and Variable Stiffness are the controllers used on the soft arm.

	Path	Door	Wipe
OSC Variable Impedance	96.67%	96.67%	100%
Joint Position Variable Impedance	93.33%	90%	96.67%
Equilibrium Low Stiffness	56.67%	73.33%	100%
Equilibrium High Stiffness	80%	90%	100%
Equilibrium Variable Stiffness	6.67%	96.67%	86.67%

5 Conclusion and Future Work

In this paper, we present a novel approach to simulating soft robots with antagonistic VSAs and test it using the GH360 robot as an example. We show that this approach works for full-task learning with a state-of-the-art deep RL algorithm in three different environments. Even though robots with soft materials are generally viewed as harder to control than their stiff counterparts, the results show that the tendon-driven robot can learn all three different tasks with relative performance increasing in contact-rich and kinematic-constrained environments. The results also suggest that equilibrium variable stiffness control is of interest for learning also on the physical robot, as it exerts the least amount of force onto the environment. We also believe that the overall stability and success rate of all control types could be further improved by optimizing the observations given to the policy and tuning the network size and training time.

Future work will look into quantifying the sim2real gap, in enabling self-identification of the tendon model, and in transferring learned behaviours onto the real robot. We are also interested in introducing Learning from Demonstrations into the system by not only tracking motions, but also using EMG sensors to demonstrate co-contraction levels during task execution.

Acknowledgement. This work was partially supported by the Research Council of Norway through grant number 28077. The authors would like to thank Fieldwork Robotics Ltd for making the design of the GH2 arm available to the project.

References

1. Akkaya, I., et al.: Solving rubik's cube with a robot hand (2019). https://doi.org/10.48550/ARXIV.1910.07113

2. Elstner, L., Tirach, R.M., Kyrkjebø, E., Stoelen, M.F.: Comparing and configuring soft tendon designs for variable stiffness actuators on a robot arm. In: 2023 IEEE International Conference on Soft Robotics (RoboSoft). IEEE (2023). https://doi.org/10.1109/robosoft55895.2023.10121971

3. Friedl, W., Hoppner, H., Petit, F., Hirzinger, G.: Wrist and forearm rotation of the dlr hand arm system: mechanical design, shape analysis and experimental validation. In: 2011 IEEE/RSJ International Conference on Intelligent Robots and Systems. IEEE (2011). https://doi.org/10.1109/iros.2011.6094616

4. Gu, S., Holly, E., Lillicrap, T., Levine, S.: Deep reinforcement learning for robotic manipulation with asynchronous off-policy updates. In: 2017 IEEE International Conference on Robotics and Automation (ICRA). IEEE (2017). https://doi.org/10.1109/icra.2017.7989385

5. Haarnoja, T., Ha, S., Zhou, A., Tan, J., Tucker, G., Levine, S.: Learning to walk via deep reinforcement learning (2019)

6. Keppler, M., Lakatos, D., Ott, C., Albu-Schaffer, A.: Elastic structure preserving (ESP) control for compliantly actuated robots. IEEE Trans. Rob. **34**(2), 317–335 (2018). https://doi.org/10.1109/tro.2017.2776314

7. Kim, S., Laschi, C., Trimmer, B.: Soft robotics: a bioinspired evolution in robotics. Trends Biotechnol. **31**(5), 287–294 (2013). https://doi.org/10.1016/j.tibtech.2013.03.002

8. Kingma, D.P., Ba, J.: Adam: a method for stochastic optimization. In: Bengio, Y., LeCun, Y. (eds.) 3rd International Conference on Learning Representations, ICLR 2015, San Diego, CA, USA, 7–9 May 2015, Conference Track Proceedings (2015). http://arxiv.org/abs/1412.6980

9. Lee, M.A., et al.: Making sense of vision and touch: Self-supervised learning of multimodal representations for contact-rich tasks. In: 2019 International Conference on Robotics and Automation (ICRA). IEEE (2019). https://doi.org/10.1109/icra.2019.8793485

10. Levine, S., Finn, C., Darrell, T., Abbeel, P.: End-to-end training of deep visuomotor policies. J. Mach. Learn. Res. **17**(39), 1–40 (2016). http://jmlr.org/papers/v17/15-522.html

11. Martin-Martin, R., Lee, M.A., Gardner, R., Savarese, S., Bohg, J., Garg, A.: Variable impedance control in end-effector space: an action space for reinforcement learning in contact-rich tasks. In: 2019 IEEE/RSJ International Conference on Intelligent Robots and Systems (IROS). IEEE (2019). https://doi.org/10.1109/iros40897.2019.8968201

12. Santina, C.D., et al.: Controlling soft robots: balancing feedback and feedforward elements. IEEE Rob. Autom. Maga. **24**(3), 75–83 (2017). https://doi.org/10.1109/mra.2016.2636360

13. Santina, C.D., Catalano, M.G., Bicchi, A.: Soft robots. In: Encyclopedia of Robotics, pp. 1–15. Springer, Heidelberg (2021)

14. Stoelen, M.F., de Azambuja, R., Rodríguez, B.L., Bonsignorio, F., Cangelosi, A.: The GummiArm project: a replicable and variable-stiffness robot arm for experiments on embodied AI. Front. Neurorob. **16** (2022). https://doi.org/10.3389/fnbot.2022.836772

15. Todorov, E., Erez, T., Tassa, Y.: MuJoCo: a physics engine for model-based control. In: 2012 IEEE/RSJ International Conference on Intelligent Robots and Systems. IEEE (2012). https://doi.org/10.1109/iros.2012.6386109
16. Vanderborght, B., Tsagarakis, N.G., Semini, C., Van Ham, R., Caldwell, D.G.: Maccepa 2.0: adjustable compliant actuator with stiffening characteristic for energy efficient hopping. In: 2009 IEEE International Conference on Robotics and Automation, pp. 544–549. IEEE (2009)
17. Zhu, Y., et al.: robosuite: a modular simulation framework and benchmark for robot learning. In: arXiv preprint arXiv:2009.12293 (2020)

Using Barometric Pressure Sensors in Soft Robots

Kedar Suthar[1]([✉]), Chapa Sirithunge[2], and Mingfeng Wang[1][ORCID]

[1] Department of Mechanical and Aerospace Engineering, Brunel University London, Uxbridge UB8 3PH, UK
kedarsuthar@outlook.com, mingfeng.wang@brunel.ac.uk
[2] Department of Engineering, University of Cambridge, Cambridge CB2 1PZ, UK
csh66@cam.ac.uk

Abstract. This paper presents a detailed methodology for calibrating barometric pressure sensors embedded in soft robots, which is crucial for ensuring accurate pressure measurement and effective control. Soft robots, with their flexible and adaptive structures, require precise internal pressure monitoring to maintain performance and functionality. We outline a comprehensive calibration process that includes the initial setup of the sensor within the soft structure, baseline measurement under unpressurised conditions, and incremental application of known pressures using a calibrated source. Data from these controlled pressure steps are recorded and used to create a calibration curve through linear regression, enabling accurate pressure readings. The calibration process is validated through additional pressure applications, ensuring minimal error and high reliability. Our methodology addresses sensor stability, environmental factors, and hysteresis, providing a robust framework for integrating barometric pressure sensors in soft robotic applications. This calibration technique ensures that the soft robot's internal pressure is accurately monitored, enhancing operational efficiency and control precision.

Keywords: Barometric Pressure Sensors · Calibration · Sensing

1 Introduction

Fluidic or barometric pressure sensors play a crucial role in the functionality and performance of soft robots. Accurate pressure monitoring is essential for maintaining the shape, movement, and functionality of these robots, as they rely heavily on precise control of internal pressures. Control and actuation are other critical areas where pressure sensors are indispensable. Soft robots often use pneumatic or hydraulic systems, and fluidic pressure sensors are an inexpensive option for regulating these systems, allowing smooth and controlled movements. In addition, these sensors can be placed outside a soft robot, which can still be connected with the help of a soft tube or a similar structure. This helps maintain the compliance of soft robots without hindering their sensing experience. An

M. N. Huda et al. (Eds.): TAROS 2024, LNAI 15052, pp. 236–241, 2025.
https://doi.org/10.1007/978-3-031-72062-8_21

example of this is shown in Fig. 1. Although barometric pressure sensors have great compliance, hence making them ideal for soft robots, signals can be noisy and challenging to calibrate. Calibration gets even more difficult when they are embedded in soft robots in space-constrained applications as shown.

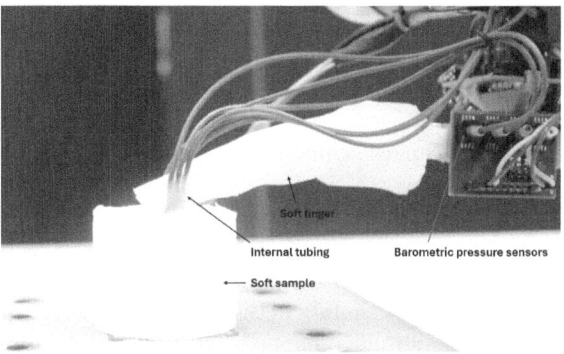

Fig. 1. An application of fluidic pressure sensors extended and embedded into a soft robotic finger for tactile discrimination of softness [2].

Adaptive response is another key benefit provided by pressure sensors. Soft robots are designed to interact with dynamic and unpredictable environments. Pressure sensors enable these robots to adapt to varying loads and external forces by adjusting their internal pressures accordingly. This capability enhances their ability to perform complex tasks effectively. However, these sensors have several disadvantages as well. The compressibility of air can absorb a portion of the applied pressure, resulting in under-recorded measurements, and the dynamic response of compressed air can introduce delays, thereby reducing the sensor's response speed.

Not only soft robotics but also anatomical information has been extracted using barometric pressure data as explained in [1]. In this work, activity-related altitude changes have been recorded with the help of barometric and inertial sensors. Masse et al. [4] provide a thorough analysis of using barometric pressure sensors for wearables. According to this work, the main considerations in using these sensors include form factor, power consumption, precision and commercial availability.

2   Related Work

Nonaka et al. used a soft finger equipped with only fluidic pressure sensors to recognise the "softness" of softer materials compared to the soft skin [2]. These sensors were placed at the fingertip and the pressure changes during the interaction of finger skin with external soft materials were recorded with the help of fluidic pressure sensors. Authors in [7] present a framework for embedding soft sensors for proprioception. This can be used as an inspiration for air chamber design for barometric pressure sensing for monitoring deformation as well as determining external forces.

Barometric pressure sensing has widely been used in mobile phones to estimate touch forces on the screen [3]. Although this technology cannot be adopted in soft robots directly, this has been made possible using air chambers as an extension from the robot to pressure sensors located outside robots [5]. This preserves the flexibility of soft robots while providing sensing capability. A tactile sensor network based on the same principle has been deployed for touch localisation in [6]. Results for the above sensor suggested that the sensor performs well for low and higher force ranges while simultaneously locating and estimating contact forces. Hence, it is worth finding out in which force ranges the sensor performs better, given the properties of elastomers used in the structure.

An advanced design of this concept can be found in [8]. This is a 3D-printed soft robotic hand comprising tactile sensing regions spread across the palm. These consist of 3D-printed soft chambers that are compressed by external forces. In this work, we present a simplified experimental setup to calibrate barometric pressure sensors to enhance their performance in similar soft robotic applications. This is an effort to promote the accurate usage of barometric pressure sensing in soft robots.

3   Experimental Setup

Calibrating a fluidic pressure sensor embedded in a soft robot or silicone layer is essential for accurate pressure measurement and effective robot control. Firstly, we ensure the sensor is properly integrated within the soft robot or silicone layer, with secure and leak-free connections to the fluidic channels. Then the sensor was connected to a stable power supply and linked to a data acquisition system, in our case, an Arduino, to record pressure readings.

We placed the silicone layer in a relaxed, unpressurised state and recorded the sensor's baseline pressure reading, which serves as the zero-pressure reference point. We recorded pressure values under stable environmental conditions, such as constant temperature and humidity, to avoid variations in sensor readings.

Next we gradually applied known pressures to the fluidic system using a calibrated pressure source: in the example case study, the sensor was calibrated with a syringe pump, as shown in Fig. 2a. The pressure was increased in small, controlled steps, allowing the system to stabilise at each pressure level before recording the sensor output. The corresponding sensor reading at each applied pressure was recorded as the next step, ensuring multiple readings are considered for any variability.

We plotted the recorded sensor readings against the known applied pressures; an example is shown in Fig. 2b. A calibration curve was fitted to the plotted data points, which can be either linear or nonlinear, depending on the sensor characteristics. Common fitting methods include linear regression or polynomial fitting. The linear calibration curve can be represented by the equation:

$$P_{applied} = a * S_{sensor} + b \tag{1}$$

where $P_{applied}$ is the applied pressure, S_{sensor} is the sensor reading, and a and b are the calibration coefficients obtained from the fit.

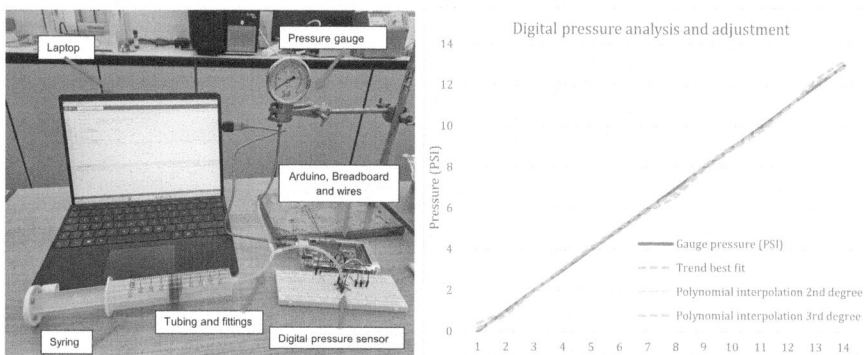

Fig. 2. Calibration in Case 1 study: (a) experimental setup and (b) linear regression or polynomial fitting (2nd and 3rd degree). The x-axis represents the known pressures, and the y-axis shows the adjusted sensor output.

A higher order polynomial in place of (1) will represent the ideal case for calibration. However, using a higher order of polynomial fitting will result in more accurate results, but will also increase computation time. Hence a second order polynomial has been used in this study to calibrate the sensor. The calibration curve was validated by applying additional known pressures and comparing the measured values to the expected values derived from the calibration curve, such as a pressure gauge used in 2. Calibration error was calculated to ensure it was within acceptable limits and, if necessary, adjust the calibration procedure or curve fitting. As the final step, the calibration curve can be implemented into the soft robot software, allowing the system to convert raw sensor readings into accurate pressure measurements. A routine calibration schedule must be maintained for proper functioning as sensor accuracy degrades over time, particularly if the soft robot operates in varying environmental conditions or undergoes significant mechanical deformation.

Important considerations include ensuring the fluidic pressure sensor has stable and repeatable output characteristics, accounting for any environmental factors that might affect sensor readings, such as temperature changes, and checking for and compensating for any hysteresis in the sensor output during pressurisation and depressurisation cycles.

3.1 Case Study: SoftPicker

In the case study of SoftPicker, a pneumatically driven soft gripper was designed for soft fruits (e.g., strawberries) harvesting. To evaluate whether the proposed soft gripper will cause any damage to the surface of a soft fruit (in this case, strawberries) during and after it is picked, the lab-based prototype is set up as shown in Fig. 3. A series of snap-picking tests have been carried out to evaluate the grasping performance under different air pressures, and an illustration of snap-picking is presented in Fig. 4.

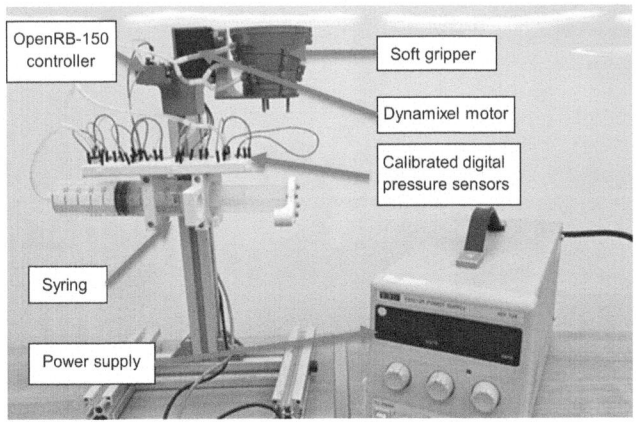

Fig. 3. Experimental setup of the SoftPicker

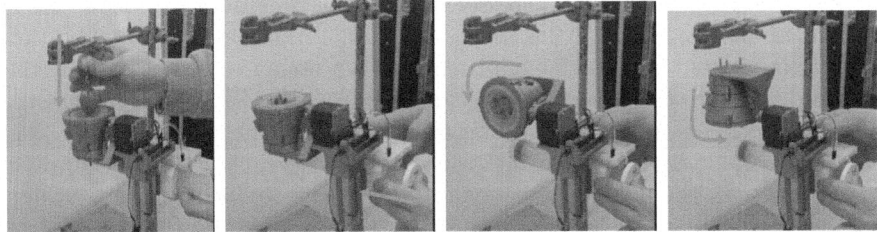

Fig. 4. Snapshots of an experimental test of snap-picking of a strawberry with the SoftPicker.

Experiments were conducted to evaluate the performance of the sensor before and after calibration. While calibration of the sensor helped in reducing error in pressure values, material properties of elastomers and tubing and air compressibility can add errors to pressure readings. However these sensors use air- which is readily available and does not contaminate the surroundings upon damage. Furthermore, refilling does not require complex mechanisms. The experiment was conducted within 1–2 hours during day time to keep the environmental conditions as constant as possible-hence minimise the effect of the environment on air compression and material properties. Performance of the sensor before and after the calibration in the fruit picking task was compared.

4   Conclusions

This study presents a comprehensive methodology for calibrating barometric pressure sensors embedded in soft robots, highlighting the importance of accurate pressure measurement for effective control. By carefully setting up the sensor, establishing a baseline measurement, and applying incremental known pressures, we created a reliable calibration curve through linear regression. This process ensures the sensor provides precise pressure readings, which is critical for the soft robot's performance and functionality. Validation through additional pressure applications confirmed the minimal error and high reliability of the calibration. Our approach also addresses key factors such as sensor stability, environmental influences, and hysteresis, offering a robust solution for integrating barometric pressure sensors in soft robotic systems. Overall, this calibration technique enhances the accuracy of internal pressure monitoring, thereby improving the operational efficiency and control precision of soft robots.

References

1. Moncada-Torres, A., Leuenberger, K., Gonzenbach, R., Luft, A., Gassert, R.: Activity classification based on inertial and barometric pressure sensors at different anatomical locations. Physiol. Meas. **35**(7), 1245 (2014)
2. Nonaka, T., Abdulali, A., Sirithunge, C., Gilday, K., Iida, F.: Soft robotic tactile perception of softer objects based on learning of spatiotemporal pressure patterns. In 2023 IEEE International Conference on Soft Robotics (RoboSoft), pp. 1–7. IEEE (2023)
3. Quinn, P.: Estimating touch force with barometric pressure sensors. In: Proceedings of the 2019 CHI Conference on Human Factors in Computing Systems, pp. 1–7 (2019)
4. Massé, F., Bourke, A.K., Chardonnens, J., Paraschiv-Ionescu, A., Aminian, K.: Suitability of commercial barometric pressure sensors to distinguish sitting and standing activities for wearable monitoring. Med. Eng. Phys. **36**(6), 739–744 (2014)
5. Gilday, K., Relandeau, L., Iida, F.: Design and characterisation of a soft barometric sensing skin for robotic manipulation. In 2022 IEEE/RSJ International Conference on Intelligent Robots and Systems (IROS), pp. 10021–10028. IEEE (2022)
6. De Clercq, T., Sianov, A., Crevecoeur, G.: A soft barometric tactile sensor to simultaneously localize contact and estimate normal force with validation to detect slip in a robotic gripper. IEEE Rob. Autom. Lett. **7**(4), 11767–11774 (2022)
7. Tapia, J., Knoop, E., Mutný, M., Otaduy, M.A., Bächer, M.: Makesense: automated sensor design for proprioceptive soft robots. Soft Rob. **7**(3), 332–345 (2020)
8. Shorthose, O., Albini, A., He, L., Maiolino, P.: Design of a 3D-printed soft robotic hand with integrated distributed tactile sensing. IEEE Rob. Autom. Lett. **7**(2), 3945–3952 (2022)

Modelling of Buoyancy Based Actuation of an Inflatable Underwater Soft Robot

Danyaal Kaleel[1]($\boxtimes$) (iD), Benoit Clement[2,3] (iD), and Kaspar Althoefer[1] (iD)

[1] Queen Mary University of London, 327 mi End Rd, London E1 4NS, UK
`d.m.kaleel@qmul.ac.uk`
[2] CNRS IRL Crossing , ENSTA Bretagne, Adelaide, Australia
[3] Flinders University, Adelaide, Australia

Abstract. This paper presents a novel actuation method to change the buoyancy of an underwater soft robot by pumping denser and less dense liquids, than the liquid the robot is immersed in, into the robot to actively change the mass of the robot and cause it to experience a change in buoyancy. The technological research gap lies in the method of pumping lighter and heavier fluids into a soft robot to cause it to experience a change in mass and depth, which has not been explored before to the best of the author's knowledge. An analysis of the forces that are placed on the robotic system and the necessary equations to determine the force produced by a solution with a particular ratio of solute to solvent are presented. Preliminary experiments were conducted to test the buoyancy-based actuation method discussed in this paper by building a two link, soft, inflatable robot arm. This robot was shown to change the floatation height of its links when denser fluid was pumped into its links.

Keywords: Depth controllable soft robot · Buoyancy controllable soft robot · Buoyancy actuated soft robot · Underwater soft robot arm

1 Introduction

1.1 Background

There are three common types of robots that are deployed underwater: Remotely Operated Vehicles (ROVs), Autonomous Underwater Vehicles (AUVs) and gliders. ROVs are robots for marine exploration, that are tethered to a support ship and are controlled by a human operator [17]. ROVs are used in a range of applications including archaeological discovery [23], inspection of oil rig platforms [19] and offshore wind farms [22], preventing marine fouling [3], biological exploration and sample collection [4], underwater mine detection and disposal [15], and search and rescue [7]. Despite their wide array of uses, these robots are limited in that they are unable to operate autonomously and rely on a highly trained human operator to control them. They also require a ship to be tethered to, which needs a crew, which is expensive. Furthermore, ROVs have problems due to visibility being limited and communication being slow underwater [9].

© The Author(s), under exclusive license to Springer Nature Switzerland AG 2025
M. N. Huda et al. (Eds.): TAROS 2024, LNAI 15052, pp. 242–253, 2025.
https://doi.org/10.1007/978-3-031-72062-8_22

This makes control especially challenging as the ROV system is working in real time and ideally feedback from its sensors need to be processed immediately as the environment can change around the ROV quickly.

Autonomous underwater vehicles (AUVs) are another type of marine robot. They are autonomously controlled and untethered [16] [5]. Gliders are a subset of AUVs that move through adjusting their buoyancy [18]. Being untethered means that a support ship is not required and being autonomously controlled enables these robots to be in operation for long periods of time as a human operator is not required. Like ROVs, AUVs and gliders are used in a range of applications including inspecting infrastructure [10], marine surveying [2], oceanography and biological sample collection [24].

Despite the large number of applications in which these current state of the art underwater robotic systems are used, they all suffer from the drawbacks of being rigid, and large in size and therefore not conformable to their environment. This means that certain underwater environments, such as those that are tight and narrow like coral reefs and underwater cave systems, are difficult for them to move through.

A potential solution to this problem could be through the use of soft robots. These robots have properties that make them desirable for underwater applications including being soft and deformable so that they have the ability to squeeze through narrow openings, but also not cause significant damage to marine structures that they come into contact with, being cheap to manufacture if degradation happens due to the marine environment or if lost at sea, being straightforward to deploy as they are light weight and can be compactly stored, and not greatly affected by pressure changes.

There are an increasing number of soft robots being designed for underwater applications. These include bio inspired robots such as fish [21], stingrays [1] and octopuses [8].

An ability that both soft and hard underwater robots should have is a method to change their buoyancy. There are many methods of achieving buoyancy control in literature.

One method of changing buoyancy is to have a chamber inside the robot that is filled with air that can be compressed to make the robot experience negative buoyancy and expanded for positive buoyancy, as seen in [12], which explored this mechanism for a soft underwater fish. A potential problem with a system such as this is that if the air chamber has a leak, such as from the degradation of sea water over time or from manufacturing defects, air could leak out preventing the robot from being able to move up and down.

Another method of changing buoyancy is to change the angle of dive planes that are located externally on the robot, like that seen in [13]. A potential problem with this mechanism is that if the robot is moving through a tight and cramped environment, such as through a coral reef, the robot's dive planes could become obstructed, preventing or limiting vertical movement.

Another method shown in literature to change buoyancy is to heat wax to cause a change of volume and thus a change in buoyancy [20]. A potential problem with this mechanism is that it could be not possible or require more energy to heat the wax in colder waters.

In this paper, a novel method of buoyancy based actuation is presented. A denser fluid than the surrounding fluid that the robot is in, is pumped into the robot to enable negative buoyancy to occur and for the robot to move downwards towards the ground and a less dense fluid than the surround fluid that the robot is in, is pumped into the robot to enable positive buoyancy to occur and for the robot to move upwards towards the surface of the water. To the best knowledge of the authors, this is the first buoyancy based system that uses fluids of different densities to enable depth change underwater. The work presented builds on our previous work [11], where we looked into a robot that uses lighter fluids to change floatation height in air.

We envisage the robot presented in this paper to be deployed either from a support ship or from an ROV as an arm like [14] and so the pumps necessary for pumping different density fluids can be contained on these vehicles. The robotic system presented in this paper is still underdevelopment, but we intend to use a single premixed denser liquid and a single premixed less dense liquid, which are both capable of applying a single amount of force. To vary the robot's position, precise modulation of these premixed liquids inside the robot is required.

1.2   Contributions

The contributions of this paper are as follows:

1. Proving mathematically that changing the composition of a fixed volume solution can result in a change of depth underwater.
2. Providing the essential equations to understand how much force a particular solution can exert based on solute and solvent compositions.
3. Developing a two link soft robotic arm in which the depth position of the links can be changed through pumping denser fluid into them.

1.3   Paper Structure

This paper is structured as follows: Sect. 2 examines the forces that the inflatable robot is under and whether certain forces are more important than others such that they do not need to be considered. Section 3 looks into the forces generated for different compositions of fluids and specifically looks into the use of salt as a solute and distilled water as the solvent for this purpose. Section 4 shows the design and build of a two link soft inflatable robotic arm for use in demonstrating the principles of the buoyancy based actuation method discussed in previous sections. Section 5 presents the results of how the link orientation of the robot arm changes when denser fluids are used to inflate the links. Section 6 concludes the paper and gives some recommendations of future work that should be performed.

2   Analysis of the Forces that Act Upon the Robot

Figure 1 shows the five different forces that act upon the robot underwater. The three larger forces are the controllable downward force $F_{solution}$, the upward force $F_{buoyant}$ and the force produced by the weight of the robot F_{robot}. There are also two other forces, namely F_{drag}, and $F_{viscous}$ that act against the robot's direction of motion. If the robot is moving downwards, F_{drag}, and $F_{viscous}$ are applied upwards, as shown in Fig. 1 and is represented mathematically by Eq. 1. If the robot is moving upwards, F_{drag}, and $F_{viscous}$ are applied downwards, as shown in Fig. 2 and is represented mathematically by Eq. 2.

2.1   Control Force

The controllable solution force, $F_{solution}$, is made up of two components, a solute such as a granular substance like sodium chloride and a solvent such as a liquid substance like water and it is the variation of these quantities of components that dictates the amount of force produced that enables the robot to either sink to greater depths or rise to lower depths.

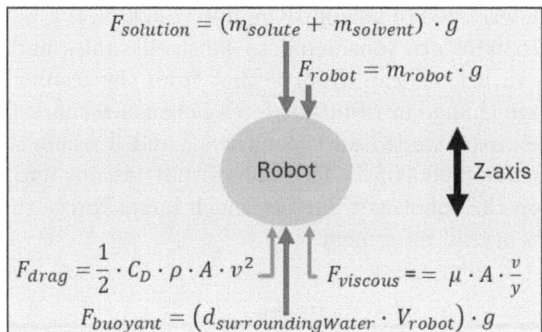

Fig. 1. Shows all the forces that are acting upon the robot when the robot is moving downwards

$$F_{resultant} = F_{buoyant} + F_{drag} + F_{viscous} - F_{solution} - F_{robot} \qquad (1)$$

$$F_{resultant} = F_{buoyant} - F_{drag} - F_{viscous} - F_{solution} - F_{robot} \qquad (2)$$

2.2   Proving Viscous and Drag Forces Are Insignificant

Mathematically, it can be proven that drag and viscous forces have negligible affect on the overall resultant force experienced by the robot as we are considering the robot to move at slow velocities such as at 0.01 m/s.

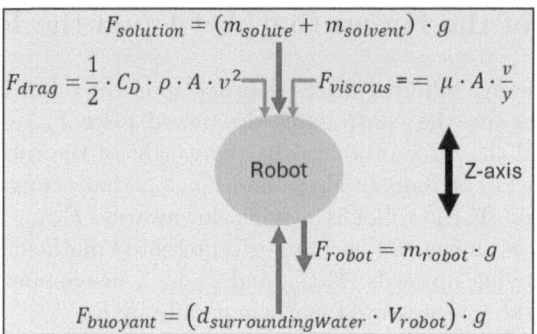

Fig. 2. Shows all the forces that are acting upon the robot when the robot is moving upwards.

Considering the first scenario where a cylindrical robot is moving downwards and constructed from PVC with the dimensions length 1000 mm and diameter 100 mm (Fig. 3), it would have a mass around 0.43 g. If we take an arbitrary mass of solute of 2500 g and mass of solvent of 7775 g, then the total downward force is approximately 100000 N. Given the dimensions of the robot, the upward buoyant force is determined to be approximately 76300 N. If $f_{solution} = 100000\,N$ and $f_{buoyant} = 76300 N$ are considered to be fixed values and drag force and viscous force are varied, we can observe that from the results seen in Table 1, that the percentage change in resultant force between scenario 1 where no drag and viscous forces are included and scenarios 2 and 3 where drag and viscous forces are included, is 0.000005%. This shows that viscous and drag force have negligible affect on the robot as there are much larger forces that play a bigger role on the robots overall movement.

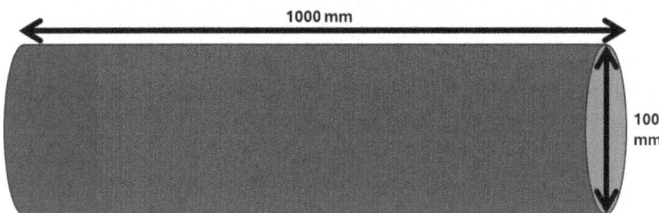

Fig. 3. Robot dimensions for proving the negligible affect of drag and viscous forces.

We can therefore not consider these forces when considering the forces that act upon the robot and consider the situation illustrated in Fig. 4 and mathematically represented in Eq. 3.

$$F_{resultant} = (d_{surroundingWater} \cdot v_{robot}) \cdot g - (m_{solute} + m_{solvent} + m_{robot}) \cdot g \quad (3)$$

Table 1. Showing the Negligible Effect of Viscous and Drag Forces

scenario	Drag Force F_{Drag}	Viscous Force $F_{Viscous}$	Resultant Force $F_{Resultant}$	Percentage change
1	0	0	-23700	0
2	0.319	0.00002	-23699.681	0.000005
3	0.319	0.0000003	-23699.681	0.000005

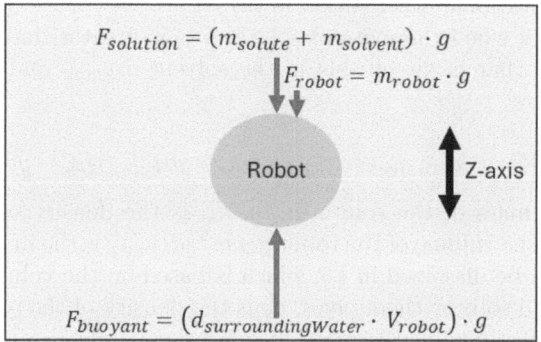

Fig. 4. Shows all the forces that make a significant difference upon the robot's motion.

3 Examining the Affect on Varying the Components of the Solution Force

As mentioned in Sect. 2, the solution force $F_{Solution}$ is formed of two components, the solute and the solvent.

In the robotic system design considered in this paper, it is intended that solutions pumped into the robot would be premixed and so the force these solutions produce would be constant. This is rather than mixing the solute and solvent in the robot itself and having a variable force that can be controlled.

However, it is important to know the force that the premixed solution pumped into the robot can generate for a number of reasons. These include applications where less force is required such as in a fragile environment, applications where more force is required such as in a pick and place task and for considering suitable controllers where change in force needs to be considered to understand how the robot will stabilise at specific depths while reducing the amount of overshooting and undershooting. This section looks into how much force a solution can generate by varying the solute and solvent quantities while maintaining constant volume.

There are two force phases that the solution goes through. The first phase, phase 1, occurs when the solute can be dissolved by the solvent and so the volume does not increase. The second phase, phase 2, occurs when the solute cannot be dissolved by the solvent and the solvent must be decreased to add more solute to maintain a constant volume. It is important to have a constant volume as the robot cannot change volume, so the solution pumped into the robot needs to be of a constant volume.

Phase 1 occurs when the mass of the solute, m_s, is less than or equal to the maximum amount of solute that is dissolvable in the solvent dis_{max} and is governed by Eq. 4.

$$F = (m_s + m_w) \cdot g \tag{4}$$

where m_s is the mass of the solute in kg, m_w is the mass of the solvent in kg and g is the gravitational acceleration in m/s^2.

Phase 2 occurs when the mass of solute, m_s, is greater than the maximum amount of solute that is dissolvable in the solvent dis_{max} and is governed by Eq. 5.

$$F = (m_s + d_w \cdot (v_{robot} - (m_s - dis_{max})/d_s) \cdot g \tag{5}$$

where m_s is the mass of the solute in kg, d_w is the density of the solvent in kg/cm^3, v_{robot} is the volume of the robot in cm^3, dis_{max} is the maximum amount of solute that can be dissolved in kg, which is based on the volume of the robot and the solute and solvent themselves, d_s is the density of the solute in kg/cm^3 and g is the gravitational acceleration in m/s^2.

3.1   Sodium Chloride Example

By taking sodium chloride as the solute and distilled water as the solvent, it can be determined through experimentation that $35.7g$ of sodium chloride can be dissolved in $100\,cm^3$ of water at $20°$ C or room temperature [6].

Then, by applying Eq. 4 and Eq. 5 for sodium chloride and distilled water, the theoretical downwards acting force, produced by a robot, which is using this fluid to change its buoyancy can be determined. Figure 5 shows how this theoretical force produced by the solution increases with sodium chloride concentration for a $100\,cm^3$ robot. An interesting observation from Fig. 5 is that the maximum dissolvable quantity of sodium chloride in phase 1, produces a greater force than the minimum undissolvable quantity of sodium chloride in phase 2. This occurs due to the reduction in distilled water at the start having a greater effect on the force than the increase in sodium chloride.

4   Case Study: Two Link Robotic Arm

We built a two link robotic arm as seen in Fig. 6 and Fig. 7 to examine the affect of our buoyancy based actuation method. Each link could be inflated with the fluids to enable it to move up and down independently of the other link.

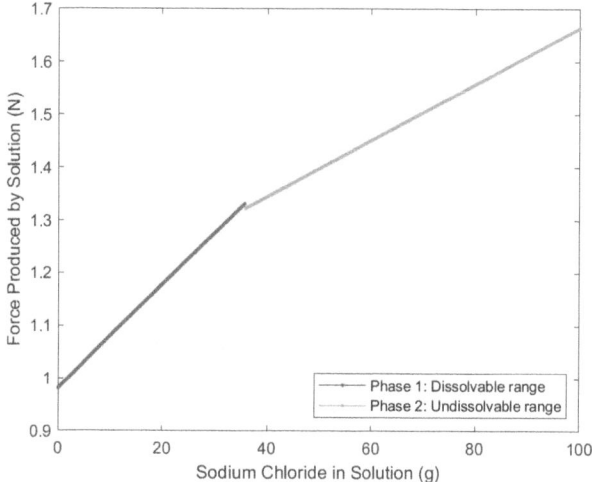

Fig. 5. Shows the theoretical force increase as the concentration of sodium chloride in a solution increases, inside a robot, with a volume of 100 cm^3 at $20\,^{\circ}$C.

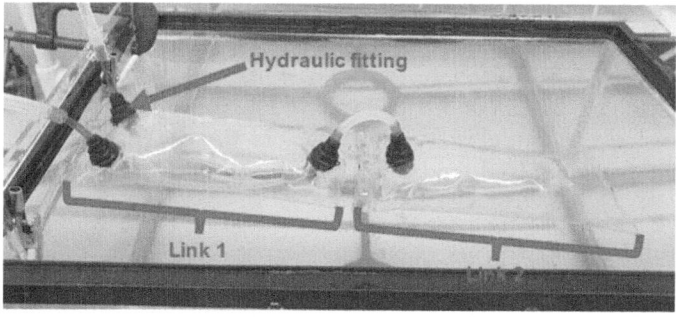

Fig. 6. Shows the robot with two links.

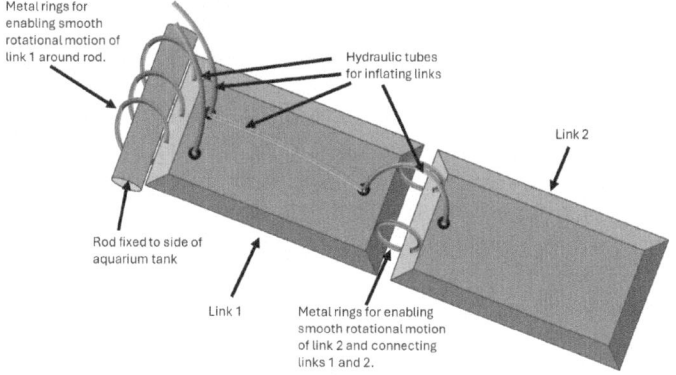

Fig. 7. 3D rendering of two link soft robot arm

4.1   Design and Build

The links were heat welded from thin low density polyethylene (LDPE) sheets using a plastic bag sealer. Small holes in the links were made for the hydraulic connectors from which hydraulic tubes could be connected to, as shown in Fig. 6. The hydraulic connectors were then screwed into the links to make water tight seals. Thin, silicone hydraulic tubes provided the means for transporting water to the links without compromising the robots ability to change bending angle. A bypass hydraulic tube to supply water to link 2 was placed through link 1, to prevent tension being placed on the links, which would prevent them from bending. Flexible metal rings were created from the rings of a spring to enable the links to be connected together and link 2 to rotate around link 1 and to connect link 1 to the rod, which enables link 1 to rotate around the rod.

5   Results

5.1   Experimental Setup

An aquarium tank was filled with distilled water with a negligible amount of sodium chloride. Sodium chloride was chosen as the solute and distilled water was chosen as the solvent in the solution that was pumped into the robot links. The salt solution used to inflate the robot consisted of approximately 36 g of sodium chloride for 200 ml of distilled water. The robot was anchored to a piece of acrylic, using the rod, close to the surface of the water. A syringe was used to pump different premixed fluid solution into and out of the robot.

5.2   Robot Link Positions

The robot started in the configuration shown in Table 2 (a), where it was parallel and in line with the surface of the water. The salt solution was then added to link 1, as shown in Table 2 (b) causing the robot to orientate link 1 downwards and as link 2 was attached to link 1, it was also oriented downwards. The premixed salt solution was then removed from link 1 and placed in link 2 causing link 2 to be orientated downwards while link 1 remained at the surface of the water, as shown in Table 2 (c). The final test determined whether the robot was able to achieve neutral buoyancy by filling both links with distilled water. The results of which are shown in Table 2 (d), where the robot appears to show itself in a neutrally buoyant position. A limitation of this test is that the robot was anchored close to the surface of the water, so it is difficult to distinguish whether the robot was experiencing neutral buoyancy or positive buoyancy. Further tests are required to determine whether the robot was experiencing positive buoyancy or neutral buoyancy. These further tests would require the robot to be anchored at a greater water depth, such that the anchoring point is submerged further underwater.

Table 2. Robot demonstrating different configurations with different solutions. (a) shows robot in initial configuration without any solution inside the links. (b) shows salt solution inside link 1. (c) shows salt solution inside link 2. (d) shows distilled water in links 1 and 2 demonstrating neutral buoyancy.

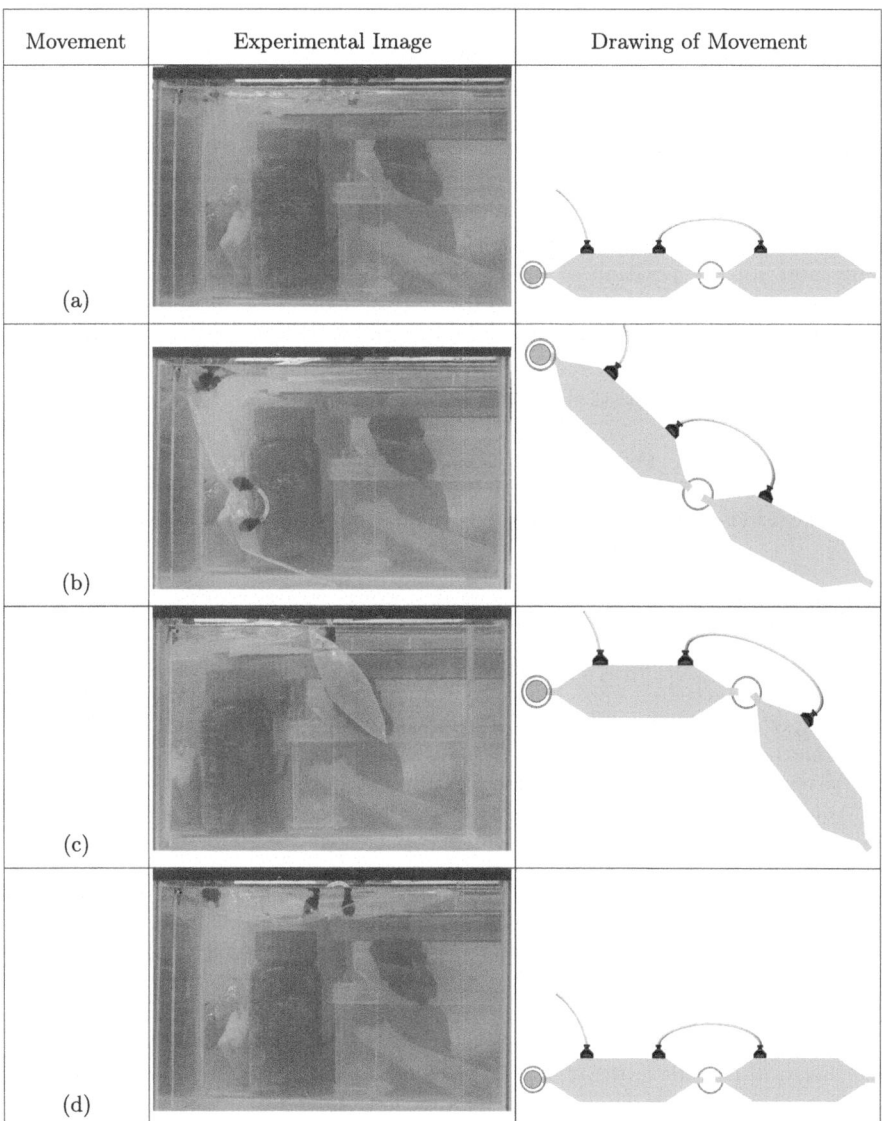

Movement	Experimental Image	Drawing of Movement
(a)		
(b)		
(c)		
(d)		

6   Conclusions and Future Work

This paper proved both mathematically and experimentally that buoyancy actuation can be achieved in a soft robotic system, through the inflation of the robot with different density fluid solutions to the fluid solution the robot is immersed in, which enables the robot to generate a specific amount of force.

There persists many challenges that need addressing to make a robot powered by this buoyancy system useful. This includes the prevention of air from going into the robot causing it to become positively buoyant to the extent that the addition of a denser fluid would not be able to counteract the upward buoyant force. Additionally, a suitable robot design is needed that has separate chambers, in which the negatively buoyant, denser solutions can be pumped in without contaminating the positively buoyant, less dense solutions through the remainder of remnants of solute after repeated actuation. Furthermore, a suitable control system is needed to pump different density fluid solutions into and out of the robot for precise position control as this paper only considers constant force values and not the position of the robot underwater. This control system could make use of peristaltic pumps to pump fluid solutions into and out of the robot. One could also consider a reinforcement learning approach with sensors that can detect depth as the control system would need to be highly adaptable due to the high dynamic variability of the underwater environment.

Acknowledgement. The work in this paper is funded by DSTL. Thank you to Dr Adrian Baker from DSTL and Dr Abu Bakar Dawood from QMUL for their guidance on this work. The authors of this paper have no competing interests.

References

1. Alvarado, P.V., Chin, S., Larson, W., Mazumdar, A., Youcef-Toumi, K.: A soft body under-actuated approach to multi degree of freedom biomimetic robots: a stingray example. In: 2010 3rd IEEE RAS & EMBS International Conference on Biomedical Robotics and Biomechatronics, pp. 473–478. IEEE (2010)
2. Barrett, N., Seiler, J., Anderson, T., Williams, S., Nichol, S., Hill, S.N.: Autonomous underwater vehicle (auv) for mapping marine biodiversity in coastal and shelf waters: implications for marine management. In: OCEANS'10 IEEE SYDNEY, pp. 1–6. IEEE (2010)
3. Bruno, F., et al.: A rov for supporting the planned maintenance in underwater archaeological sites. In: Oceans 2015-Genova, pp. 1–7. IEEE (2015)
4. Chaloux, N., Phillips, B.T., Gruber, D.F., Schelly, R.C., Sparks, J.S.: A novel fish sampling system for rovs. Deep Sea Res. Part I **167**, 103428 (2021)
5. Collins, K.: Untethered auv's can reduce costs for offshore inspection jobs. In: Offshore Technology Conference. pp. OTC–7114. OTC (1993)
6. Duerst, M.D.: The critical impact of nacl on human history and development. In: Chemistry's Role in Food Production and Sustainability: Past and Present, pp. 47–62. ACS Publications (2019)
7. Fattah, S., Abedin, F., Ansary, M., Rokib, M., Saha, N., Shahnaz, C.: R3diver: remote robotic rescue diver for rapid underwater search and rescue operation. In: 2016 IEEE Region 10 Conference (TENCON). pp. 3280–3283. IEEE (2016)
8. Fras, J., Noh, Y., Macias, M., Wurdemann, H., Althoefer, K.: Bio-inspired octopus robot based on novel soft fluidic actuator. In: 2018 IEEE International Conference on Robotics and Automation (ICRA), pp. 1583–1588. IEEE (2018)

9. Ho, G., Pavlovic, N., Arrabito, R.: Human factors issues with operating unmanned underwater vehicles. In: Proceedings of the Human Factors and Ergonomics Society Annual Meeting, vol. 55, pp. 429–433. SAGE Publications, Los Angeles (2011)

10. Inzartsev, A., Pavin, A.: Auv application for inspection of underwater communications. Underwater Vehicles () (2009)

11. Kaleel, D., Clement, B., Althoefer, K.: A framework to design and build a height controllable eversion robot. In: 2023 11th International Conference on Control, Mechatronics and Automation (ICCMA), pp. 239–244. IEEE (2023)

12. Katzschmann, R.K., DelPreto, J., MacCurdy, R., Rus, D.: Exploration of underwater life with an acoustically controlled soft robotic fish. Sci. Rob. **3**(16), eaar3449 (2018)

13. Katzschmann, R.K., Marchese, A.D., Rus, D.: Hydraulic autonomous soft robotic fish for 3d swimming. In: Experimental Robotics: The 14th International Symposium on Experimental Robotics, pp. 405–420. Springer, Heidelberg (2015)

14. Koch, J., Leichty, J.: Development of a robotic arm for mini-class rov dexterous manipulation. In: OCEANS 2019 MTS/IEEE SEATTLE, pp. 1–5. IEEE (2019)

15. Matika, D., Koroman, V., (CROATIA), M.O.D.Z.: Undersea detection of sea mines. MINISTRY OF DEFENCE ZAGREB (CROATIA) (2001)

16. Nicholson, J., Healey, A.: The present state of autonomous underwater vehicle (auv) applications and technologies. Mar. Technol. Soc. J. **42**(1), 44–51 (2008)

17. Petillot, Y.R., Antonelli, G., Casalino, G., Ferreira, F.: Underwater robots: from remotely operated vehicles to intervention-autonomous underwater vehicles. IEEE Rob. Autom. Maga. **26**(2), 94–101 (2019)

18. Rudnick, D.L., Davis, R.E., Eriksen, C.C., Fratantoni, D.M., Perry, M.J.: Underwater gliders for ocean research. Mar. Technol. Soc. J. **38**(2), 73–84 (2004)

19. Salgado-Jimenez, T., Gonzalez-Lopez, J., Martinez-Soto, L., Olguin-Lopez, E., Resendiz-Gonzalez, P., Bandala-Sanchez, M.: Deep water rov design for the mexican oil industry. In: OCEANS'10 IEEE SYDNEY, pp. 1–6. IEEE (2010)

20. Shibuya, K., Kawai, K.: Development of a new buoyancy control device for underwater vehicles inspired by the sperm whale hypothesis. Adv. Robot. **23**(7–8), 831–846 (2009)

21. Shintake, J., Cacucciolo, V., Shea, H., Floreano, D.: Soft biomimetic fish robot made of dielectric elastomer actuators. Soft Rob. **5**(4), 466–474 (2018)

22. Trslić, P., Rossi, M., Sivčev, S., Dooly, G., Coleman, J., Omerdić, E., Toal, D.: Long term, inspection class rov deployment approach for remote monitoring and inspection. In: OCEANS 2018 MTS/IEEE Charleston, pp. 1–6. IEEE (2018)

23. Warren, D.J., Church, R.A., Cullimore, R., Johnston, L.: Rov investigations of the dkm u-166 shipwreck site to document the archaeological and biological aspects of the wreck site: final performance report. US department of commerce, national oceanic and atmospheric administration, office of ocean exploration. Silver Spring, Maryland (2004)

24. Wood, S., Allen, T., Kuhn, S., Caldwell, J.: The development of an autonomous underwater powered glider for deep-sea biological, chemical and physical oceanography. In: OCEANS 2007-Europe, pp. 1–6. IEEE (2007)

Designing and Manufacturing Low-Cost, Tendon-Driven Soft Robots

Ted Winter-Glasgow and Pablo Borja$^{(\boxtimes)}$ (iD)

University of Plymouth, Plymouth PL4 8AA, UK
stead.winter-glasgow@students.plymouth.ac.uk,
pablo.borjarosales@plymouth.ac.uk

Abstract. Soft robots have the potential to solve many problems in areas like robotics surgery, human-robot interaction, unmanned exploration tasks, and food harvesting. However, several aspects of soft robotics are still in development, e.g., efficient designs, actuation methods, modeling, and control. This paper provides some designs of tendon-driven soft robots that can be used for educational and research purposes, boosting the understanding of these systems. The proposed designs focus on five features: low manufacturing costs, accessible fabrication methods, simple assembly processes, an open-source approach, and durability.

Keywords: Soft robots · Low-cost prototyping · Tendon actuation

1 Introduction

Soft robots are robots containing compliant components in their mechanical structure [6]. In recent years, these systems have fostered a considerable interest among researchers in different disciplines because of their potential in areas such as the food industry, robotic surgery, human-robot interaction, and exploration of dangerous and unstructured environments [6,7]. Most of these systems are inspired by nature—see, for instance, the robots developed in [16,17]—where rigid behaviors are rare in animals and other living beings. Illustrative examples of this are elephant trunks, starfish bodies, chameleon tails, and octopus arms. Some specific features of soft robots are their potentially efficiency to perform a task—enabled by their compliant nature—and their capacity to adapt to unexpected environmental changes. While these properties are appealing, they have not reached their full potential in current applications, as soft robotics is still a relatively new area concerning the well-established theory and methods for rigid robots [19]. Some fundamental aspects of soft robotics that are still in development are the design, actuation approaches, modeling, and control of these systems [7,19].

This paper presents several low-cost, tendon-driven soft robot designs. The overarching aim of this work is to help bridge the current gaps in soft robotics

---

This research was funded by UKRI. Grant Reference EP/X525789/1.

M. N. Huda et al. (Eds.): TAROS 2024, LNAI 15052, pp. 254–265, 2025.
https://doi.org/10.1007/978-3-031-72062-8_23

by providing accessible prototype designs that can be used for educational and research purposes for a broader audience. To this end, the proposed designs focus on the following aspects:

(i) Low manufacturing costs. Existing prototypes often employ high-end materials, components, and fabrication processes.

(ii) Accessible fabrication methods. The proposed designs avoid highly specialized machinery and employ low-cost off-the-shelf components wherever possible.

(iii) Simple assembly processes. We minimize the necessity of uncommon specialized tools and propose an assembly process that only requires basic skills such as screwing bolts and tying knots. Additionally, this process can be carried out by one person.

(iv) An open-source approach. The repository [3] provides all the documentation to (re)produce the units. In addition, the design files can be modified to create new prototypes tailored to the user's needs.

(v) Durability and easy maintenance. The robot units resulting from the proposed designs must be durable under normal operation circumstances and withstand reasonably vigorous control actuation, where broken or worn-off pieces must be easy to replace.

The outline of this document is as follows: Section 2 is devoted to the mechanical design of the prototypes. The manufacturing process is described in Sect. 3. Section 4 illustrates the functionality of the units. Finally, Sect. 5 discusses the conclusions and some proposed future research directions.

2   Mechanical Design

This section discusses the mechanical design of the robotic systems, focusing on the five design aspects highlighted in Sect. 1.

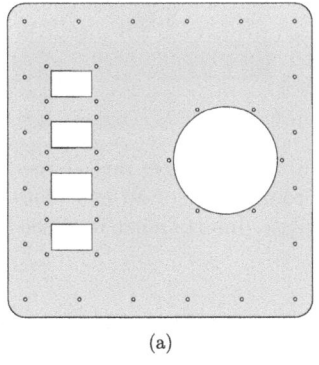
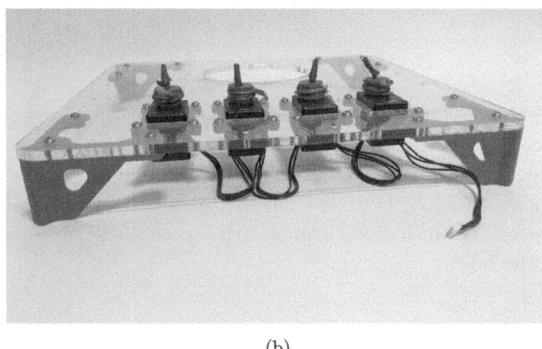

(a)   (b)

Fig. 1. Universal chassis: (a) Schematic. (b) Chassis made of acrylic with legs and servomotors.

2.1 Universal Chassis

We propose a universal chassis for all the soft robot variants—see Fig. 1(a). This structure includes four slots for servomotors and strategic holes to place legs, the compliant structure (i.e., the body of the robot), and other devices (e.g., a microcontroller).

Given its lightweight, low cost, and durability, we suggest using acrylic for the main board of the chassis. Then, by adding four 3D-printed legs, one in each corner, we elevate the board from the surface, permitting placing the servomotors and other devices (e.g., a microcontroller to operate the actuators) under the main board. Similarly, the chassis includes four slots with dimensions 31 [mm] by 48 [mm], where 3D-printed brackets can be attached with bolts to hold servomotors. The dimensions and design of such brackets depend on the chosen servomotors. We provide the design of brackets for two Dynamixel servomotor models—namely, XH430-W350-T and XL330-M288-T—for illustration purposes. Nevertheless, cheaper servomotors can be used to operate the robots. Figure 1(b) depicts an assembled set of chassis, legs, brackets, and servomotors.

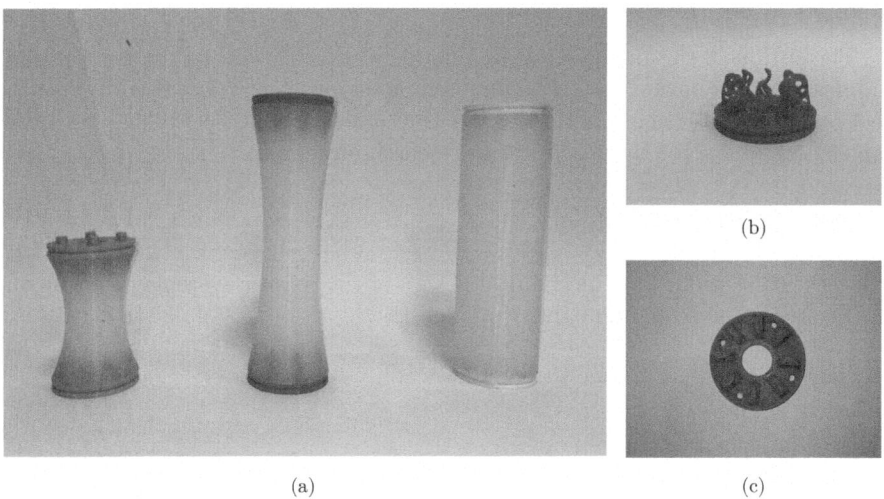

(a)   (b)

                                   (c)

Fig. 2. Compliant structures: (a) Silicone segments with different sizes and shapes. Notice the dowels at the tip of the short segment. (b) Frontal view of an end plate. Notice the fin-like structures. (c) Upper view of an end plate. Inserts must be placed in the holes.

2.2 Compliant Structure

Each robot's body consists of one or several silicone segments, endowing the proposed designs with the flexibility distinctive of soft robotic systems. Such silicone-made components can have different sizes, shapes, and stiffness, as shown in Fig. 2(a). Silicone is one of the most expensive components of the proposed designs; nevertheless, it is not used in large quantities. We suggest using Smooth-On silicones—e.g., Dragon Skin or Ecoflex—to create the segments because they are relatively easy to handle and skin-safe. Moreover, these silicones are commonly available online.

Both ends of each silicone segment consist of a 3D-printed plate with fin-like structures that help maintain the silicone attached to the 3D-printed material—see Fig. 2(b). Four metallic threaded inserts are added to each plate so the silicone structure can be bolted to the rest of the robotic system—the holes where for the metallic inserts are shown in Fig. 2(c). Then, one end of the robot's body is bolted to a 3D-printed base, while the other is bolted to a universal top plate as illustrated in Fig. 3(a). The design of the 3D-printed base depends on the number of actuators, where three or four are suggested to ensure that the robot's tip can bend in any direction—The configuration of the bases for three and four actuators is shown in Fig. 3(b). We recommend using acrylic for the universal top plate—only one acrylic sheet with dimensions 360[mm] × 360[mm] × 6[mm] is needed to manufacture a robotic unit, including the chassis and the top plate. Furthermore, the top plate design is called universal because it permits the installation of two, three, four actuators. The schematic of the top plate is provided in Fig. 3(c).

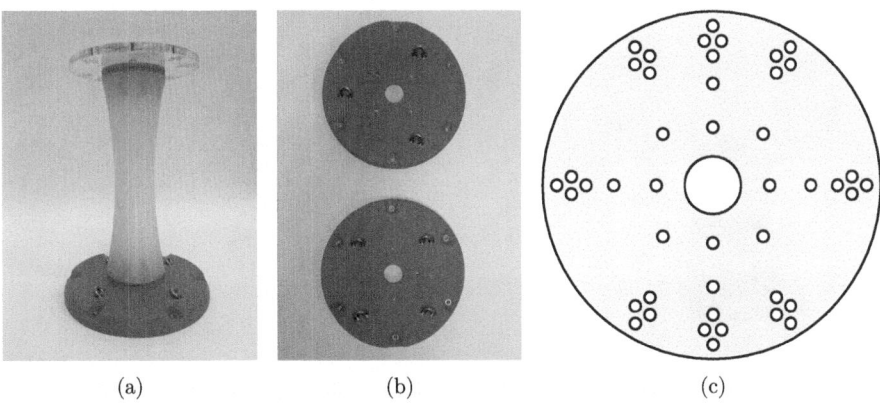

(a)   (b)   (c)

Fig. 3. Robot's body: (a) Silicone segment bolted to a top plate and a base for four tendons. (b) Upper view of bases for three and four tendons. (c) Schematic of the top plate.

2.3   Actuation

Among the different actuation approaches for soft robots, we find the so-called tendon-driven robots. As the name indicates, this actuation approach is based

on a system of tendons—e.g., wires or threads—which are often driven by servomotors or linear actuators. Then, the tendons connected to a specific part of the
robot to modify its shape by modifying the tension, i.e., using the tendons to
pull a specific part of the robot. Some examples of these systems are reported in
[1,8–10,12,13,16,17]. However, the robotic systems developed in those references
require more expensive components, more specialized tools and machinery, and
more involved mechanical designs, making their reproduction more complex.

For the actuation system using servomotors, we 3D-print capstans so that
the motors can pull, release, and spool the tendons, as shown in Fig. 4(a). The
tendons are connected from the capstans to the top plate using U-bolts placed on
the 3D-printed base plate as guides—see Fig. 4(b). Then, the end of the tendons
are secured (tied up) to the top plate using a bolt and a nut as depicted in
Fig. 4(c).

For the tendons, we recommend using fishing or kite lines, as they are cheap,
easy to find, strong, and exhibit low friction. However, other materials, e.g.,
metallic thread, can be used. Concerning the guiding system, U-bolts are also
cheap, low-friction, and off-the-shelf components. We provide the design for 3D-
printed bases with three of four tendon guides (U-bolts). Such guides are separated by 120° in the former design and by 90° in the latter. If needed, the design
can be easily adapted to two guides, in which case the robot bends only on a
plane—similar to the planar prototype developed in [16]. In all cases, six metallic threaded inserts are added to the 3D-printed base to bolt it to the universal
chassis.

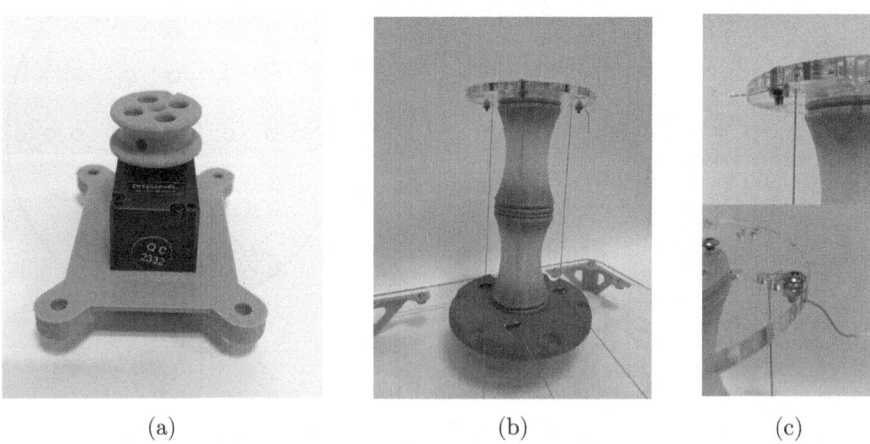

(a)   (b)   (c)

Fig. 4. Actuation system: (a) Assembly of servomotor Dynamixel XL330-M288-T, 3D-
printed bracket, and 3-printed capstan. (b) Example of the proposed actuation system
for three servomotors. Notice the U-bolts acting as guides for the tendons. (c) Example
of how to secure a tendon to the top plate. Notice how the bolt and nut hold the tendon.

3   Manufacturing Process

This section provides an overview of the robots' manufacturing process. The estimated manufacturing and assembling process is 48 h, considering no machinery needs to be outsourced. To assemble the units, 12 [mm] bolts, metallic threaded inserts, and nuts—all M4—are required. For more details on assembling the units, please consult [3].

3.1   Universal Chassis, Top Plate, and 3D-Printed Components

The CAD files corresponding to the universal chassis and top plate are provided in [3]. If acrylic is used, as suggested in Sect. 2, these components can be laser-cut in less than one hour. The top plate can be 3D printed if the application does not require vigorous actuation or heavy loads. Similarly, wood or cardboard can be used for the chassis if they are stiff enough to stand the robot body's weight without deforming.

The files for the 3D-printed components are also available in [3]. A soldering iron can be used to add the metallic inserts to the 3D-printed parts (base, plates, legs, and brackets)—we suggest dedicating a conical tip to this purpose. The U-bolts can be inserted into the base using a hammer, and the capstans need to be screwed to the motor shafts—most servomotor models include the screws. Regarding the filament specifications, we recommend using PLA filaments due to their low price and low health risks. The required time to print all the components may vary according to the 3D printer(s) and number of segments. However, a standard 3D printer requires between 25 and 35 h to produce the components.

The laser cutter and/or 3D printer can be outsourced—these services are typically available online. However, this may increase the manufacturing costs.

3.2   Silicone Segments

We stress that the metallic threaded inserts must be added to the plates before casting the silicone. The segments are manufactured by pouring silicone into 3D-printed molds with different shapes. This paper illustrates the use of hourglass-shaped and cylindrical silicone segments—see Figs. 5(a) and 5(b)—but further options are discussed in Sect. 5. The molds consist of two parts that are bolted or clamped during the casting process—see Fig. 5(a). Before pouring the silicone, both end plates of the segment must be placed between the mold parts as illustrated in Fig. 5(c). Then, the central hole of one end plate must be covered with a 3D-printed plug (design provided in [3]), and the silicone must be poured through the central hole of the other end plate. We recommend using molding clay to seal the molds to avoid silicone leakages—this material is cheap and can be reutilized. Similarly, we suggest letting the silicone cure for at least 24 h before using the unit. Because of the segments' thickness and shapes, vacuum degassing is not necessary, although most suppliers recommend it.

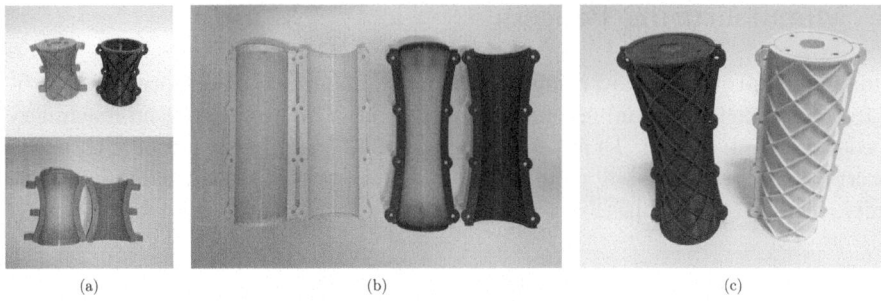

Fig. 5. Molds for silicone casting: (a) Molds for 100 [mm] hourglass-shaped segments. The blue mold is fastened with clamps, while the black mold is fastened with bolts. (b) Open molds for 300 [mm] cylindrical (white) and hourglass-shaped (grey) segments. (c) Closed (not fastened) molds with silicone segments inside. The silicone must be poured in the shown position, sealing the bottom part to avoid leakages. In addition to the bolts, it is recommended that the union of both parts of the mold be sealed with modeling clay. (Color figure online)

3.3   List of Manufactured Pieces

Table 1 provides a list of pieces to be manufactured. A list of (suggested) off-the-shelf components can be found in [3].

4   Functional Units and Suggested Operation

We explore two actuation configurations to test the functionality of the mechanical designs proposed in Sect. 2: quad actuation (four servomotors) and triad actuation (three servomotors). While the workspace of robots with triad and quad actuation is the same, the latter is more intuitive if the robots need to be manipulated manually. We also test units with three different compliant bodies: a single cylindrical segment of 300 [mm] made with Ecoflex 00-30, a single hourglass-shaped segment of 300 [mm] made with Dragon Skin 20, and two hourglass-shaped segments of 100 [mm] made with Ecoflex 00-30. The two hourglass-shaped segments are connected via 3D-printed dowels in the multi-segment case. The mentioned dowels can be observed in Figs. 2(a) and 5(a). Different combinations of actuation and compliant bodies are shown in Fig. 6. Note that any silicone column can be attached to the same base and top plate. Hence, one set consisting of a chassis, a top plate, and the actuation system can be used to test segments with different properties—e.g., stiffness or shape. We remark that hourglass-shaped silicone segments are more flexible than cylindrical ones. Therefore, using stiffer silicone is recommended for long hourglass-shaped segments.

Table 1. Manufactured components. The symbol n represents the number of segments.

Description	Quantity per robot	Suggested material
Universal chassis	1	Acrylic
Top plate	1	Acrylic
Legs	4	PLA filament
Base	1	PLA filament
Capstans	1 per motor	PLA filament
Motor brackets	1 per motor	PLA filament
End plates	$2 \times n$	PLA filament
Mold	1 (two pieces)	PLA filament
Clamps	6	PLA filament
End plug	1	PLA filament
Funnel	1 (optional)	PLA filament
Dowels	$(4 \times n) - 1$	PLA filament
Compliant segments	n	Silicone

The proposed design permits stacking multiple silicone segments. However, including more segments results in a less stable column due to the effect of gravity. If the robot's body consists of three or more hourglass-shaped segments, stiff silicone (e.g., Dragon Skin 20) is recommended. Another alternative is to opt for cylindrical segments. Note that a multi-segment robot can have different stiffness properties along the body. For instance, the segments closer to the base can be stiffer than those attached to the top plate. Combinations of shapes are also possible.

The operation of the units depends on the application. For instance, a desired trajectory can be defined, and the corresponding commands can be sent to the motors using a microcontroller. A control system is necessary if the desired trajectory is unknown or the robot needs to interact with the environment in real-time. Here, we provide two simple open-loop control alternatives to test the units: using two joysticks to control a unit with quad actuation—see Fig. 7(a)—and an inertial measurement unit (IMU) to control a robot with triad actuation—see Fig. 7(b). For the robot with quad actuation, we use Dynamixel XH430-W350-T motors with an OPEN-RB 150 board. Similarly, we employ Dynamixel XL330-M288-T servomotors with a Dynamixel motor shield and an Arduino Uno for the robot with triad actuation. Videos of these robots in action can be found in [3].

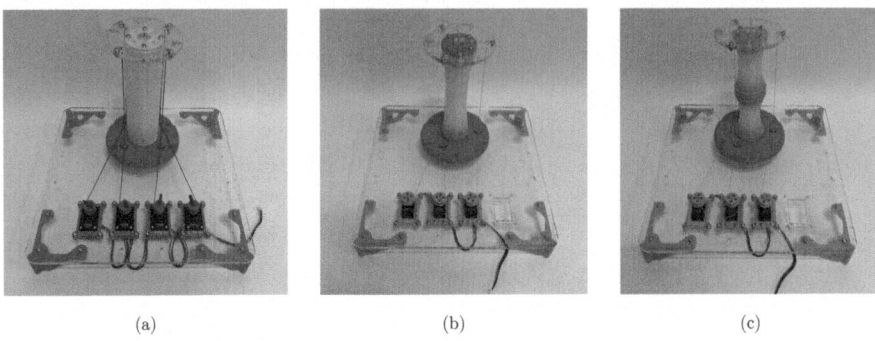

(a)　　　　　　　(b)　　　　　　　(c)

Fig. 6. Examples of robotic units: (a) Robot with quad actuation and a 300 [mm] length compliant body. The cylindrical silicone segment is made of Ecoflex 00-30. The actuators are Dynamixel XH430-W350-T. The tendons are made of 1.5 [mm] diameter Dyneema kite line. (b) Robot with triad actuation and a 300 [mm] length compliant body. The hourglass-shaped silicone segment is made of Dragon Skin 20. The actuators are Dynamixel XL330-M288-T. The tendons are made of 0.8 [mm] diameter Dyneema kite line. (c) Robotic unit with a multi-segment compliant body. Both hourglass-shaped silicone segments have a length of 100 [mm]. The actuation configuration, tendons, silicone specifications, and actuators are the same as those of the robot shown in (b).

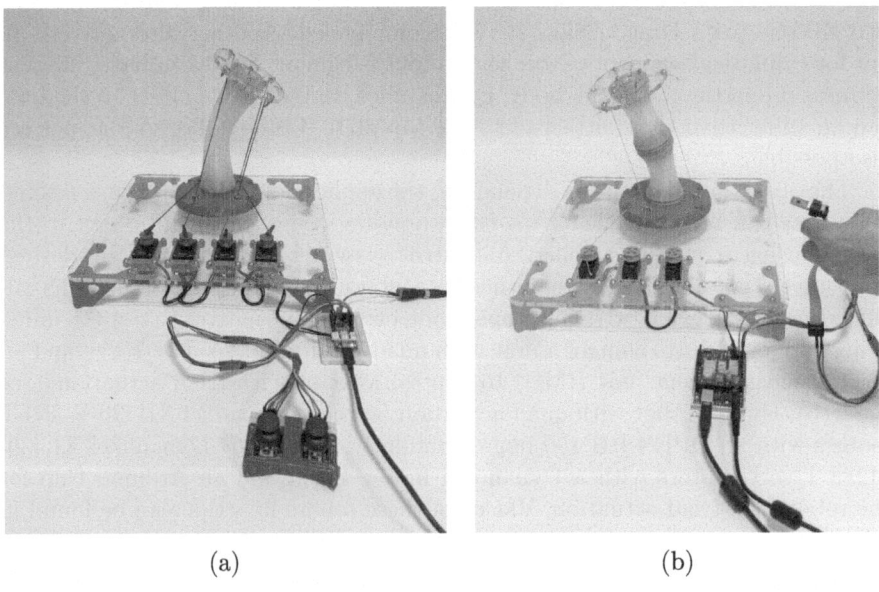

(a)　　　　　　　　　　　　(b)

Fig. 7. Operation of the robotic units: (a) Robotic unit with quad actuation operated with two joysticks. Each joystick axis controls a servomotor. (b) Robotic unit with triad actuation operated with an IMU. The servomotors are controlled by changing the IMU's roll, pitch, and yaw angles.

5   Concluding Remarks and Future Research

The proposed mechanical designs are based on low-cost materials, off-the-shelf components, standard tools, and commonly available machinery that can be out-sourced whenever unavailable. Additionally, the assembly process requires basic skills and can be carried out by a single person. From an education viewpoint, the mentioned features make the robotic units ideal for prototyping, mechanical design, robotics, and embedded systems courses. From a research perspective, the proposed units offer low-cost alternatives for control algorithm tests, vibration analysis, modeling validation, and robotics-related proof of concept. Consequently, these units are an attractive, low-cost, and easy-to-manufacture alternative to exploring the use of soft robots for research and education purposes.

5.1   Suggested Design Variants

The current design requires the following M4 bolts: three per leg, four per motor bracket, one to secure each tendon, eight to fasten the robot's body, and six to attach the base to the chassis. Therefore, 41 and 46 M4 bolts are necessary to assemble a unit with triad and quad actuation, respectively. However, the design can be modified to reduce the number of bolts. In particular, the design can be modified to use only three bolts and inserts per end plate of the robot's body and four bolts instead of six to attach the base to the chassis.

Concerning actuation, linear actuators can be used instead of servo motors. The central holes at each side of the chassis can be used to place the linear actuators. Then, pulleys can be placed on the chassis corners to guide the tendons to the U-bolts on the base—see Fig. 8(a). We remark that linear actuators are often slower but more powerful than servomotors. Another variant is to adapt the base plate to operate the motor with only two actuators. Note that the workspace of this configuration is more limited because the robot can bend only on a plane. Similarly, the design can be adapted to have two independent actuation system for a two-segment robot.

In this paper, we propose two particular shapes for the silicone segments. Nevertheless, the molds (see Fig. 5) can be easily modified to achieve further shapes, such as helix, convex, and conical. These shapes can be used to study the elastic and stability properties of various structures.

The top plate has extra holes that can be used for different purposes, e.g., to add sensors, actuators, or another structure. Some of these holes are highlighted by blue squares and rectangles in Fig. 8(b). By fastening a gripper to the top plate, the robots can be used for pick-and-place routines. Similarly, by adding the corresponding structures, the robots can transform into a humanoid neck or a ball-on-plate system.

If the robots are intended for closed-loop control applications, sensors are required. We recommend attaching a low-cost IMU to the top plate to obtain the end effector's position and velocity. If available, a motion capture system can measure the curvature and position of different points of the robot's body.

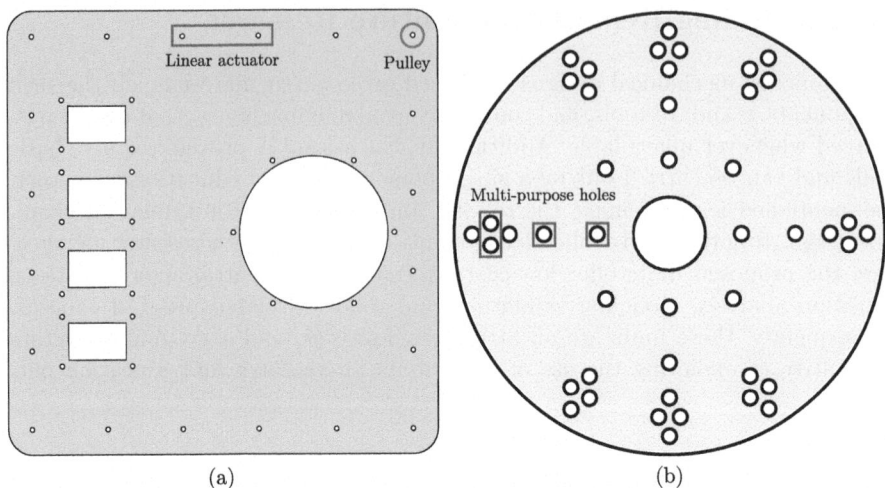

Fig. 8. Variants to use linear actuators and attach components to the top plate: (a) Linear actuators can be fastened to the chassis using the holes highlighted by the green rectangle. The current design allows placing up to four linear actuators. The potential place for the pulley is highlighted in a red circle. (b) The top plate design includes multi-purpose holes, highlighted in blue. Depending on the actuation (triad or quad), more holes are available to attach components to the robot's body. (Color figure online)

Furthermore, a solid tube can be inserted through the central holes of the end plates—see Fig. 2(c)—during the casting process to create an inner chamber in the robot's body, allowing an endoscope camera to be added to the robotic unit.

5.2 Future Research

We suggest three research directions using the robotics units:

1. Validate dynamic models for soft robots. In particular, the model apporaches reported in [5, 7, 15, 17–19].
2. Proof of concept of optimized body structures. The robot's body can be optimized for specific purposes using tools such as the ones provided in [4, 11].
3. To validate model-based control algorithms. For instance, the controllers proposed in [2, 14, 20].

Acknowledgement. This research was funded by UKRI. Grant Reference EP/X525789/1.

References

1. Amanov, E., Nguyen, T.D., Burgner-Kahrs, J.: Tendon-driven continuum robots with extensible sections–a model-based evaluation of path-following motions. Int. J. Robot. Res. **40**(1), 7–23 (2021)

2. Borja, P., Dabiri, A., Della Santina, C.: Energy-based shape regulation of soft robots with unactuated dynamics dominated by elasticity. In: 2022 IEEE 5th International Conference on Soft Robotics (RoboSoft), pp. 396–402. IEEE (2022)
3. Borja, P.: Low-cost tendon-driven SR (2024). https://github.com/PabloBorja/Low-cost-tendon-driven-SR
4. Caasenbrood, B.J., Pogromsky, A.Y., Nijmeijer, H.: Sorotoki: a Matlab toolkit for design, modeling, and control of soft robots. IEEE Access **12**, 17604–17638 (2024)
5. Camarillo, D.B., Milne, C.F., Carlson, C.R., Zinn, M.R., Salisbury, J.K.: Mechanics modeling of tendon-driven continuum manipulators. IEEE Trans. Rob. **24**(6), 1262–1273 (2008)
6. Della Santina, C., Catalano, M.G., Bicchi, A., Ang, M., Khatib, O., Siciliano, B.: Soft robots. Encycl. Robot. **489**, 1–15 (2020)
7. Della Santina, C., Duriez, C., Rus, D.: Model-based control of soft robots: a survey of the state of the art and open challenges. IEEE Control Syst. Mag. **43**(3), 30–65 (2023)
8. Deutschmann, B., Reinecke, J., Dietrich, A.: Open source tendon-driven continuum mechanism: a platform for research in soft robotics. In: 2022 IEEE 5th International Conference on Soft Robotics (RoboSoft), pp. 54–61. IEEE (2022)
9. Li, M., Kang, R., Geng, S., Guglielmino, E.: Design and control of a tendon-driven continuum robot. Trans. Inst. Meas. Control. **40**(11), 3263–3272 (2018)
10. Liu, Y., Alambeigi, F.: Effect of external and internal loads on tension loss of tendon-driven continuum manipulators. IEEE Robot. Autom. Lett. **6**(2), 1606–1613 (2021)
11. Mathew, A.T., Hmida, I.B., Armanini, C., Boyer, F., Renda, F.: SoRoSim: a matlab toolbox for hybrid rigid-soft robots based on the geometric variable-strain approach. IEEE Robot. Autom. Mag. **30**(3), 106–122 (2022)
12. Nguyen, T.D., Burgner-Kahrs, J.: A tendon-driven continuum robot with extensible sections. In: 2015 IEEE/RSJ International Conference on Intelligent Robots and Systems (IROS), pp. 2130–2135. IEEE (2015)
13. Norouzi-Ghazbi, S., Mehrkish, A., Fallah, M.M.H., Janabi-Sharifi, F.: Constrained visual predictive control of tendon-driven continuum robots. Robot. Auton. Syst. **145**, 103856 (2021)
14. Pustina, P., Borja, P., Della Santina, C., De Luca, A.: P-satl-D shape regulation of soft robots. IEEE Robot. Autom. Lett. **8**(1), 1–8 (2022)
15. Rao, P., Peyron, Q., Lilge, S., Burgner-Kahrs, J.: How to model tendon-driven continuum robots and benchmark modelling performance. Front. Robot. AI **7**, 630245 (2021)
16. Reinecke, J., Deutschmann, B., Fehrenbach, D.: A structurally flexible humanoid spine based on a tendon-driven elastic continuum. In: 2016 IEEE International Conference on Robotics and Automation (ICRA), pp. 4714–4721. IEEE (2016)
17. Renda, F., Cianchetti, M., Giorelli, M., Arienti, A., Laschi, C.: A 3D steady-state model of a tendon-driven continuum soft manipulator inspired by the octopus arm. Bioinspiration Biomimetics **7**(2), 025006 (2012)
18. Renda, F., Giorelli, M., Calisti, M., Cianchetti, M., Laschi, C.: Dynamic model of a multibending soft robot arm driven by cables. IEEE Trans. Rob. **30**(5), 1109–1122 (2014)
19. Russo, M., et al.: Continuum robots: an overview. Adv. Intell. Syst. **5**(5), 2200367 (2023)
20. Shao, X., et al.: Model-based control for soft robots with system uncertainties and input saturation. IEEE Trans. Ind. Electr. **71**, 7435–7444 (2023)

Bio-Inspired Soft Pneumatic Gripper for Agriculture Harvesting

Alex Clark, Liam Goodsell-Carpenter, Pia Buckow, Daniel Hewett, Francis White, Adil Imam⏺, Nabila Naz⏺, Breeshea Robinson⏺, Soumya K. Manna$^{(\boxtimes)}$ ⏺, and Abdullahi Ahmed⏺

Canterbury Christ Church University, Canterbury CT1 1QU, UK
{a.clark1303,l.goodsellcarpenter612,p.buckow554,d.hewett609,
f.white369,adil.imam,nabila.naz,breeshea.robinson,
soumyakanti.manna,abdullahi.ahmed}@canterbury.ac.uk

Abstract. For agricultural harvesting, handling fruits and vegetables are important processes which need a flexible but firm grip with dexterity. Due to the existing limitations of standard traditional grippers, a diverse range of soft grippers has been developed utilising elastomer, silicon, flexible plastic material etc. The manufacturing process includes mainly moulding or 3D printing and is actuated by pneumatic/ hydraulic drives, tendon drives or electrical signals. Still, there is a gap in the current market of soft grippers in terms of creating the required holding force with flexibility and bending to handle delicate foods. To overcome these existing limitations, a bio-inspired soft gripper is developed from 3D printing using Ninjaflex material. The proposed pneumatic controlled gripper is integrated with a hair-like fine structure and DIP-inspired feature to provide better grip to surfaces along with electrostatic attraction and suction. A comprehensive FEA simulation has been performed on material selection to find lower stress, longer lifespan, and higher bending capacity. Three basic tests are conducted to evaluate its performance: static friction, block force and radius of curvature. Besides, a simple grasping experiment is also conducted for holding tomatoes and coffee cups. Considering these results, the developed gripper has the potential to be acceptable in the current agricultural industry for handling objects.

Keywords: Soft gripper · Harvesting · 3D printing · Pneumatic controlled

1 Introduction

Each year, the United Kingdom's farming industry wastes 2.9 million tons of edible food during harvesting, equivalent to 6.9 billion meals [1]. One of the main reasons is the lack of affordable manual labour, leading to crops not being harvested in time. To tackle those issues, automation and robotics are diversely utilized for various operations of agriculture harvesting where different types of robotic grippers play an important part in handling different processes [2]. Mainly three operations are commonly used in autonomous agricultural robots for harvesting plants: indirect harvesting consists of

M. N. Huda et al. (Eds.): TAROS 2024, LNAI 15052, pp. 266–277, 2025.
https://doi.org/10.1007/978-3-031-72062-8_24

applying a force to the plant, such as shaking the tree, which causes the food to detach from the stem, as seen with olive harvesting [3]; direct harvesting uses twisting, pulling or bending to detach the fruit from the stem [4]; direct harvesting with cutting actuation involves cutting the stalk of the food to remove it from the rest of the plant [5]).

Several major factors are considered for designing a gripper, such as grip/suction force, gripper stroke, precision, weight, size, controller, and energy consumption. Traditional grippers can be mechanical, pneumatic, magnetic or electrical which are mainly electromechanical devices used in industrial applications for grasping and lifting, holding, rotating, and placing objects into set locations [6]. Although those types of grippers can provide linear and high force, those materials are equipped with limited adaptability in handling uneven delicate objects and limiting damage to fruit during contact. Considering the limitations of traditional grippers for handling delicate objects, soft robotic grippers appear to be a potential resource for harvesting crops [7]. The low elastic moduli present in the material enables high degrees of strain to occur when stress is applied during operation [8]. Overall, this inherent material compliance enables soft-bodied grippers to manipulate their structure to comply with their surroundings, for example, when grasping an object, it can deform around the object. The application includes harvesting strawberries [9], tomatoes [10], iceberg lettuce [11], mushrooms [12] etc. Soft robotic grippers have also been designed to handle ingredients post-harvesting [13]. Based on the structural configurations and controlling mechanism, the grippers can be divided into four categories: Pneumatic/Hydraulic network bending gripper [14] (Fig. 1(a)), Smart memory alloy driven gripper [15] (Fig. 1(b)), Elastomer-based gripper [16] (Fig. 1(c)), and Tendon-driven gripper [13] ((Fig. 1(d))).

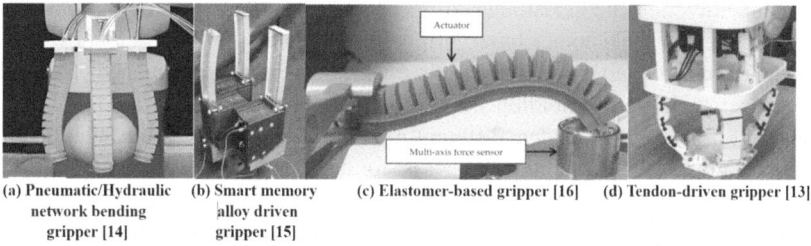

(a) Pneumatic/Hydraulic network bending gripper [14]   (b) Smart memory alloy driven gripper [15]   (c) Elastomer-based gripper [16]   (d) Tendon-driven gripper [13]

Fig. 1. Different types of soft grippers

Different types of materials and manufacturing techniques have been used in soft robotic grippers to handle food products. For example, Birrell et al., 2019 [11] utilized an elastic polymer gripper, actuated by a pneumatic signal, to make contact and grasp the iceberg lettuce while the stem was cut using an automated cutting tool. Fluid/pneumatic actuation achieves bending deformation by creating a pressure difference between the chamber's internal and external environment. The structure usually has an asymmetrical cross-section by including materials with different Young's moduli [17]. Due to the presence of a positive pressure difference within the chambers compared to the external environment, bending occurs in the direction of the material with the greater Young's modulus; the opposite is true when a negative pressure difference is present [18]. Overall, hydraulic and pneumatic-powered soft actuators are deemed as propitious solutions due to the systems being easy to fabricate having a desirable power-to-weight ratio, low material cost, and a lightweight structural body.

Soft grippers can also be actuated by a small stroke of Shape memory alloy (SMA) wire embedded within the polymeric matrix. To increase the bending angle and bending force, free-sliding SMA wires [15] are used as tendons for soft actuation by decoupling the length of the matrix and SMA while driving SMA wires. Dielectric elastomer-based actuation methods are mainly comprised of two compliant electrodes that are fitted on either side of an elastomeric membrane [19]. When an external voltage is passed through the actuator's materials, Maxwell stress occurs resulting in the flexible electrodes compressing and providing linear extension around the actuator about a fixed frame [20], however, the inclusion of a rigid reference frame can result in a reduction in overall system flexibility. Tendon actuation, inspired by tendons in the biological world, provides actuation by shortening the length of a cable or thread controlled using a servo motor, which causes the hinges to rotate [13]. To ensure the actuator returns to its original state once the cable or thread has been lengthened, elastic hinges are used. Unlike hydraulic and pneumatic-powered grippers, tendon-driven grippers are not constructed from fully compliant materials due to the presence of motors and solid components. Gafer et al., 2020 [13] designed and tested a quad-spatula gripper consisting of four tendon-driven actuators and a cradling system. The function of this soft robotic gripper was to cradle the object by providing a base for the object to rest on, which reduced the pressure exerted on objects when compared to a grasping device that used hyperplastic pneumatic actuators to apply pressure and friction to secure the product.

However, there are some limitations in the existing models such as the active length of the SMA wire is limited [15], hence the bending angle will be restricted as well. Similarly pneumatic-controlled elastomer gripper has non-linear control with a possibility of leakage. Dielectric elastomer-based grippers possess major drawbacks [19] such as low payload capacity and slow response time which affect its performance. The tendon driver gripper is limited to holding flat surfaces and needs free surroundings without obstruction for grasping. Besides, different materials with hardness for manufacturing soft grippers could be other inhibiting factors for developing soft grippers.

Considering the current limitations of the existing gripper design, the purpose of this project is to design and develop a bio-inspired soft gripper through the 3D printing technique. The proposed gripper is controlled through a pneumatic signal (0 bar to 4 bar). Upon investigation of the existing experimental testing methodologies for soft pneumatic grippers, 3D printing [21] and traditional moulding techniques [22] are largely used. After going through several testing scenarios such as blocked force and the bend angle experiments, it appears that 3D-printed soft grippers can perform better than or equal to their moulded counterparts while being more reliable and robust [23]. It also shows that 3D printing may support the design and fabrication of more complex soft grippers. Another research focused on strawberry picking [24] assessed critical metrics such as bending behaviours, deflections, holding force, friction etc. for effective fruit retrieval. The bending behaviour and deflection of soft pneumatic actuators [25] ensure the gripper achieves an adequate angle for fruit encircling. Combining these deflection criteria with force performance evaluations lays a strong foundation for comparing their effectiveness. Besides, the average static friction coefficients can also be calculated [26] to avoid any slippage between food items and gripper contact. Therefore, the functional properties of the developed soft pneumatic gripper are evaluated through three experiments: analyzing the friction strength of the material, the block force strength, and the curvature of the radius of the gripper. The test data have been used to determine the effectiveness of the gripper.

2 Design and Manufacture of the Pneumatic Soft Gripper

2.1 Soft Gripper Design

The proposed pneumatic soft gripper is integrated with two bio-inspired features: gecko's foot [27] and human finger [28], as shown in Fig. 2. The design of the gripper is inspired by the delicate, hair-like structure found on a gecko's toe pad, as depicted in Fig. 2(a). This intricate design [27] enables adhesive contact with surfaces, elucidating the impressive grip of gecko feet. Another bio-inspired feature that is also integrated into the design is the Distal Phalangeal (DIP) joint of the human finger, illustrated in Fig. 2(b). Incorporating this feature [28] could extend the range of motion of the gripper, making it easier to pick up objects. The proposed model consists of three legs (Fig. 2(d)) and each leg consists of eighteen inflatable rectangular air chambers. Compressed air can be introduced into these chambers through a pneumatic access port (Fig. 2(c)) which causes the legs to open and close to grasp objects. During the design process, several key features such as wall thickness, gaps between chambers, and cross-sectional geometry are considered. The present design is inspired by the research of Hu et.al, 2018 [16] focusing on enhancing the actuator's performance by optimizing its structure, wall thickness, chamber gaps, and cross-sectional geometry (Table 1).

Table 1. Design comparison between the standard criteria and the proposed model

Design parameter	Standard criteria [16]	Proposed Design
Wall thickness	1 mm	1 mm
Bottom layer thickness	4.5 mm	1 mm
Gaps between chambers	1 mm (higher bending angle but more stress on joints)	0.5 mm
Geometry of cross-section	Rectangular	Rectangular

In the proposed model, the bottom layer thickness and the gaps between chambers have been changed to improve its performance. The bottom layer thickness is decreased to maximize the bending angle of the gripper and to expedite prototype manufacturing. Thinner walls speed up the 3D printing process and reduce material shrinkage. Additionally, the gaps between chambers are reduced to 0.5mm to enhance flexibility.

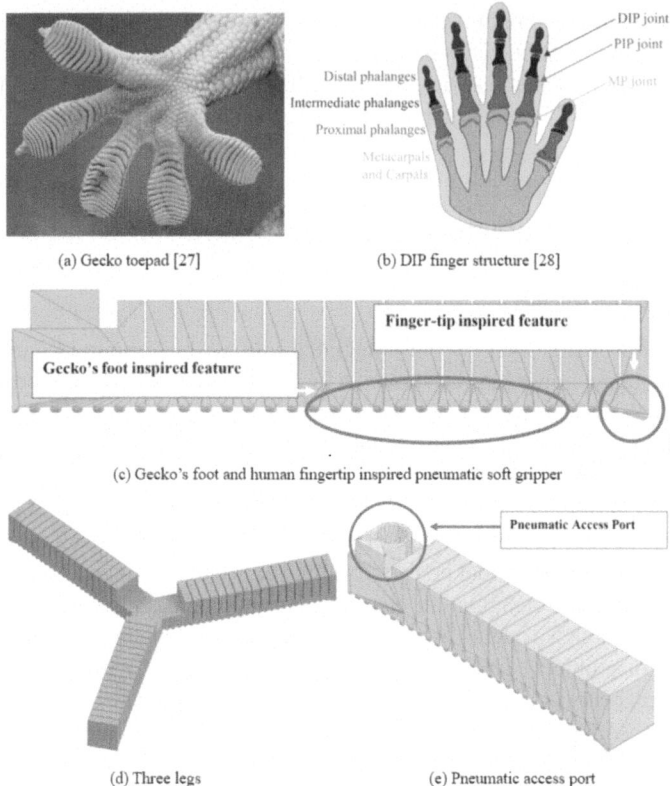

(a) Gecko toepad [27]

(b) DIP finger structure [28]

(c) Gecko's foot and human fingertip inspired pneumatic soft gripper

(d) Three legs

(e) Pneumatic access port

Fig. 2. Structural framework of the proposed gripper

2.2   Manufacturing

The gripper is manufactured using an Ultimaker S5 Fusion Deposition Model (FDM) 3D printer with 0.2mm layer height and 100% infill density. To ensure the flexibility of the gripper, NinjaFlex 85A is chosen which is a Thermoplastic polyurethane (TPU) with shore hardness 85A. NinjaFlex 85A offers high elongation and flexibility that enables the gripper to adapt to various shapes and sizes of fruits, ensuring gentle and secure handling. Good abrasion and chemical resistance of TPU ensure the gripper's durability, reliability, and longevity in a farming environment, enhancing its suitability for the agricultural industry (Fig. 3). A key aspect of FDM printing is its layer-by-layer approach. Anticipating potential air leakage due to the printing process, 3D-printed components have been coated with PVA glue and R Pro 10 silicone. This dual-coating treatment enhances airtightness.

2.3   Finite Element Analysis (FEA)

FEA is conducted on individual legs to analyse the bending behaviour of the gripper. NinjaFlex 85A is selected for simulation study with a minimum mesh element size of

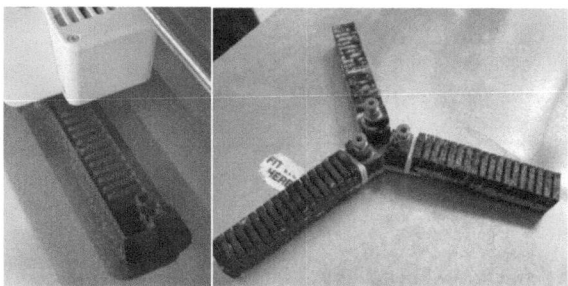

Fig. 3. Final setup of pneumatic soft gripper with channel attachments

0.2 mm and internal pressure on all internal faces of 5 bar with a fixed constraint on the vertical face of the design. The results obtained are shown in Table 2.

Table 2. FEA results for the proposed gripper

Parameters	Values
Mass (g)	23.14
Von Mises Stress (MPa) (Yield)	77.48
1st Principal Stress (MPa) (Tensile)	116.52
3rd Principal Stress (MPa) (Compressive)	34.54
Displacement (mm)	1879.18
Safety Factor	15

3   Experimental Set-Ups for Performance Analysis

3.1   Friction Testing

ES8-experiment friction tool and an inclined plane are used for the friction test (Fig. 4). The test bed was set to a horizontal level using an angle pointer and a spirit level to validate the angle pointer's figure. To identify the coefficient of friction for a test material, a known material was placed on the sledge and another on the test bed's surface. The materials mentioned above were interchanged enabling an average coefficient of friction to be calculated. The test material of the soft gripper was placed between the two materials. The test material was circular with a diameter of 20 mm. During the test, 1 g mass was added gradually until the sledge moved; this weight was recorded to calculate the friction coefficient between the sample and the materials.

The static coefficient of friction (μ) can be calculated as

$$\mu = F/N \tag{1}$$

Fig. 4. Friction test setup (TECQUIPMENT)

$$F = \mathrm{m}/1000 * 9.81 \tag{2}$$

$$N = m * g \tag{3}$$

F = Frictional force, N = Normal force, m = Mass needed to cause the sledge to move, and g is gravitational potential energy.

On average, the coefficients of friction are 0.32 and 0.32 for two types of test materials plastic and rubber which are close to the texture of fruits (Table 3). The friction force was decreased when the rubber material was utilised instead of plastic. The coefficient of friction is still low for a food product so a solution of coating the gripper in silicone may aid in increasing the coefficient of friction.

Table 3. Material friction test results

Sledge material	Surface material	Weight (including the weight hanger) to Move	Friction Force (N)	Static coefficient of Friction
Plastic	Plastic	45	0.44	0.32
Plastic	Rubber	38	0.37	0.27

3.2 Block Force Testing

For the experiment, the gripper was mounted into a clamp and controlled through a Festo pressure rig from 1 bar to 4 bar and the weight is measured (Fig. 5). The set-up would actuate towards the scales, by setting the clamp height so that the actuator contacts the scales but not applying any force on the scales. Both samples (Silicone-coated and PVA-coated) followed the same trend of increasing block force with the input pressure. The silicone-coated gripper performed better than the PVA-coated gripper with over double the force produced at 1 bar where the silicone-coated gripper produced 2.37 N and the PVA-coated gripper produced only 1.01 N (Fig. 6). The pressure is increased as the silicone-coated gripper produced 3.11 N at 4 bar whereas the PVA-coated gripper produced 2.7 N. The data plateaued around 4 bar as the silicone-coated gripper only increased by 0.01 N from 3 to 4 bar.

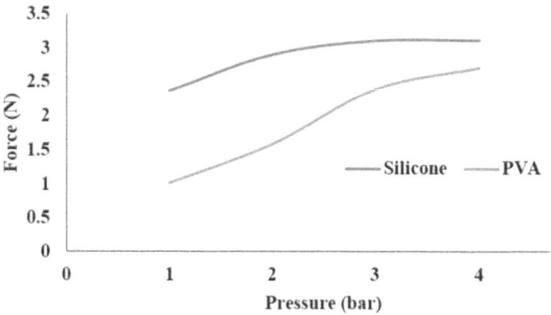

Fig. 5. Block force test results

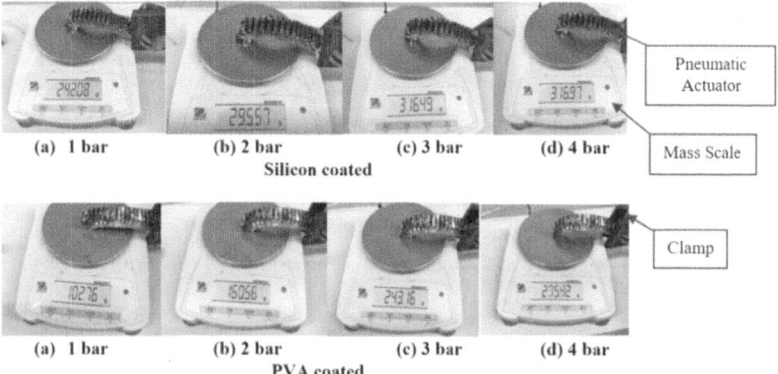

Fig. 6. Comparative analysis of block force from Silicone-coated and PVA-coated gripper

3.3 Radius of Curvature Testing

The set-up was arranged to measure the bending angle of the gripper at different pressures. The gripper was clamped and allowed to bend in its natural form by gravity. Behind the gripper, a grid paper is used for comparison of the subsequent angles. A large tripod with a camera was levelled with the gripper (Fig. 7) which reduces the risk of inaccurate measurements. To ensure consistency in results, a photo was taken before any pressure was introduced just in case the gripper did not move to its original position after flexing.

Stable cyclic loading was assessed, and the angle of curvature was recorded before and after loading up to 3 bars (Fig. 8). It has been found that the gripper returned to its original location of 36° each time.

Furthermore, the silicone coating on the gripper produced a higher change in angle than the PVA coating (Fig. 9) at lower pressure. At 1 bar, the silicone-coated gripper rotated up to 125° whereas the PVA-coated gripper rotated up to 94°. The change of curvature angle was increased for both samples as the pressure increased. The silicone-coated sample plateaued around 3 bar with an angle of 132° whereas the PVA-coated gripper reached an angle of 129° at 4 bar.

Fig. 7. Radius of curvature test setup

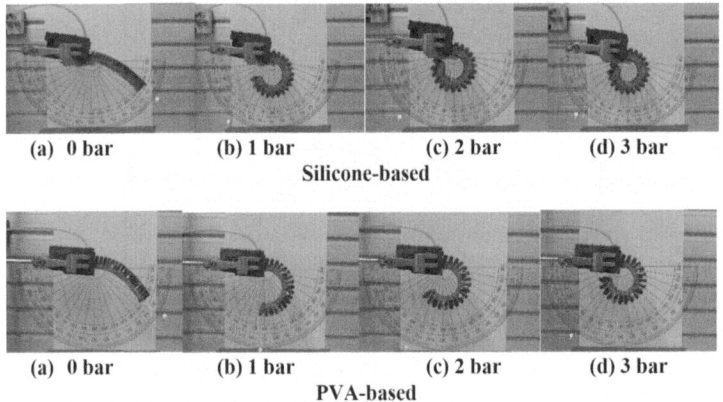

(a) 0 bar   (b) 1 bar   (c) 2 bar   (d) 3 bar

Silicone-based

(a) 0 bar   (b) 1 bar   (c) 2 bar   (d) 3 bar

PVA-based

Fig. 8. Radius of curvature for both samples at respective pressures

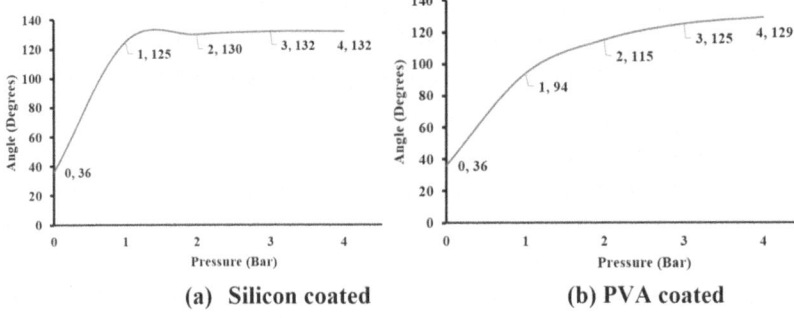

(a) Silicon coated   (b) PVA coated

Fig. 9. Comparative analysis of curvature angle from Silicon-coated and PVA-coated samples

Compared to the existing designs where the bending angle can be limited up to 90° [14] and a constant block force of 2.75 at all pressures [16], the proposed gripper shows a capacity of bending angle of up to 132° and block force of up to 3.11 N, mostly a consistent behaviour with the changing of pressure. The flexibility is enhanced due to the implementation of Gecko foot and DIP structure and leaking was prevented due to coating. For the gripping experiment, a salad tomato and coffee cup weighing 80 g and 20 g respectively were used. With both grippers producing a minimum of 100g at 1 bar, it is predicted that the gripper only needs 1 bar to hold those objects. To check this theory, a joint was manufactured to connect three of the grippers (two PVA-coated and one silicone-coated). All three grippers were connected to a Festo pressure rig where the gripper could hold the tomato or the coffee cup with 1 bar of pressure (Fig. 10).

Fig. 10. Three grippers holding a tomato and a coffee mug

4   Conclusions

The article describes the design and development of a novel bio-inspired soft pneumatic gripper. The effectiveness of its use in agricultural harvesting applications has been evaluated through several qualitative experiments. The friction test highlighted that the chosen material NinjaFlex does not have high coefficient of friction and was low with rubber-type materials. Hence, it is recommended to coat with a material having a higher coefficient of friction such as silicone. After adding the silicone and PVA coatings, the samples were tested and compared. The block force test identified that the gripper design performed substantially better when coated in silicone as opposed to PVA, especially at lower pressures. The radius of curvature test echoed similar results with the block force test where the silicone-coated sample outperformed the PVA gripper. Regarding future scope, the frictional test needs to be done with coating and an additional gripping test needs to be conducted to find the accuracy of the developed grippers. It is still not determined whether three legs will be optimal to hold a diverse range of objects.

Disclosure of Interests..   The authors acknowledge no conflict of interest.

References

1. WWF-UK.: Hidden waste: The scale and impact of food waste in primary production. WWF (2022). https://www.wwf.org.uk/our-reports/hidden-waste
2. Rose, D.C., Bhattacharya, M.: Adoption of autonomous robots in the soft fruit sector: grower perspectives in the UK. Smart Agric. Technol. **3**, 100118 (2023)

3. Nasini, L., Proietti, P.: Olive harvesting. The Extra-Virgin Olive Oil Handbook, pp. 87–105. John Wiley & Sons, Hoboken (2014)
4. Dimeas, F., Sako, D.V., Moulianitis, V.C., Aspragathos, N.A.: Design and fuzzy control of a robotic gripper for efficient strawberry harvesting. Robotica **33**(5), 1085–1098 (2015)
5. Muscato, G., Prestifilippo, M., Abbate, N., Rizzuto, I.: A prototype of an orange picking robot: past history, the new robot and experimental results. Ind. Rob. Int. J. **32**(2), 128–138 (2005)
6. Birglen, L., Schlicht, T.: A statistical review of industrial robotic grippers. Rob. Comput.-Integrat. Manuf. **49**, 88–97 (2018)
7. Mazzolai, B., et al.: Roadmap on soft robotics: multifunctionality, adaptability and growth without borders. Multifunct. Mater. **5**(3), 032001 (2022)
8. Polygerinos, P., et al.: Soft robotics: review of fluid-driven intrinsically soft devices; manufacturing, sensing, control, and applications in human-robot interaction. Adv. Eng. Mater. **19**(12), 1700016 (2017)
9. Hayashi, S., et al.: Evaluation of a strawberry-harvesting robot in a field test. Biosys. Eng. **105**(2), 160–171 (2010)
10. Yaguchi, H., Nagahama, K., Hasegawa, T., Inaba, M.: Development of an autonomous tomato harvesting robot with rotational plucking gripper. In: IEEE/RSJ International Conference on Intelligent Robots and Systems, pp. 652–657. IEEE, Daejeon (2016)
11. Birrell, S., Hughes, J., Cai, J.Y., Iida, F.: A field-tested robotic harvesting system for iceberg lettuce. J. Field Rob. **37**(2), 225–245 (2020)
12. Galley, A., Knopf, G.K., Kashkoush, M.: Pneumatic hyperelastic actuators for grasping curved organic objects. Actuators **8**(4), 76 (2019)
13. Gafer, A., Heymans, D., Prattichizzo, D., Salvietti, G.: The quad-spatula gripper: a novel soft-rigid gripper for food handling. In: 2020 3rd IEEE International Conference on Soft Robotics, pp. 39–45. IEEE, New Haven (2020)
14. Wang, Z., Or, K., Hirai, S.: A dual-mode soft gripper for food packaging. Robot. Auton. Syst. **125**, 103427 (2020)
15. Lee, J.H., Chung, Y.S., Rodrigue, H.: Long shape memory alloy tendon-based soft robotic actuators and implementation as a soft gripper. Sci. Rep. **9**(1), 11251 (2019)
16. Hu, W., Mutlu, R., Li, W., Alici, G.: A structural optimisation method for a soft pneumatic actuator. Robotics **7**(2), 24 (2018)
17. Hegde, C., Su, J., Tan, J.M.R., He, K., Chen, X., Magdassi, S.: Sensing in soft robotics. ACS Nano **17**(16), 15277–15307 (2023)
18. Walker, J., et al.: Soft robotics: A review of recent developments of pneumatic soft actuators. Actuators **9**(1), 3 (2020)
19. Youn, J.H., et al.: Dielectric elastomer actuator for soft robotics applications and challenges. Appl. Sci. **10**(2), 640 (2020)
20. Plante, J.S., Dubowsky, S.: On the performance mechanisms of dielectric elastomer actuators. Sens. Actu. A **137**(1), 96–109 (2007)
21. Tawk, C., Gillett, A., in het Panhuis, M., Spinks, G. M., Alici, G.: A 3D-printed omni-purpose soft gripper. IEEE Trans. Rob. **35**(5), 1268–1275 (2019)
22. Venter, D., Dirven, S.: Self morphing soft-robotic gripper for handling and manipulation of delicate produce in horticultural applications. In: 24th International Conference on Mechatronics and Machine Vision in Practice, pp. 1–6. IEEE, Auckland (2017)
23. Yirmibeşoğlu, O.D., Oshiro, T., Olson, G., Palmer, C., Mengüç, Y.: Evaluation of 3D printed soft robots in radiation environments and comparison with molded counterparts. Front. Rob. AI **6**, 40 (2019)
24. Ranasinghe, H.N., Kawshan, C., Himaruwan, S., Kulasekera, A.L., Dassanayake, P.: Soft pneumatic grippers for reducing fruit damage during strawberry harvesting. In: 2022 Moratuwa Engineering Research Conference, pp. 1–6. IEEE, Moratuwa (2022)

25. Alici, G., Canty, T., Mutlu, R., Hu, W., Sencadas, V.: Modeling and experimental evaluation of bending behavior of soft pneumatic actuators made of discrete actuation chambers. Soft Rob. **5**(1), 24–35 (2018)
26. Jahanbakhshi, A., Rasooli Sharabiani, V., Heidarbeigi, K., Kaveh, M., Taghinezhad, E.: Evaluation of engineering properties for waste control of tomato during harvesting and postharvesting. Food Sci. Nutr. **7**(4), 1473–1481 (2019)
27. Rasmussen, M.H., et al.: Evidence that gecko setae are coated with an ordered nanometre-thin lipid film. Biol. Let. **18**(7), 20220093 (2022)
28. Stojiljković, D., et al.: Simulation, analysis, and experimentation of the compliant finger as a part of hand-compliant mechanism development. Appl. Sci. **13**(4), 2490 (2023)

Swarms and Multi-Agent Systems

A Leader-Follower Collective Motion in Robotic Swarms

Mazen Bahaidarah[1,2], Ognjen Marjanovic[1], Fatemeh Rekabi-bana[3], and Farshad Arvin[3(✉)]

[1] Department Electrical and Electronic Engineering, University of Manchester, Manchester, UK
mazen.bahaidarah@manchester.ac.uk
[2] King Abdulaziz University, Jeddah, Saudi Arabia
[3] Department of Computer Science, Durham University, Durham, UK
farshad.arvin@durham.ac.uk

Abstract. Collective Motion (CM) is a basic phenomenon observed in nature, such as in birds, insects, and schooling fish. In swarm robotics, virtual links among the swarm members generate attractive and repulsive forces to attain self-organised CM behaviour. However, their manoeuvre in a cluttered environment can be challenging. Therefore, this study demonstrated the implementation of an Optimised Collective Motion (OCM) model with a Leader-Follower method to steer the swarm towards a desired position and improve its mobility. Two environmental configurations were designed to evaluate the swarm navigation while avoiding obstacles. Numerical simulations on MATLAB showed a significant performance, attaining 99% of the swarm rapidly converge while maintaining the formation cohesiveness.

Keywords: Swarm Robotics · Leader-Follower · Collective Motion

1 Introduction

Collective Motion (CM) is one of the fundamental phenomena that are frequently observed in organisms ranging from molecules to massive biological systems, such as insect swarms, fish schools, and bird flocks [1]. The collective coordination improves the efficiency of the species' decision-making, allowing them to carry out complex tasks [2]. This fascinating feature of nature has encouraged scientists to develop many mathematical models to address real-world problems in the field of swarm robotics, which studies the ability of multiple miniature and identical robots working together to accomplish tasks that exceed the capabilities of a single robot [3]. Adaptability, scalability, and robustness are some of the benefits of swarm robotics that make it applicable for practical applications.

In swarm robotics, CM has been effectively used to tackle a wide range of real-world problems, producing remarkable outcomes in domains such as agriculture [4], environment exploration [5] and search and rescue [6]. In order to maintain coordinated movement, robots in these swarms usually share important information, including orientation, speed, and position. Many studies inspired by natural systems have used the Leader-Follower (L-F) method to improve CM performance. For instance, the energy-efficient V-shaped formations observed in flocks of birds led by a leader to control the group movement [7]. Thus, the L-F method proves beneficial for complex tasks, including agricultural operations [8], tracking formations [9], and seabed mapping [10].

Reynolds presented the basic model for artificial flocking behaviours in an early study [11]. The model emphasised robot cohesion through attractive forces, collision avoidance through repulsive forces, and uniform movement direction of the swarm through alignment. The Self-Propelled Particles (SPP) concept, inspired by the features of statistical physics, has greatly impacted CM algorithms. The Standard Vicsek Model (SVM) [12] is a widely used implementation of SPP, mainly examining the effects of noise and particle size on the transition between disordered and ordered states. To achieve CM, the SVM modifies each particle's trajectory to point in the direction of its neighbours' average heading using a velocity alignment criterion. Subsequently, this model has inspired numerous research in swarm robotics, as demonstrated by practical experiments that use e-puck robots to illustrate CM [13]. Although the SVM is simple, it requires sharing orientation data, which poses difficulties for smaller robots with limited processing and sensory capabilities.

Several studies have explored the use of position-based models for CM, which avoids the need to exchange orientation data. In [14], Model Predictive Control (MPC) is used with a position-based CM approach to minimise communication overhead and promote quick convergence. In [15], a CM method that uses positional data to steer a robot swarm towards a stable flocking pattern while maintaining network connectivity was implemented. Furthermore, in [16], Ferrante et al. have created a complex CM model called the Active Elastic Sheet (AES) that includes elastic interactions depending on the robots' relative positions, establishing a new standard in the discipline. Subsequent studies have investigated the AES model from diverse angles. Research has been conducted to assess the impact of network architecture on CM [17], the model's performance with measurement noise [18], and the effects of external forces to steer the swarm [19]. In our recent study [20], we introduced the Optimised Collective Motion (OCM) model, which integrates virtual viscoelastic links between robots to enhance the efficiency of CM. Further investigations have addressed the CM behaviour in various challenges, such as obstacle avoidance [21], the fine-tuning of control parameters to optimise CM [22,23], and scalability of robotic systems [24]. However, the impact of geometric formations on the swarm movement was not addressed in the aforementioned studies.

Many strategies have been investigated to enhance swarm robotics performance regarding coordinated motion and preserving shape integrity. The L-F method played a key role in this context, as it categorises robots into two roles: leaders, who are in charge of establishing the main desired behaviour, and followers, who maintain a specific distance from others and perform simple tasks.

The L-F approach is applied in [25] to keep UAV robot swarms together while they move around in the environment. In [26], the L-F method was applied to indoor exploration and navigation tasks, using fiducial markers to assist in coordination. The study highlighted a notable reduction in communication delays among robots. Additionally, in [27], a cluster formation containment framework is developed for networked robots that incorporates a two-layer formation control strategy using the L-F method. This approach has proven to enhance fault tolerance, stabilise formations, and decrease the communication burden among the robots.

This study integrates the OCM model with the L-F approach to improve swarm trajectory management in complex environments. This integration is particularly suitable for practical scenarios, such as precision agriculture. The OCM model leverages viscoelastic interactions among the robots to quickly reduce fluctuations within the swarm during geometric formation changes, significantly enhancing the stability of the swarm formation. The main contributions of this paper are as follows: (i) an L-F swarm control mechanism is established to optimise the behaviours of CM, and (ii) the Monte Carlo method is utilised to evaluate the robustness and efficiency of the suggested method against measurement noise.

2   Collective Motion

2.1   Optimised Collective Motion (OCM)

A virtual viscoelastic mechanism is introduced by the OCM model [20] to reduce unwanted fluctuations among swarm members. As shown in Fig. 1, a viscoelastic connection between two robots is made up of a damping element and a spring placed in parallel with coefficient c and stiffness k, respectively. Virtual forces $\boldsymbol{F}$, which result from interactions between the robot and its neighbours $\mathcal{N}_i$, control the swarm's collective motion.

At the start of an experiment, N robots are arranged in a specific formation in a two-dimensional arena. The virtual force acting on the ith robot is the total effect of all viscoelastic links with $\mathcal{N}_i$ adjacent robots. Only the central leader robot and its immediate two neighbours are visible to each robot in any formation configuration considered in this paper. For instance, as Fig. 1 illustrates, the robots r_l, r_{f5}, and r_{f2} are the neighbours of the first follower r_{f1}. Here, r denotes the position of the robot as $[x, y]^T$. Therefore, the leader has $(N - 1)$ links, whereas each follower in the swarm has three links.

Each robot's position $\boldsymbol{x}_i$ and direction θ_i are updated at discrete intervals throughout the experiment. These updates, which are determined by the cumulative virtual force operating on the ith robot, are calculated as follows:

$$\boldsymbol{x}_i = v_0 \, \hat{n}_i + \alpha[(\boldsymbol{F} + D_r \hat{\xi}_r) \, . \, \hat{n}_i]\hat{n}_i \, , \tag{1}$$

$$\theta_i = \beta[(\boldsymbol{F} + D_r \hat{\xi}_r) \, . \, \hat{n}_i^\perp] + D_\theta \xi_\theta \, , \tag{2}$$

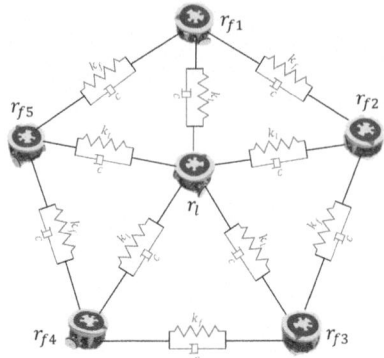

Fig. 1. A swarm of six MONA [28] robots fully connected by viscoelastic links. The swarm is formed in a pentagon shape, where the leader is at the centre, and the followers are at the corner of the pentagon.

$$\hat{n}_i = \begin{bmatrix} \cos(\theta_i) \\ \sin(\theta_i) \end{bmatrix}. \tag{3}$$

In Eqs. (1) and (2), the inverse transitional and rotational parameters, denoted by the parameters α and β, respectively, determine the linear and angular velocities of the ith robot. The symbol v_0 indicates the constant speed that every robot maintains. While $\hat{n}_i^\perp$ is a unit vector perpendicular to $\hat{n}_i$, the unit vector $\hat{n}_i$ aligns with the heading direction of robot i. Two different types of stochastic processes are integrated into the model. $D_r \hat{\xi}_r$ denotes measurement noise, which results from errors in force measurement caused by inaccurate positional sensing. Actuation disturbances introduce fluctuations into the robot's motion, mainly affecting its angular orientation θ_i, as denoted by $D_\theta \xi_\theta$. The scaling factors D_θ and D_r modify the magnitude of the corresponding sounds and disturbances.

2.2   Leader-Follower Mecahnism

In this paper, the swarm structure mainly focuses on the leader positioned at the centre of the geometry shape and the followers at the edge of a shape. The leader robot can be the same type as the followers (homogeneous swarm) or potentially a more advanced robot (heterogeneous swarm). As noted earlier, these connections are facilitated by viscoelastic links to mitigate the fluctuations among the robots. Algorithm 1 contains the pseudocode that depicts the mechanism of the suggested work. The virtual force F_i exerted on a follower robot is expressed through the following mathematical formulation:

$$F_i = \sum_{j \in \mathcal{N}_i} -k_f(\|r_{ij}\| - l_{ij}) \frac{r_{ij}}{\|r_{ij}\|} - (c\, v_{ij}), \; i \in \{2, ..., N\} \tag{4}$$

where, l_{ij} is defined as the natural length connecting robots i and j, and $\mathcal{N}_i$ identifies the neighbours of the ith robot. k_f is the spring constant and c is the damper coefficient. $\boldsymbol{r}_{ij}$ quantifies the distance between i^{th} and j^{th} robots. The variable $\boldsymbol{v}_{ij}$ captures the relative velocity difference between robot i and its neighbour j.

Furthermore, the virtual force $\boldsymbol{F}_l$ exerted by the leader robot, centrally positioned within the swarm formation, is mathematically defined as follows:

$$\boldsymbol{F}_l = w \sum_{j \in \mathcal{N}_l} -k_l(\|\boldsymbol{r}_{lj}\| - l_{lj}) \frac{\boldsymbol{r}_{lj}}{\|\boldsymbol{r}_{lj}\|} - (c\,\boldsymbol{v}_{lj}) , \tag{5}$$

where, $\mathcal{N}_l$ represents the neighbours of the leader robot and l_{lj} is the natural length between the leader and all the followers. w is a parameter that scales the value of the leader's virtual force to be in the range of the followers' forces. To update the position and heading using (1) and (2), the virtual force ($\boldsymbol{F} = \boldsymbol{F}_l$) for the leader and ($\boldsymbol{F} = \boldsymbol{F}_i$) for the follower.

This work introduces an adaptive approach to spring stiffness [29] between the robots' links while maintaining a constant damper coefficient across all links. The spring constants k_f between followers are dynamically adjusted and inversely proportional to the square root of $\boldsymbol{r}_{ij}$. In contrast, the spring constant for the leader, k_l, remains constant throughout the simulation. The equations for calculating these dynamic and fixed spring constants are presented below.

$$k_f = \frac{k}{\sqrt{\|\boldsymbol{r}_{ij}\|}}, \quad k_l = 2k. \tag{6}$$

2.3 External Forces of OCM Model

External virtual forces can be added to the swarm system to move or direct the individuals to specific paths. The exerted force can be an attractive force that attracts the swarm members to a desired location. In contrast, a repulsive force allows the swarm members to avoid colliding with obstacles or neighbour robots.

Attraction Force: This force $\boldsymbol{F}_g$ is used to attract the swarm towards a desired location and only exerted on the leader. The leader of the swarm knows the desired location and is responsible for guiding the swarm towards that location. Therefore, the goal force $\boldsymbol{F}_g$ can be calculated as:

$$\boldsymbol{F}_g = k_g \frac{\boldsymbol{r}_{lg}}{\|\boldsymbol{r}_{lg}\|} , \tag{7}$$

where r_{lg} is the Euclidean distance between the leader and the desired location. k_g is a constant that determines the strength of the attractive force.

Repulsion Force: The Artificial Potential Field (APF) approach is utilised to generate $\boldsymbol{F}_{obs}$ to endeavour the swarm away from an obstacle. This force is exerted only on the followers that are responsible for exploring the environment, detecting obstacles, and avoiding them. The force originates based on the distance between the follower robot and the obstacle. Apparently, $\boldsymbol{F}_{obs} = 0$ as the

---

Algorithm 1: The generation of target and obstacle forces in the OCM Model using the L-F approach.

Result: Force F and Alignment ψ.
Data: Initialisation of N, L, η, l_{ij} and T_{sim}.
Set robots' positions in a particular formation and random headings θ.
while $t \neq T_{sim}$ **do**
   for $i \leftarrow 1$ **to** N **do**
      if $i == 1$ **then**
         Calculate F_g using Eq. (7);
         Calculate F_l using Eq. (5);
      else
         Calculate F_{obs} using Eq. (8) ;
         Calculate F_i using Eq. (4) ;
      end
   end
   Update new position x_i using Eq. (1) ;
   Update new heading θ_i using Eq. (2);
end

---

distance between the robot and the obstacle is larger than an obstacle-sensing radius R_o. Hence, the repulsive force can be calculated as:

$$
F_{obs} = \begin{cases} \left(\dfrac{k_o}{\|r_{io}\|^2} \right) \dfrac{r_{io}}{\|r_{io}\|}, & \text{if } \|r_{io}\| \leq R_o. \\ 0, & \text{otherwise.} \end{cases} \tag{8}
$$

Here, r_{io} is the Euclidean distance between the follower and the obstacle. k_o is a constant that determines the strength of the repulsive force.

3 Simulation Setup

A two-dimensional arena with a size $L \times L$ was built on MATLAB to conduct extensive numerical simulations and investigate the feasibility of the proposed work. A swarm of $N = 15$ robots are positioned in predefined locations in a particular geometry shape and random orientations. Accordingly, the robot's status in the arena is represented as $[x, y, \theta]^T$. The radius of the swarm shape is $10\,\text{m}$, and each robot is linked with its neighbours based on the sensing radius R_s and also linked with the leader as displayed in Fig. 1. The optimised control parameters of the collective motion behaviour α, β, k, and c are defined based on our previous study [20]. Table 1 lists the rest of the simulation parameters. At the beginning of the simulation, the natural lengths denoted as l_{ij} are calculated, which are the initial distances between the robots before they move. Regarding the simulation time, each iteration in the simulation lasts for 1 ms. The parameter w is empirically selected to keep the leader's force value proportionate to the followers' forces. Furthermore, a measurement noise $\eta = 0.1$ was added in order to evaluate how resilient the CM behaviour was to noise. To evaluate the work in this paper, a Monte Carlo method that conducts 50 simulations is used. Two experiments (Case 1 and Case 2) are designed to test the performance of the L-F mechanism with the OCM model: (i) avoiding one obstacle and (ii) navigating through a maze towards a desired location.

Table 1. List of simulation parameters used in this work

Parameters	Description	Values
N	Number of robots	15
α	Inverse transitional parameter	0.0262
β	Inverse rotational parameter	0.5627
k	Spring stiffness	1.0
c	Damping coefficient	1.7503
L	The side length of the arena	140 [m]
v_0	Constant speed	0.075
R_s	Robot's sensing radius	4.5 [m]
T_{sim}	Simulation time	2300 [ms]
w	Coefficient of the leader's force	0.09
η	Noise level	0.1

3.1   Case 1: Swarm Avoiding an Obstacle

Simulations in Case 1 examine the impact of having an external repulsive force of an obstacle and an attractive force of a target position on the OCM model. An obstacle in a square shape is located between the swarm and the target position. The leader at the swarm's centre aims to guide the robots while the followers attempt to detect obstacles within the obstacle sensing range R_o.

3.2   Case 2: Swarm Navigation Through a Maze

After the swarm shows promising performance in avoiding obstacles, a maze is designed to evaluate the CM performance by increasing the complexity of the environment. Based on the current size of the arena, three walls are set to build the maze. As mentioned previously, the leader only knows the desired position to guide the swarm towards it, whereas the followers inspect for obstacles to protect and maintain the structure of the swarm in the maze.

3.3   Performance Metric

To measure the performance of the CM behaviour, the alignment ψ was defined as a primary metric. The degree of alignment ψ is crucial for evaluating the consensus of the robots' headings, which is essential for the stability of CM. It is defined as follows:

$$\psi = \frac{1}{N} \left\| \sum_{i=1}^{N} \hat{n}_i \right\|, \tag{9}$$

where $\hat{n}_i$ represents the heading direction of the ith robot. A high degree of alignment $\psi \approx 1$, when all swarm's members are oriented to the same angle. In contrast, the disordered state can be indicated by $\psi \approx 0$.

4   Results and Discussion

In this section, the results of the simulations are analysed by studying the total force and the degree of alignment of the swarm. The utilisation of the L-F method with the OCM model has notably improved by (i) keeping the swarm structure, (ii) increasing alignment convergence, and (iii) navigating towards the target position. The simulations correspond to a swarm in a ring formation and a size of $N = 15$ robots. Since the OCM model is resilient against the measurement noise, the noise intensity $\eta = 0.1$ is examined. The results of 50 simulations for both experiments (Case 1 and Case 2) are depicted as shaded plots where the middle line is the median, and the lower and upper lines represent the first and third quartiles, respectively. A sample run of Case 1 is shown in Fig. 2, and Case 2 is shown in Fig. 3.

4.1   Avoiding an Obstacle

As shown at $t = 0$ ms in Fig. 2, the robots are formed in a ring formation and randomly oriented. The swarm's members rapidly converge to the same headings and move towards the target position. At $t = 700$ ms, the followers detected the obstacle when the distance between them and the obstacle was less than R_o. Then, the swarm decided to turn right at $t = 900$ ms. It is worth mentioning that the swarm randomly selects the turning direction to avoid an obstacle based on the virtual total force. At $t = 1100$ ms, the swarm returned again to the order state and moved to the target position, as depicted at $t = 1500$ ms. The magenta line illustrates the leader's path from the starting point to the final point.

Figure 4(a) shows the force and alignment results of the 50 simulations. The rapid convergence is remarkably noted at $t = 10$ ms approximately $\psi = 0.99$. Moreover, the narrow bounds of the shaded plot prove the consistency of the simulations, taking into account the stochasticity of the swarm movement in the arena. The swarm was in the order state $\psi = 0.99$, and its force value became stable. Then, at $t = 700$ ms, there was a drop in the alignment value $\psi = 0.85$ due to avoiding the obstacle.

4.2   Navigation Through a Maze

At $t = 0$ ms in Fig. 3, the robots are formed in a ring formation and randomly oriented at the starting position. The walls and target position are specified

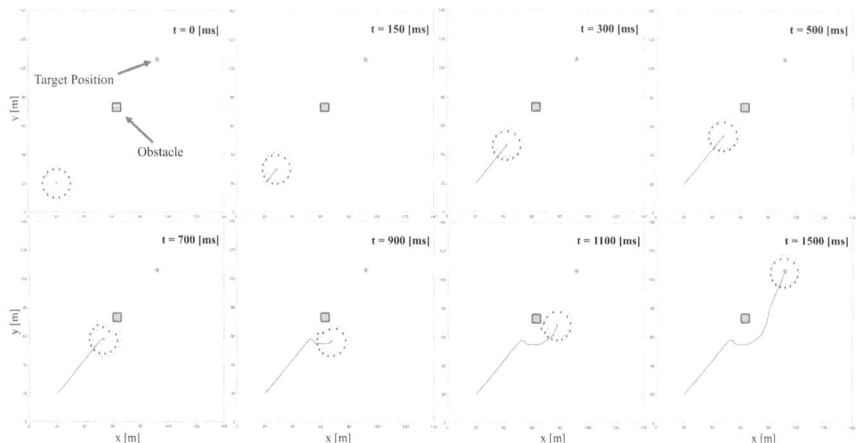

Fig. 2. Multiple snapshots of $N = 15$ robots in a ring formation utilising the OCM model to bypass an obstacle while maintaining the formation to arrive at the target location. The leader's path is shown by the magenta line.

in the maze. At $t = 150$ ms, the followers detected the first wall and turned right. Although the swarm lost a bit of cohesiveness until $t = 600$ ms, the swarm recovered its original formation at $t = 900$ ms. That's because the spring stiffness's dynamic helps the swarm to maintain its structure. At approximately $t = 1400$ ms, the swarm detected another wall and turned left to bypass it at $t = 1500$ ms. Finally, at $t = 2200$ ms, the swarm arrived at the target position while maintaining the shape stability. The magenta line shows the leader's path from the starting point throughout the simulation.

Figure 4(b) shows the force and alignment results of the 50 simulations. The slight drop in the alignment $\psi = 0.89$ happened early as the wall was close to the swarm. Then, at $t = 1600$ ms, another drop in alignment and force values happened due to the detection of the second wall. Then, at $t = 1900$ ms, the swarm became in the order state until the end of the simulation. Similarly, the narrow bounds of the shaded plot prove the consistency of the simulations. In Figs. 4(a) and 4(b), the high total force at $t = 0$ ms indicates that the robots are initially not in their desired relative position. As they adjust and align over time, the positional errors decrease, reducing the total force.

4.3   Discussion

The performance of the proposed work was extensively evaluated via sensitivity analysis to observe the degree of alignment metric. First, multiple noise values are tested $\eta \in \{0, 0.1, 0.2, 0.3\}$. As expected, results showed that as the noise increases, the performance decreases slightly. Second, different swarm shapes, such as square, pentagon, and hexagon, are tested. The performance of these shapes in both experiments (Case 1 and Case 2) is approximately similar. Third,

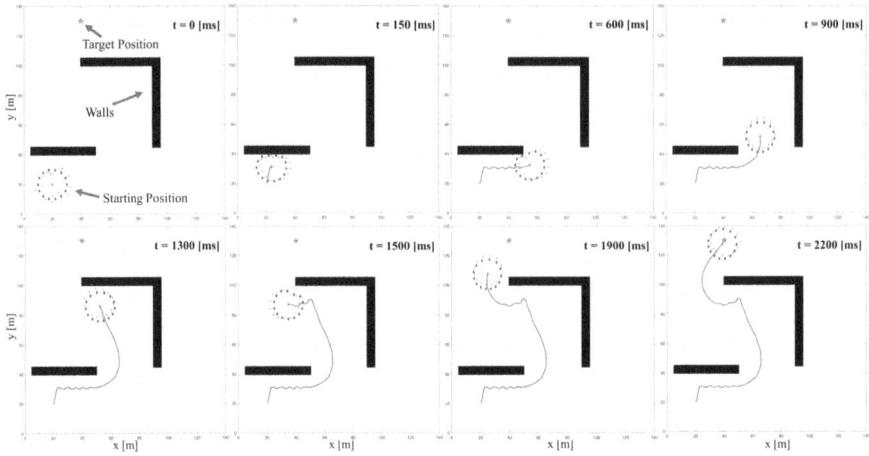

Fig. 3. Multiple snapshots of $N = 15$ robots in a ring formation (one leader and 14 followers) utilising the OCM model to navigate a maze to reach the target location by avoiding the walls. The magenta line denotes the path of the leader.

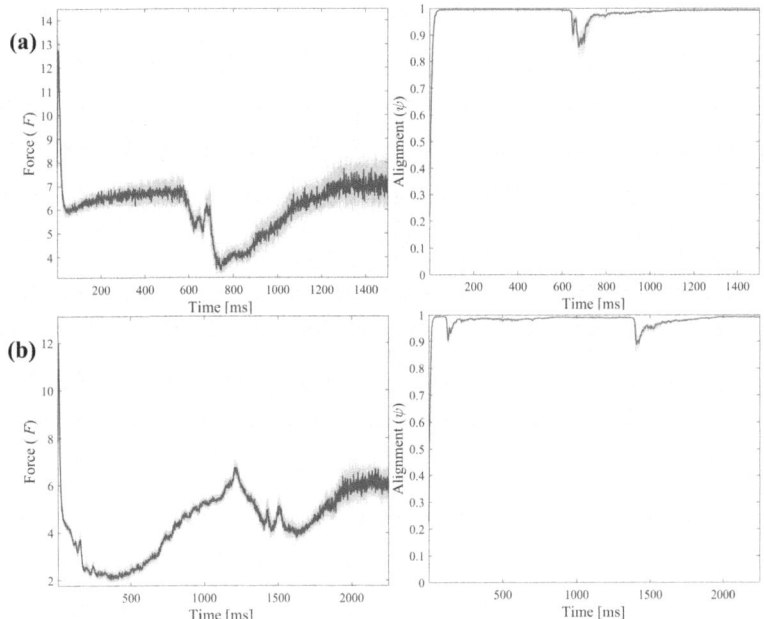

Fig. 4. Total force and degree of alignment of the swarm repeated 50 simulations in (a) Case 1: avoiding one obstacle and (b) Case 2: navigating through a maze.

the performance is tested by positioning the swarm in different locations in the arena. The swarm arrived successfully at the target position. The number of robots in this work is set to $N = 15$ robots to prove the concept. However, since we used the OCM model, which supports the scalability feature, a larger swarm size can be used with the need for a larger arena. Although the proposed work yields promising results, it is important to recognise its limitations. The swarm formation cannot compress to pass between two obstacles when the distance between them is less than the radius of the swarm shape. Moreover, if the obstacle between the swarm and the target position is a wall corner, the swarm might get stuck there.

5   Conclusion

This study demonstrated the implementation of the Optimised Collective Motion (OCM) model with the Leader-Follower (L-F) method to improve swarm trajectory. By including external forces, the swarm could navigate and avoid obstacles in the arena. Numerical simulations on MATLAB showed significant performance, attaining 99% of the swarm's members rapidly converge while maintaining the formation cohesiveness. In the future, the proposed work will be tested in real-world applications, considering a heterogeneous swarm in which the leader is a sophisticated robot and the followers are simple.

Acknowledgement. This work was partially supported by EU H2020-FET RoboRoyale (964492). Additionally, the first author wants to thank King Abdulaziz University (KAU) for supporting the work.

References

1. Camazine, S., Deneubourg, J.L., Franks, N.R., Sneyd, J., Theraula, G., Bonabeau, E.: Self-organization in biological systems. In: Self-Organization in Biological Systems. Princeton University Press (2020)
2. Hemelrijk, C.K., Hildenbrandt, H.: Schools of fish and flocks of birds: their shape and internal structure by self-organization. Interface focus **2**(6), 726–737 (2012)
3. Schranz, M., et al.: Swarm intelligence and cyber-physical systems: concepts, challenges and future trends. Swarm Evol. Comput. **60**, 100762 (2021)
4. Albani, D., IJsselmuiden, J., Haken, R., Trianni, V.: Monitoring and mapping with robot swarms for agricultural applications. In: IEEE International Conference on Advanced Video and Signal Based Surveillance, pp. 1–6 (2017)
5. Hu, J., Turgut, A.E., Lennox, B., Arvin, F.: Robust formation coordination of robot swarms with nonlinear dynamics and unknown disturbances: design and experiments. IEEE Trans. Circuits Syst. II Express Briefs **69**(1), 114–118 (2021)
6. Din, A., Jabeen, M., Zia, K., Khalid, A., Saini, D.K.: Behavior-based swarm robotic search and rescue using fuzzy controller. Comput. Electr. Eng. **70**, 53–65 (2018)

7. Mirzaeinia, A., Heppner, F., Hassanalian, M.: An analytical study on leader and follower switching in v-shaped Canada goose flocks for energy management purposes. Swarm Intell. **14**, 117–141 (2020)
8. Ju, C., Son, H.I.: Modeling and control of heterogeneous agricultural field robots based on Ramadge-Wonham theory. IEEE Robot. Autom. Lett. **5**(1), 48–55 (2019)
9. Wu, K., Hu, J., Ding, Z., Arvin, F.: Finite-time fault-tolerant formation control for distributed multi-vehicle networks with bearing measurements. IEEE Trans. Autom. Sci. Eng. **21**, 1346–1357 (2023)
10. Lee, J.H., et al.: Unmanned surface vehicle using a leader-follower swarm control algorithm. Appl. Sci. **13**(5), 3120 (2023)
11. Reynolds, C.W.: Flocks, herds and schools: a distributed behavioral model. In: Proceedings of the 14th Annual Conference on Computer Graphics and Interactive Techniques, pp. 25–34 (1987)
12. Vicsek, T., Czirók, A., Ben-Jacob, E., et al.: Novel type of phase transition in a system of self-driven particles. Phys. Rev. Lett. **75**(6), 1226 (1995)
13. Turgut, A.E., Çelikkanat, H., Gökçe, F., Şahin, E.: Self-organized flocking in mobile robot swarms. Swarm Intell. **2**, 97–120 (2008)
14. Zhan, J., Li, X.: Flocking of multi-agent systems via model predictive control based on position-only measurements. IEEE Trans. Industr. Inf. **9**(1), 377–385 (2012)
15. Su, H., Wang, X., Chen, G.: A connectivity-preserving flocking algorithm for multi-agent systems based only on position measurements. Int. J. Control **82**(7), 1334–1343 (2009)
16. Ferrante, E., Turgut, A.E., Dorigo, M., Huepe, C.: Collective motion dynamics of active solids and active crystals. New J. Phys. **15**(9), 095011 (2013)
17. Turgut, A.E., Boz, İ.C., Okay, İ.E., Ferrante, E., Huepe, C.: Interaction network effects on position-and velocity-based models of collective motion. J. Royal Soc. Interface **17**(169), 20200165 (2020)
18. Zheng, Y., Huepe, C., Han, Z.: Experimental capabilities and limitations of a position-based control algorithm for swarm robotics. Adapt. Behav. **30**(1), 19–35 (2022)
19. Raoufi, M., Turgut, A.E., Arvin, F.: Self-organized collective motion with a simulated real robot swarm. In: Annual Conference Towards Autonomous Robotic Systems, pp. 263–274 (2019)
20. Bahaidarah, M., Rekabi-Bana, F., Marjanovic, O., Arvin, F.: Swarm flocking using optimisation for a self-organised collective motion. Swarm Evol. Comput. **86**, 101491 (2024)
21. Ban, Z., Hu, J., Lennox, B., Arvin, F.: Self-organised collision-free flocking mechanism in heterogeneous robot swarms. Mob. Netw. Appl. **26**(6), 2461–2471 (2021)
22. Bahaidarah, M., Bana, F.R., Turgut, A.E., Marjanovic, O., Arvin, F.: Optimization of a self-organized collective motion in a robotic swarm. In: International Conference on Swarm Intelligence, pp. 341–349 (2022)
23. Bahaidarah, M., Marjanovic, O., Rekabi-Bana, F., Arvin, F.: An optimised robot swarm flocking with genetic algorithm. In: 2023 IEEE International Conference on Mechatronics and Automation (ICMA), pp. 1823–1828 (2023)
24. Michel, O.: Cyberbotics ltd. webots™: professional mobile robot simulation. Int. J. Adv. Robot. Syst. **1**(1), 5 (2004)
25. Quesada, W.O., et al.: Leader-follower formation for UAV robot swarm based on fuzzy logic theory. In: AI & Soft Computing, pp. 740–751 (2018)
26. Evangeliou, N., Chaikalis, D., Tsoukalas, A., Tzes, A.: Visual collaboration leader-follower UAV-formation for indoor exploration. Front. Robot. AI **8**, 777535 (2022)

27. Hu, J., Bhowmick, P., Jang, I., Arvin, F., Lanzon, A.: A decentralized cluster formation containment framework for multirobot systems. IEEE Trans. Rob. **37**(6), 1936–1955 (2021)
28. Arvin, F., Espinosa, J., Bird, B., West, A., Watson, S., Lennox, B.: Mona: an affordable open-source mobile robot for education and research. J. Intell. Robot. Syst. **94**(3), 761–775 (2019)
29. Khaldi, B., Cherif, F.: A virtual viscoelastic based aggregation model for self-organization of swarm robots system. In: Towards Autonomous Robotic Systems, pp. 202–213 (2016)

Duckling Platooning - Safety Guarantees Through Controlled Information Disclosure

James R. Heselden(✉)⬤ and Gautham P. Das⬤

Lincoln Institute of Agricultural Technology, University of Lincoln, Lincoln, Lincolnshire, UK
15591313@students.lincoln.ac.uk

Abstract. Autonomous robots are expected to increase drastically across many sectors over the coming years. It is commonplace in deployments for robots to traverse large spaces. This is often approached through mapping and explicit route planning. Under these approaches, problems such as scalability and deadlocks can cause serious issues. Alongside this, for an MRPP system to consider all of the vehicle dynamics of many robots moving in close proximity, it requires a large amount of processing. To better facilitate robots navigating along in shared spaces and in close proximity, platooning offers suitable approaches. In platooning approaches, robots are grouped together and planning is completed by the collective.

In this approach, we place safety as the primary focus. We address safety guarantees through controlled information disclosure and construct a framework around this concept to facilitate safe, robust and efficient platooning. In this, we break down the approach to platooning into four systems which each focus on removing the potential for safety risks by limiting the information each subsequent robot is allowed. This approach is tested in practical experiments with multiple robots in an indoor environment. The experiments focus on the validation of the safety systems whilst unit-testing is used to validate the scalability of the approach and its robust nature. The results showed effective platooning despite dynamic navigation problems attributed to occasional localisation drift.

Keywords: Platooning · Highway Platooning · Multi-Robot Systems

1 Introduction

The number of robots used in sectors such as agriculture, warehouse logistics, and connected autonomous vehicles is set to increase in the coming years. To manage the shared navigational spaces they will operate within, it is imperative to improve the navigational reliability of multiple robots sharing the same space and routes within. Often, a group of robots may have to move together from one

© The Author(s), under exclusive license to Springer Nature Switzerland AG 2025
M. N. Huda et al. (Eds.): TAROS 2024, LNAI 15052, pp. 294–306, 2025.
https://doi.org/10.1007/978-3-031-72062-8_26

region of an environment to another region, even when they have independent targets (in tasks and space). This is similar to a challenge in autonomous vehicles going from one region of a city to another region, where they will have to use the same route.

Although different multi-robot path planning (MRPP) algorithms exist [3] to tackle such scenarios, route assignments for individual robots or autonomous vehicles will result in inefficient executions of the paths often resulting in navigation deadlocks or large delays. Vehicle platooning is a coordination strategy which works to simplify this problem in a scalable manner through collective management. In this approach, the group of independent autonomous vehicles sharing the route are coordinated as a single entity [8], reducing the computational complexity associated with the planning. It is key to both improving navigational throughput, while also maintaining the explicit control architectures required for safety validation. A challenge in such approaches, however, is ensuring the navigational safety within the platoon formation which may have to consider the complexities of the dynamic navigational constraints and the geometrical shapes of the multi-robot formation.

In this paper, we propose a platooning framework for such robotic fleets, which focuses on transparent guaranteeing safety through validation-capable system design. Specifically, the approach utilises controlled information disclosure to limit information that if used by systems in error, could be at risk to safety concerns. The system is designed based on the navigational strategy found in biological systems such as duckling follower behaviour (Fig. 1), where each duckling will follow the duckling right in front of it. As explained later, this modular approach is highly flexible in maintaining the formation and using controlled information disclosure to enable safety.

(a) AgileX LIMO robots (b) Ducks following closely
following with 0.25m safety in river.
radius.

Fig. 1. Representation of the parity between this approach and biological platooning found in nature.

The proposed platooning framework is designed of a collection of systems which, establish the platoon, define paths, facilitate safety validation, and execute navigation. Depending on the position of the robot within the platoon, different systems are active, this is detailed in Sect. 3.

Section 4, details the process for generating new paths and how to optimise the spawning method for different scenarios. The auction system is outlined in Sect. 5, which facilitates formation of the platoon, and enables both automatic attachment from clusters and self-healing. In Sect. 6, the safety filtering system based on controlled information disclosure is detailed. Sect. 7, overviews the navigation approach and how follower-based safety works along with management of path re-interception.

We assess the developed approach in Sect. 8, in which we validate the approach through both physical experiments with AgileX LIMO robots and unit testing. Finally in Sect. 9, we summarise our findings.

2   Related Works

Multi-robot formations have been studied in literature both in multi-robot and swam robotic use cases. Platooning is one of the different multi-robot formations, with a clear linear arrangement of the involved robots.

There are many approaches to platooning, with specific areas of focus. Intersection Management [7] and Roundabout Management [2], explore approaches in specific junctions. Whilst the areas of merging at on-ramps [11,13], highway platooning [4], and lane changing [5], focus on joining, maintaining, and leaving the platoon in single-direction highways [6]. Platooning approaches in this manner can be described as leader-follower approaches, in which a leader is assigned to the formation and other members of the platoon are considered followers [12]. Further, this is broken down into three groups; static leader, where the leader is one fixed agent, dynamic leader, where the leader changes based on the formation, and virtual leader, in which the leader is a virtual construct or moving point being tracked.

In [15], the coupling between the information flow and vehicle dynamics is explored. In this, the authors detail and categorise the different information-sharing topologies based on the direction of information sharing, and how many agents down the platoon such information is shared with.

Many approaches to maintaining the liveness of highway formations, such as in [9], utilise control theory to handle the distancing between agents. These approaches validate their approaches through discrete modelling and simulated instances rather than scaled models or in-situ experiments. However, this is mostly due to the complexity of scaled deployments, and the repeatability and scalability able to be achieved through simulations [6].

As detailed in [1], for these approaches, there is limited systematic safety analysis using well-established methods, with most mention of safety analysis being on hardware redundancy. These approaches thus become less transparent and harder to validate for safety guarantees due to their complexity. Despite this trend, in approaches such as in [14], safety validation is considered within the system design, in which gap time and minimum clearance are used as validation metrics to verify the safe distance between agents within the platoon.

To address these limitations in transparent, safety considerations for platooning approaches, we are proposing a platooning framework which utilises systemic software safety with guarantees through controlled disclosure of information.

3   Positional Systems

There are three types of positions in the proposed platoon, as illustrated in Fig. 2. The first robot denoted as R_1 is labelled the HEAD, and leads the direction of the fleet through manual (where safety is a critical concern) or autonomous operation (where safety systems are validated), and the TAIL position R_n is the last robot in the platoon. All other positions between the HEAD and the TAIL are classified as the BODY and denoted by $R_{2...n-1}$.

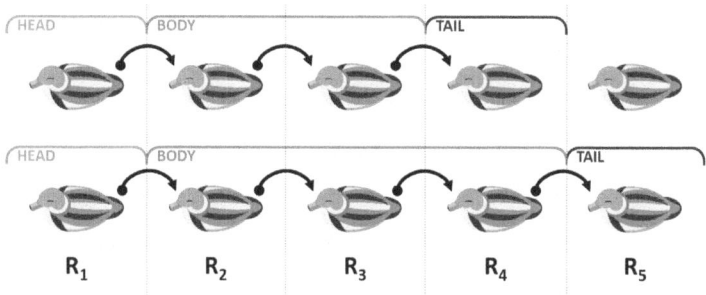

Fig. 2. Arrangement of positional groups within the platoon wherein the platoon is moving towards the left. The top diagram shows positions of a platoon of four, while the bottom shows a platoon including a fifth member. Arrows indicate the movement of information.

In this approach, a hybrid information flow topology is used in which single-step bidirectional information flow is used in the formation of the platoon. However, this is restricted to a predecessor-following topology during the platooning stage, in which information only travels to the next agent in the platoon. This is represented in Fig. 2 by the directed arrows.

To clarify terminology, the HEAD refers to R_1, however the term "leader" indicates the leader of a pair of connected agents anywhere in the platoon. For instance, R_{i+1} is the follower with respect to R_i yet the leader with respect to R_{i+2}.

Depending on the position within the platoon, different sub-systems are active. In this work, we assume that it is manually driven for explicit safety validation, however this is equally functional with the HEAD acting autonomous.

HEAD. The HEAD robot runs three sub-systems responsible for (i) spawning, (ii) filtering, and (iii) finding a follower. The spawning system (see Sect. 4) is responsible for periodically dropping new way-points based on the distance the robot has moved. The filtering system (see Sect. 6), is responsible for classifying list of way-points as safe or unsafe, based on proximity to robot. The HEAD also runs an auction (see Sect. 5), in which robots bid to become its direct follower.

BODY. Each robot in the BODY runs three sub-systems responsible for, (i) finding a follower, (ii) filtering, (iii) and path-following. The auctioneer system (see Sect. 5) is triggered on award of a follower position and functions the same as on the HEAD. Similarly, the filtering system (see Sect. 6), also works the same as on the HEAD. The path-following (see Sect. 7) works to drive the robot, following along the list of waypoints.

TAIL. As the platoon forms, each robot except the HEAD is initially the TAIL. While the behaviour of the TAIL is ultimately the same as that of each member of the BODY, it can be identified by the distinction that it will always have an open auction searching for new robots to join the platoon.

4   Way-Point Spawning

As the HEAD moves, it periodically spawns new way-points. These way-points are what dictate the path which each subsequent robot within the platoon must follow.

In this, let the position of the last spawned node be $\mathbf{p}_{\text{node}}$ and created at time t_{spawn}, and let the position of the robot be $\mathbf{p}_{\text{robot}}$. The conditions for automated spawning of an additional node include (i) exceeding a distance threshold ($r_{\text{threshold}}$) from the last node or (ii) exceeding the time threshold ($t_{\text{threshold}}$) since spawning the last node.

$$\|\mathbf{p}_{\text{node}} - \mathbf{p}_{\text{robot}}\| \geq r_{\text{threshold}} \tag{1}$$

This spawning radius is a configurable parameter with a value that depends on the size of the robot and the possible obstacles in the environment. For instance, if the robot is to move around an obstacle only 0.5 m wide, and a node is spawned on each side the path may not capture the correct size of the obstacle and its actual navigated distance around it. However, a spawning radius of 0.2 m may be appropriate for this. This is illustrated in Fig. 3.

The system can also be configured to use a time delay with minimum displacement for spawning new nodes. In this, a new way-point will be spawned after a certain timeout is reached so long as this new way-point is beyond a given radius of the previously spawned way-point. This minimum displacement, also configurable at runtime, is included to ensure the time delay does not spawn many nodes if the HEAD robot is sitting still for a while. This radius is recommended to be fairly small, at around a third of the robot's length.

$$(t_{\text{now}} - t_{\text{spawn}}) \geq t_{\text{threshold}} \quad \text{and} \quad \|\mathbf{p}_{\text{node}} - \mathbf{p}_{\text{robot}}\| \geq r_{\text{min}} \tag{2}$$

Additionally, the system offers manual node-spawning in which operators can spawn additional nodes with the press of a button on the controller. If spawning

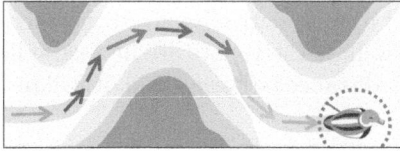 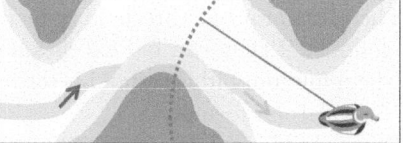

(a) Path created with small spawning radius. (b) Path created with large spawning radius.

Fig. 3. Illustration of how spawning radius can impact follower paths. Displayed with small and large radii. Arrows show the locations of each spawned way-point.

fails to initialise way-points effectively, it could cause problems under certain conditions. For this reason, the system has been designed with each spawning mechanism encapsulated. Allowing safety testing to be completed via unit tests for continuous integration.

5 Follower-Auction

To facilitate the formation of the platoon, this approach uses an auction system. It is made up of the auctioneer which is active on the TAIL, and the bidder which is active on all robots not in the platoon. The auctioneer starts by advertising it's position. This is proceeded with an open stage where any bidder can bid to become the new TAIL. Once the completion criteria has been met, bidders are notified of which robot won the position. The winner then connects to the auctioneer as the new TAIL, and triggers its own auction for the chain to continue.

Let the set of robots be R. Each robot r_i bids to be a follower based on the path length $d(r_i, A)$ to the auctioneer A. The robot r_i with the minimum path length is selected as the next follower:

$$r_{\text{closest}} = \arg \min_{r_i \in R} d(r_i, L) \tag{3}$$

In this manner, the platoon formation enables flexible management in a way which is transparent, explainable, and modular. To enable utility in different use-cases, options for constrained customisation have also been included. For the auctioneer, this customisation is available in the auction time limit, total bids cap, and minimum bid threshold.

The auction time limit is used to ensure the auction is completed within a certain amount of time. The total time of platoon formation with this can be calculated as $t_{\text{total}} \leq t_{\text{limit}} \times n$. To reduce this time, additional completion criteria are included.

The total bids cap completes the auction when a given number of total bids have been received. In situations where many agents are distributed sparsely, this is useful for limiting the time of the auction. If followers cannot be found due to high-density clustering, this may be increased.

The minimum bid threshold is included to automatically award the auction to a follower if they can bid exceptionally well. If the robots are in a close line, and the minimum bid threshold is set to the distance between them, it will result in a fast platoon formation.

On the bidder side of the auction, customisation is available in the form of the bid calculation. By default, the bid uses the path length, that being the length of the shortest navigable path from the bidder's pose, to the auctioneer's pose. As the bidding is encapsulated for unit-test safety validation, this can also be replaced with alternate bid value functions making more consideration of situational parameters. For instance, when the robots are constrained by computing power, the Euclidean distance between the pose of the TAIL and the robot may be considered as the bid value for low computational costs.

Automatic Attachment from Cluster. As illustrated in Fig. 4, through the utilisation of the path length for bidding, formations can be made with the initial positions of each robot being clustered together. Note how in (a) the bidder of 7, despite being closer to the TAIL than the bidder of 5, scores less as its path requires a turning radius.

Some robots may not initially find a path to the leader due to obstacles or other constraints. This can be denoted as $d(r_i, L) = \infty$. However, as illustrated in Fig. 4, when using this approach agents unable to attach will get new opportunities as the formation forms: $\lim_{t \to \infty} d(r_i, L) < \infty$

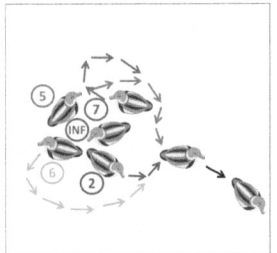 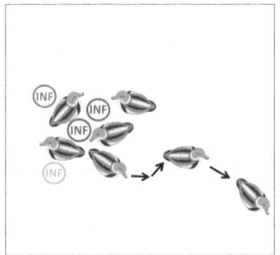 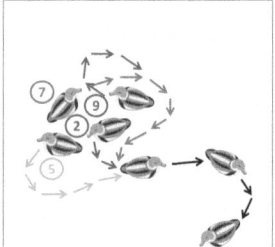

(a) Five ducks bid to follow.   (b) No ducks can make bids.   (c) Blocked in duck can now bid.

Fig. 4. Illustration showing how the formation can be formed when the robots are initially in a cluster. Temporal changes in the cluster formation through the platoon moving, new ducks can make bids. Note how the duck in the centre, despite initially being trapped is able to join the platoon as it starts moving.

Self-healing Platoon. As robots within the platoon arrive at their destinations, a functional requirement is the ability to detach. In this approach, it is considered that the HEAD is unable to autonomously detach as it is being manually controlled. For every other member of the platoon, its follower (if it has one) must re-open the auction to search for a new leader, and its leader must begin searching for a new follower. This results in self-healing of the gap in the platoon and is illustrated in 5.

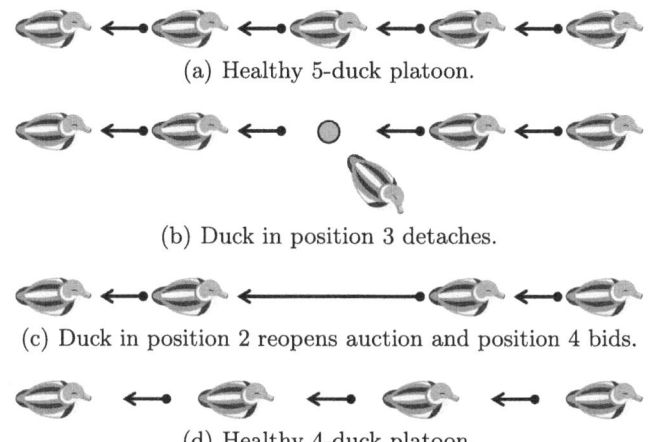

(a) Healthy 5-duck platoon.

(b) Duck in position 3 detaches.

(c) Duck in position 2 reopens auction and position 4 bids.

(d) Healthy 4-duck platoon.

Fig. 5. Illustration showing how the auction system can heal the platoon when agents detach.

6   Node Safety Filter

To ensure the safety of all robots in the platoon, a node safety filter is applied to each robot. This works to mark each node as safe or unsafe based on a given safety radius, before sending the list of safe nodes to the next robot. This approach ensures that each subsequent robot only has knowledge of the nodes it is capable of navigating to safely.

In this, let r_i denote the i-th robot in the platoon R. Also let W_1 denote the list of waypoints generated by the r_1 of the platoon: $W = \{w_1, w_2, \ldots, w_n\}$.

The $i + 1$-th robot (r_{i+1}) receives a filtered list of nodes from r_i under (4), where $d(w_j, r_i)$ is the distance between way-point w_j and r_i:

$$W_i = \{w_j \in W_{i-1} \mid d(w_j, r_i) > r_{\text{radius}}\} \tag{4}$$

Through this restricting of the full list of nodes, the system applies concepts of controlled information disclosure. In practice, this works so each robot has only a subset of the full list of nodes as shown in Fig. 6 where each subsequent robot has only knowledge of the nodes deemed safe by its leader.

Safety Radius. The default calculation for determining this safety is a straight-forward Euclidean distance from the robot as the Safety Radius. To improve flexibility, the implementation is designed so this can be replaced with a customised

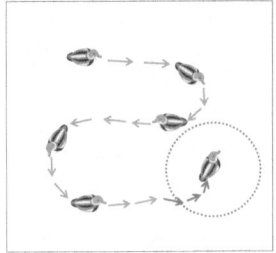 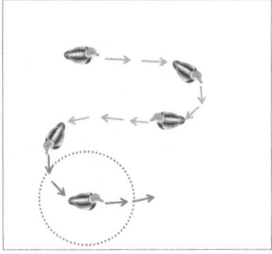 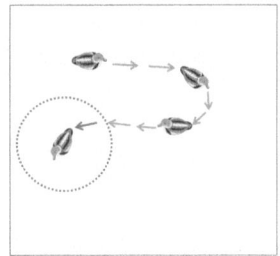

(a) HEAD filters 4 as un- (b) Filters 5 as unsafe and (c) Filters 2 as unsafe and
safe and sends 16.   sends 11.   sends 9.

Fig. 6. How controlled information disclosure works to reduce the length of the list for
each subsequent duck by removing unsafe nodes at each stage.

version such as using the path distance or some combination. The impact of
this safety radius can be seen in Fig. 6 where around the leader in each frame,
is a dotted ring, with the unsafe nodes marked in red. For additional safety
validation, this section of the system is self-enclosed for use with unit-tests and
continuous integration.

7   Follower Navigation

The intended use for this approach is for the HEAD of the platoon to be driven
with safety checking and navigational avoidance either by a manual operator or
a higher-specification safety system. This means the safety validation over the
route is completed explicitly for the HEAD without additional steps, and the
safety validation for the platoon is completed implicitly. However, this requires
followers must execute correct follower-navigation. To ensure the safety of the
fleet, the navigation system utilises a navigate-through poses system to ensure
the robot paths through each pose along the route.

Path Re-Interception. As the HEAD is controlled manually, it is key that
operators are trained to not re-intercept the path of the platoon. However, consideration has also been made to such a situation regardless.

Whilst the node filtering classifies nodes as unsafe if robots ahead in the
platoon are within their proximity. Figure 7, illustrates a situation wherein the
HEAD has reentered the path of the platoon. In this situation, the filtering
marks the additional nodes as unsafe. When the next robot to enter this unsafe
portion receives the list, its path-following would jump straight to the other side,
ignoring the safety issue and potential for collision.

To compensate for this, the follower runs its safety filtering on the path. In
this, it builds its own path only using the connecting nodes up until the point
where the re-interception occurs and ensuring safe navigation.

(a) TAIL has route of 9 way-points. (b) TAIL's route is intercepted by the HEAD. (c) TAIL's route is opened as the HEAD moves past.

Fig. 7. Demonstration of instances of path re-interception in which the duck at the top constructs its permissible route as the safe nodes up to the next obstruction.

8   Experiments

In our experiments, we aimed to validate the safety guarantees of the approach. The experiments focused on a physical assessment of the approach deployed on real robots in an indoor setting. In this, a series of manually piloted trials with two AgileX LIMO robots platooning along a corridor. The spawning radius was set to 0.25 m whilst the safety radius was also set to 0.25 m.

The robots are each equipped with a Jetson Nano. They run ROS2 through a docker container which communicates with the base driver for motion control. The robots communicate over MQTT through a VPN. They both run AMCL for localisation over a map generated using SLAM and tailored by hand.

In the assessment, we looked for risk situations, in which the robots were behaving in a categorically unsafe manner. Our key metric for observation was in their separation distance, and in the manner the robots navigate around one another for extreme conditions of unsafe driving by the operator.

In this experiment, we found the robots were able to behave within acceptable parameters. The follower did drift occasionally beyond the path laid out by the leader, however, this was limited in frequency over around an hour of platooning. Of these instances, around 80% were caused by localisation drift, and occurred at the same location in the navigational space, while the remaining times were caused by the driver entering the path of the follower and causing an obstruction or the path of the leader being too tight to spawn an adequate number of nodes.

Whilst this localisation drift is a valid concern for the deployment of this approach, it is a responsibility for the robot's navigation autonomy rather than the platooning system. This will be factored into further experimentation and can be removed entirely with the use of external localisation such as in [10].

Figure 8 shows a sample of the platooning in action during the practical experiments. In this series of images, the robots can be seen platooning around an open space. The arrows indicate the positions of the nodes as the leader was driven, whilst the path shows its navigation around the space, i.e. the path the follower would ideally follow through. Notable is in the fifth frame, in which the follower is outside the path, this is due to a localisation error.

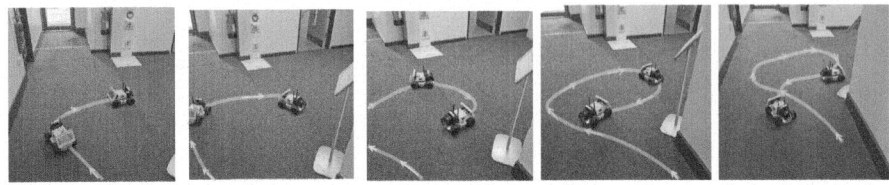

Fig. 8. Demonstration of path following and localisation drift problems.

In Fig. 9, the path taken by the robots is a long straight with slight meandering, followed by a tight 180° turn and back from where it came. Notable in this instance, is a unique situation wherein the path outlined by the leader was cut tightly by the follower. This was caused by the radius of the turn being too tight to adequately spawn an additional node towards the tip, a situation detailed in Fig. 3. Also notable is how in frame 3 the follower stops and waits at the third node. Only when the leader begins to move from the fourth node does the leader begin moving. The follower effectively maintains the safety distance from the leader as it comes into its proximity by not proceeding to the next node.

Fig. 9. Demonstration of importance of safety distances and spawning distances.

9   Conclusion

In this work, we have explored the viability of controlled information disclosure as a means for verifiable fleet safety in multi-robot platooning. Our approach utilises, to achieve this, customisable auction systems to enable fleet formation from clusters, and safety filtering on the leader through controlled information disclosure, and on the follower through re-interception filtering to ensure the platoon is incapable of traversing shared unsafe spaces.

Practical experimentation with two physical AgileX LIMO robots has been used to demonstrate the utility of controlled information disclosure for safety management and the feasibility of the follower auction. As the approach is modular the platoon can be extended by repeating the same systems, this has been further validated through the use of unit-testing. In this, additional virtual instances of robots can be deployed and used for safety validation through continuous integration.

Further work in this approach will include larger experimentation in both simulation and with over 15 physical robots. It will also aim to further functionality through the use of alternative formation patterns and more dynamic supporting safety systems.

Acknowledgements. This work was supported by AgriFoRwArdS CDT, under the Engineering and Physical Sciences Research Council [EP/S023917/1].

References

1. Axelsson, J.: Safety in vehicle platooning: a systematic literature review. IEEE Trans. Intell. Transp. Syst. **18**(5), 1033–1045 (2016)
2. Azimi, R., Bhatia, G., Rajkumar, R., Mudalige, P.: V2V-intersection management at roundabouts. SAE Int. J. Passenger Cars-Mech. Syst. **6**(2), 681–690 (2013)
3. Heselden, J.R., Das, G.P.: Heuristics and rescheduling in prioritised multi-robot path planning: a literature review. Machines **11**(11), 1033 (2023)
4. Hu, M., Zhao, X., Hui, F., Tian, B., Xu, Z., Zhang, X.: Modeling and analysis on minimum safe distance for platooning vehicles based on field test of communication delay. J. Adv. Transp. **2021**(1), 5543114 (2021)
5. Karbalaieali, S., Osman, O.A., Ishak, S.: A dynamic adaptive algorithm for merging into platoons in connected automated environments. IEEE Trans. Intell. Transp. Syst. **21**(10), 4111–4122 (2019)
6. Khayatian, M., et al.: A survey on intersection management of connected autonomous vehicles. ACM Trans. Cyber Phys. Syst. **4**(4), 1–27 (2020)
7. Lin, S.H., Ho, T.Y.: Autonomous vehicle routing in multiple intersections. In: Proceedings of the 24th Asia and South Pacific Design Automation Conference, pp. 585–590 (2019)
8. Mariani, S., Cabri, G., Zambonelli, F.: Coordination of autonomous vehicles: taxonomy and survey. ACM Comput. Surv. (CSUR) **54**(1), 1–33 (2021)
9. Moode, S.N., Soriguera, F.: Microscopic modeling of connected autonomous vehicles platooning: stability and safety analysis. Transp. Res. Procedia **71**, 260–267 (2023)
10. Nitsche, M., Krajník, T., Čížek, P., Mejail, M., Duckett, T.: WhyCon: an efficient, marker-based localization system. In: IROS Workshop on Open Source Aerial Robotics (2015)
11. Rios-Torres, J., Malikopoulos, A.A.: A survey on the coordination of connected and automated vehicles at intersections and merging at highway on-ramps. IEEE Trans. Intell. Transp. Syst. **18**(5), 1066–1077 (2016)
12. Soni, A., Hu, H.: Formation control for a fleet of autonomous ground vehicles: a survey. Robotics **7**(4), 67 (2018)

13. Xue, Y., Ding, C., Yu, B., Wang, W.: A platoon-based hierarchical merging control for on-ramp vehicles under connected environment. IEEE Trans. Intell. Transp. Syst. **23**(11), 21821–21832 (2022)
14. Yoo, Y., Chen, T., Jang, K., Kim, I.: Impact on platooning of autonomous vehicles using macro to micro simulations. Procedia Comput. Sci. **220**, 235–242 (2023)
15. Zheng, Y., Li, S.E., Wang, J., Cao, D., Li, K.: Stability and scalability of homogeneous vehicular platoon: study on the influence of information flow topologies. IEEE Trans. Intell. Transp. Syst. **17**(1), 14–26 (2015)

Robust Mitigation Strategy for Misleading Pheromone Trails in Foraging Robot Swarms

Ryan Luna🄳 and Qi Lu$^{(\boxtimes)}$🄳

Department of Computer Science, The University of Texas Rio Grande Valley,
Edinburg, TX 78539, USA
{ryan.luna01,qi.lu}@utrgv.edu

Abstract. This study advances the security of swarm robotics by examining the resilience of stigmergic communication in foraging robot swarms against deceptive strategies. We specifically investigate the swarm's vulnerability to attacks via misleading pheromone trails laid by detractor robots, which significantly hinder foraging performance. Through simulations, we evaluated the adverse effects of such attacks on resource collection and forager capture rates, highlighting a notable decline as the percentage of detractors increases. To counter these threats, we implement a robust defense mechanism utilizing DBSCAN for density-based clustering of pheromone trails, complemented by a cluster grouping method that effectively isolates batches of detractors early in the simulation. This approach incorporates an adaptive timing mechanism to discern and counteract misleading trails, substantially mitigating forager captures and enhancing swarm foraging efficiency. Furthermore, we extend our analysis by introducing obstacles in the simulation environment to test the defense's robustness under varied and complex conditions. These experiments demonstrate that our defense strategy remains effective, maintaining operational stability even when faced with additional environmental challenges. This research not only underscores critical security vulnerabilities in pheromone-based foraging algorithms but also sets the foundation for developing more secure and resilient swarm robotics systems for real-world applications where robustness against both deceptive strategies and environmental complexities is essential.

Keywords: Swarm Robotics · Intrusion Detection · Autonomous Robots

1 Introduction

Drawing inspiration from natural systems like ants, termites, and birds, current research in swarm robotics spans behaviors such as self-organization [13,19],

This work is supported by the GAANN program (P200A210144 - 22) from the U.S. Department of Education. The authors would also like to acknowledge the partial funding provided by the CREST MECIS and the MSI programs through NSF Award No. 2112650 and NSF Award No. 2318682, and DHS Award No. 21STSLA00009-01-00.

M. N. Huda et al. (Eds.): TAROS 2024, LNAI 15052, pp. 307–319, 2025.
https://doi.org/10.1007/978-3-031-72062-8_27

aggregation [3,4], object sorting [31,32], and foraging [14,15,18,21,25]. However, these tasks are all optimized for benign environments, assuming no malicious activity. This paper aims to address the less explored aspects of security and reliability [16], particularly focusing on the vulnerabilities of pheromone-based foraging robot swarms to security threats like the ant mill phenomenon [8,9] and intrusion attacks that exploit virtual pheromone trails to deceive robots [5,27]. These vulnerabilities are crucial as they impact real-world applications in environmental monitoring, search and rescue, disaster recovery, and military operations, demanding more robust and reliable swarm robotics technologies [2,20,22,36]. Our research dives into stigmergic communication, where the environment is a medium for interaction, enhancing collective adaptability but also exposing inherent vulnerabilities [5,27,29,35]. We investigate the impact of detractors-hijacked robots laying misleading pheromone trails leading to forager capture and removal, which hampers foraging efficiency. Our proposed defense employs an enhanced Density-Based Spatial Clustering of Applications with Noise (DBSCAN) alongside an isolation strategy to effectively remove multiple detractors from the environment at once [12]. This novel application of DBSCAN not only mitigates specific vulnerabilities but also broadens the security protocols in swarm robotics and stigmergic communication.

The paper proceeds to detail past work on swarm robotics vulnerabilities (Sect. 2), the central-placed foraging model (Sect. 3), the methodology for addressing pheromone trail exploits (Sect. 4), our experimental setup (Sect. 5), and the evaluation of our experiments (Sect. 6). We conclude with a summary of our contributions and directions for future work (Sect. 7).

2   Related Work

Insect colonies utilize pheromone-based communication for coordination, which, while effective, has vulnerabilities that various organisms exploit, leading to evolutionary arms races [1,6,7,11,17,30,35]. A notable example is the "ant mill" or "army ant syndrome," where ants are trapped in a deadly loop [8]. Despite evolved defensive strategies, ants remain susceptible due to their inability to override instinctual responses to pheromone trails, highlighting inherent fragilities in stigmergic communication [26,28,33].

Similarly, virtual pheromone trails in robotic swarms facilitate efficient coordination but introduce vulnerabilities exploitable through signal manipulation [14,15,18,21,23,25]. Emergent behaviors like the ant mill can undermine foraging algorithms, suggesting that current systems are prone to dysfunction and manipulation by malicious entities, known as "detractors," who mislead robots with indistinguishable false trails [5,9,27].

This paper builds on previous work by exploring how malicious robots may disrupt stigmergic communication, assessing the broader implications for swarm robotics security and proposing strategies to mitigate these risks. This focus addresses the gap in research where the resilience of pheromone trails and their susceptibility to interference in artificial systems have been minimally explored.

3   Central-Placed Foraging

Central-placed foraging (CPF) is a well-established model where a nest acts as a central collection zone within the search space, and robots exhibit four major behavioral states [14]:

Departing: Robots leave the center, searching randomly or returning to previously successful locations using site fidelity or pheromone waypoints. Upon reaching a target, robots transition to *Searching*.

Searching: Robots search using random walk [10]. Successful searches lead to *Surveying*, while unsuccessful ones may end with a return to the center, determined by a probability p_r.

Surveying: Robots assess local resource density within a search radius r_{search} (Table 1), recording the count k of resources.

Returning: Robots carry resources back to the center. The density of resources λ_{lp} is taken into account to generate a probability (see Eq. 1) of laying a new pheromone waypoint. The robots then restart the cycle from *Departing*.

Behavioral parameters like λ_{sf} (site fidelity rate) and λ_{lp} (pheromone waypoint rate) are managed by a Poisson Cumulative Distribution Function (CDF) [15]:

$$\text{POIS}(k, \lambda) = e^{-\lambda} \sum_{i=0}^{[k]} \frac{\lambda^i}{i!} \tag{1}$$

Decisions to act are triggered if $\text{POIS}(c, \lambda_{sf}) > \mathcal{U}(0, 1)$ or $\text{POIS}(c, \lambda_{lp}) > \mathcal{U}(0, 1)$, leading to the use of site fidelity or the laying of a new pheromone waypoint.

The pheromone waypoints have an initial strength of 1 and decay exponentially over time by $w = e^{-t\lambda_{pd}}$. Waypoints below a threshold γ are removed, maintaining only the most relevant paths.

4   Methodology

Attack Strategy. We introduce a scenario where malicious agents, known as detractors, disrupt the foraging process by distributing misleading pheromone trails. These trails lead benign foragers to designated capture sites, effectively removing them from the foraging task as shown in Fig. 1. Detractors mimic benign foragers to avoid detection, but instead of reporting actual resource locations, they report coordinates within the vicinity of capture sites. This deceptive strategy is inspired by historical hunting techniques such as buffalo jumping, where the prey is pushed to fatal traps [34]. The attack's effectiveness is measured by tracking the number of foragers misled to capture sites, and the rate of their capture, providing insights into the impact of these tactics on overall swarm efficiency.

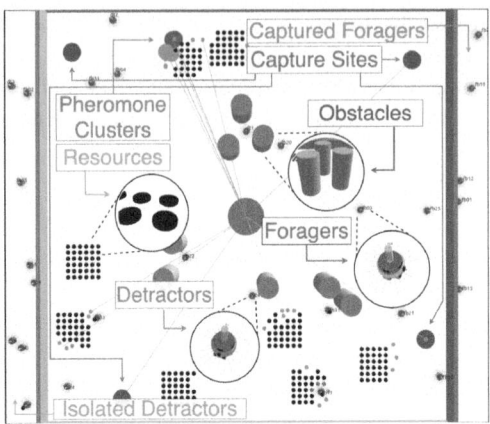

Fig. 1. A mid-simulation phase depicting the foraging scenario in ARGoS.

4.1 Defense Strategy

To counteract threats from detractors, our defense employs the DBSCAN (Density-Based Spatial Clustering of Applications with Noise) algorithm to analyze and cluster pheromone trails. This method helps identify clusters formed by similar pheromone trail patterns, distinguishing between those laid by genuine foragers and detractors.

DBSCAN Implementation: We apply DBSCAN to the spatial coordinates given by the pheromone trail waypopints. We configure DBSCAN with a search radius ϵ equal to the robots' search radius for spatial relevance. The minimum points threshold $minPts$ is set to 2, due to the expected data sparsity in small-scale swarm scenarios.

Cluster Grouping: Post-DBSCAN clustering, we construct a graph to link clusters based on common creator IDs. The nodes represent individual clusters, and the edges are formed between the nodes if their associated pheromone trails share at least one common creator ID. Using this graph model depicted in Fig. 2, we create larger cluster groupings to facilitate the identification robots who are exhibiting similar trail laying behavior.

Travel Time Estimation and Adjustment: In order to detect and flag misleading trails, we incorporate a travel-time estimation strategy. Each pheromone trail laid by robots includes an estimated travel time calculated based on the distance to the destination and the robot's average speed. When a robot follows a trail, the actual time taken to travel is recorded and compared against the estimated time. Significant deviations suggest potential misleading trails, triggering an isolation event for the trail creator and other robot's associated with the flagged trail's cluster group.

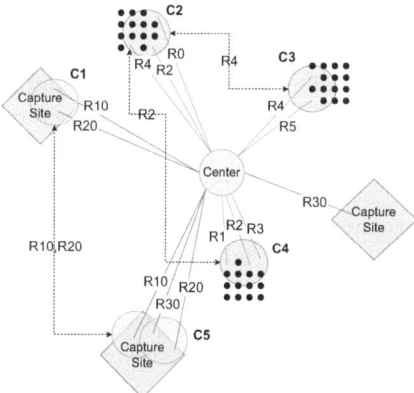

Fig. 2. A graphical representation of cluster connectivity based on common creator IDs.

This strategy involves updating our understanding of trail legitimacy continuously: If the actual travel time consistently exceeds the estimated by a significant margin, it suggests the presence of a misleading trail. However, obstacles and collisions can affect travel time and must be taken into account. Therefore, we update the estimated travel-time using a simple feedback loop taking into consideration the measured travel time when a robot returns. The estimate is only updated when a robot returns after the current estimated travel time of the trail.

Isolation and Reintegration: A 'strike' is given to all robots associated with a cluster group when a robot does not return within the estimated travel time of one of the trails within the group. Accumulating a predefined threshold of five strikes (determined empirically) results in the robot's temporary isolation from the swarm to prevent further disruption of foraging activities. This threshold ensures a balance between prompt response to threats and the minimization of false positives, essential for maintaining swarm integrity.

However, if a forager returns on a trail within a flagged cluster group after its estimated travel time, the trail creator and other robots associated with the trail's cluster group are reintegrated (if isolated) or have a 'strike' removed. This addresses false positives, allowing dynamic adaptation to new information and maintaining operational efficiency.

5   Experiment Setup

We utilize the ARGoS multi-robot simulator to conduct our experimental studies [24]. The experiments are designed to evaluate the impact of detractor behaviors and the effectiveness of our proposed defense mechanisms under various scenarios, with experimental parameters detailed in Table 1. Conducted under controlled settings with consistent arena sizes, resource clusters, and robot numbers, our experiments ensure repeatability and reliability. Key assumptions include uniform robot capabilities, randomized resource distribution, and specific detractor behaviors to isolate the effects of variable parameters on the foraging efficiency and defense effectiveness.

Experiment 1 - Pheromone Trail Dynamics: The first experiment investigates how the rate of laying pheromone trails, denoted as λ_{fg} for foragers and λ_{dt} for detractors, affects the foraging performance and susceptibility to attacks. We aim to determine the most impactful settings on resource collection efficiency and forager capture rates, setting the stage for robust testing of our defense strategies in subsequent experiments.

Experiment 2 - Detractor Impact Analysis: In Experiment 2, we assess the resilience of our defense strategy by varying the number of detractors (n_{dt}) and adjusting the pheromone lay rates for both foragers and detractors. This setup allows us to observe the direct effects of increased detractor presence on the foraging process and the efficacy of our defense mechanisms under escalated threat conditions.

Experiment 3 - Obstacle Dynamics: Additionally, a third experiment introduces environmental obstacles to evaluate the robustness of our defense under more complex and realistic conditions. We manipulate the obstacle density (n_{obs}) and distribution patterns (d_{obs}), either randomly or in an annular configuration, to test how physical barriers affect the foraging algorithms and the defense system's capability to adapt and maintain effectiveness.

Table 1. Consolidated Experimental Parameters

Parameter	Exp. 1	Exp. 2	Exp. 3	Description
D_{arena}	(10,10,1)			Foraging area dimensions (x, y, z)
$D_{cluster}$	(6,6)			Resource cluster size (l, w)
n_{fg}	24			# of foragers
n_{dt}	25%	10%,20%,30%,40%,50%	25%	# of detractors based on n_{fg}
n_{cl}	8			# of resource clusters
n_{obs}	N/A		0, 4, 8, 12, 16	Density of obstacles
r_{center}	0.25			Radius of the center
$r_{resource}$	0.05			Radius of resource
r_{search}	$4 \cdot r_{resource}$			Foot-bot search radius
r_{cs}	$r_{center}/2$			Capture site radius
λ_{fg}	1, 4, 8, 12	4		Lay rate of pheromone trails for foragers
λ_{dt}	1, 4, 8, 12	1		Lay rate of pheromone trails for detractors
d_{obs}	N/A		Random, Annular	Obstacle & resource distribution
ϵ	r_{search}			DBSCAN cluster neighborhood radius
$minPts$	2			DBSCAN minimum points per cluster

6   Results

6.1   Experiment 1: Evaluation of Pheromone Trail Dynamics

The first experiment focused on varying the rates of laying pheromone trails, λ_{fg} and λ_{dt}, and their effects on foraging performance. Figure 3a illustrates the impact on resource collection efficiency and forager capture rates over 50 runs. The tested rates were 1, 4, 8, and 12, with the smaller values indicating a higher frequency of trail laying.

Key findings include: 1) At $\lambda_{fg}, \lambda_{dt} = 1$, resource collection was at 61.61%, and the forager capture rate was high at 88.75%. 2) Increasing the rate to 4 resulted in a slight increase in resource collection to 61.82% and the highest capture rate of 92.50%. 3) At a rate of 8, the resource collection decreased to 54.71%, and the forager capture rate decreased to 72.92%. 3) The 12 rate showed an increase in resource collection to 73.19% and a significant reduction in forager captures to 14.31%.

These results underscored the risk associated with higher trail laying frequencies, with a rate of 4 selected for further testing in Experiment 2 due to its pronounced impact on forager captures.

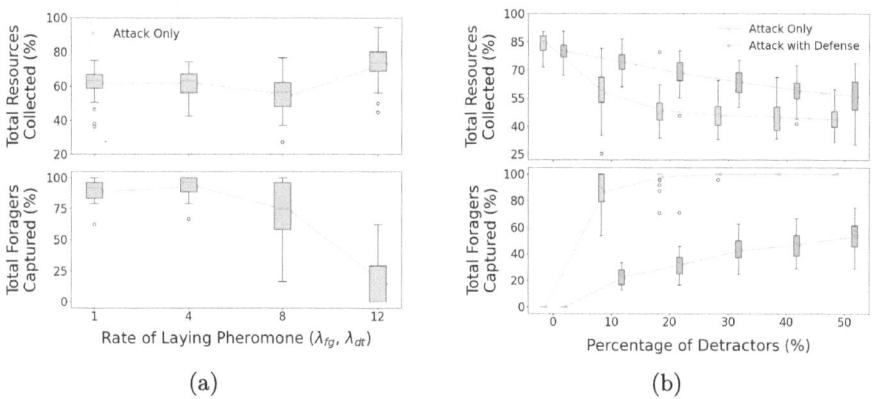

Fig. 3. (a) The foraging performance and foragers captured across different rates of laying pheromones while under attack. (b) Detractor Isolation Rates.

6.2   Experiment 2: Evaluation of the Defense Strategy

Experiment 2 assessed the defense strategy against varying levels of detractors, from 10% to 50% of the swarm. Figure 3b presents the performance of the foraging system with and without the defense mechanism.

The findings of Experiment 2 include: 1) Without detractors, the baseline foraging efficiency was approximately 85%. 2) With detractors, efficiency decreased to 57.92% at 10% detractors and further to 43.66% at 50% detractors. 3) The defense improved foraging performance by 16.06% at 10% detractors and by 12.33% at 50% detractors. 4) Capture rates reduced significantly with the defense, especially noticeable at higher detractor levels.

The defense showed substantial effectiveness in mitigating detractor impacts, confirming its capability to adapt and counteract increasing threats.

Forager Capture Dynamics: Figure 4a presents the fluctuating rates of forager captures per minute under various detractor levels, highlighting the defense mechanism's responsiveness. Notable points include: 1) Forager captures peak early but reduces to near-zero levels beyond minute 16, indicating effective threat neutralization. 2) At 10% detractors, the capture rate peaks at 0.866 per minute by minute 4, then decreases steadily to 0.033 by minute 15. 3) At 50% detractors, the capture rate reaches a higher peak of 2.6 per minute by minute 4, then drops sharply to nearly 0.0 by minute 16.

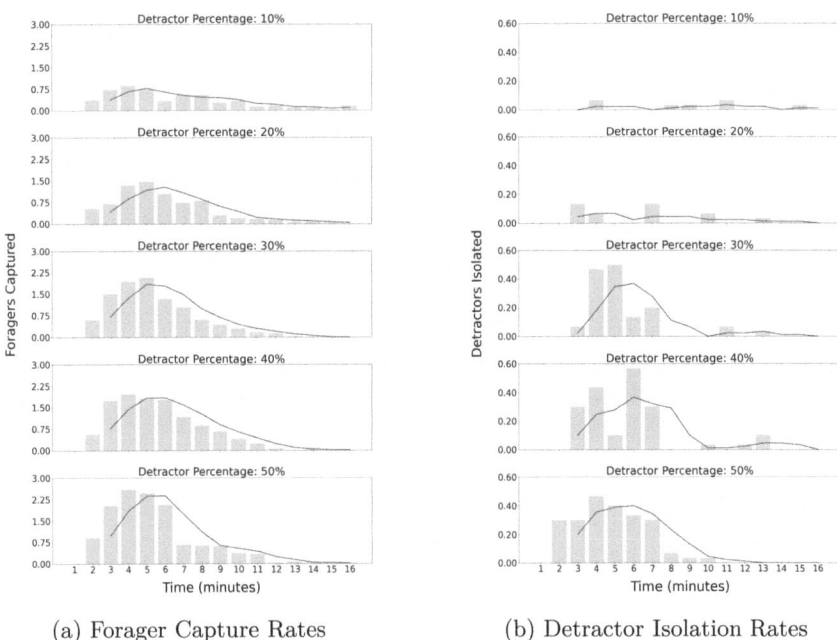

(a) Forager Capture Rates   (b) Detractor Isolation Rates

Fig. 4. Dynamics of foragers captured and detractors isolated over time in Experiment 2.

Detractor Isolation Dynamics: Figure 4b shows the frequency of detractor isolation events per minute, demonstrating the precision and efficiency of the defense strategy: 1) Isolation events decrease to virtually none after minute 16, showcasing the prompt efficacy of the defense. 2) At 10% detractors, isolation rates are minimal, peaking only at 0.066 at minute 11. 3) Increased percentages of detractors lead to higher isolation rates, with 20% peaking at 0.133 by minute 7 and 30% reaching a peak of 0.5 by minute 5, indicating a scaled response to greater threats.

6.3   Experiment 3: Robustness of Defense in Varied Environments

In Experiment 3, the robustness of the defense was challenged by introducing environmental obstacles, which tested the system's adaptability to physical changes that could impact navigation and the effectiveness of pheromone trails. The defense mechanism proved resilient across various obstacle densities, effectively maintaining foraging efficiency and minimizing forager captures.

The performance in environments with different obstacle densities showed no significant degradation in the defense's ability to protect the foragers. This indicates that the system is not only effective under direct attacks but also adaptable to complex operational conditions that include physical barriers. This resilience is crucial for real-world applications where environmental unpredictability is common.

These findings emphasize the defense's proficiency in maintaining operational stability and highlight its potential for deployment in dynamic and challenging real-world scenarios where both security threats and environmental obstacles are prevalent.

Figure 5 shows 1) a peak in "Total Resources Collected" at 66.26% with a standard deviation of 7.78 when faced with sixteen obstacles, suggesting a resilient adaptation to navigational challenges. 2) "Total Foragers Captured" remained relatively stable despite increased obstacle density, with minimal mean capture rates fluctuating slightly, demonstrating the defense's effectiveness under varied conditions. 3) The "False Positives Detected" metric was stable across different obstacle densities, indicating that the defense system maintained accuracy in threat detection even with environmental complexity. 3) The 'final Univelocity' exhibited minor variations, reflecting the system's ability to adapt to increased navigational challenges posed by obstacles.

These findings underscore the defense mechanism's robustness, not only in terms of managing detractor threats but also in handling environmental factors that might complicate navigation and communication within the swarm. The detailed results of Experiment 3 provide critical insight into defense capabilities, strengthening its effectiveness and adaptability under complex and dynamically changing conditions, confirming the readiness of the system for deployment in real-world scenarios where natural and artificial obstacles are present.

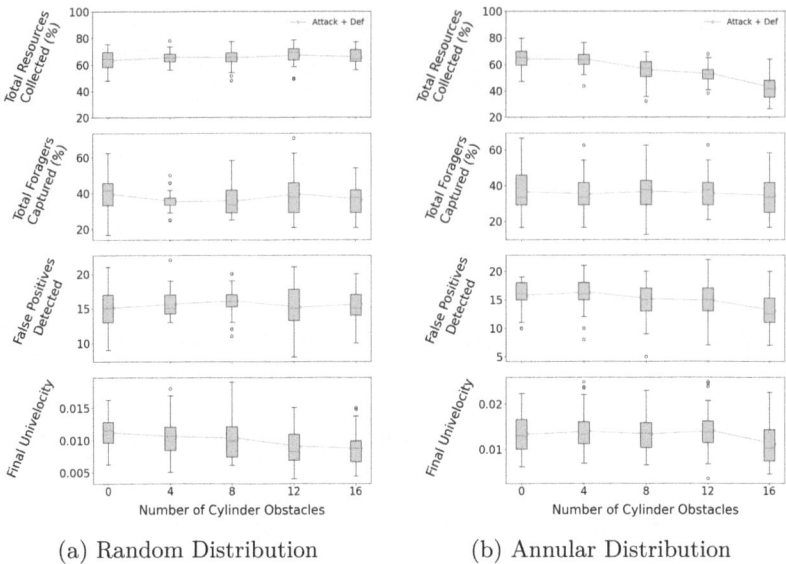

(a) Random Distribution   (b) Annular Distribution

Fig. 5. Performance metrics under different obstacle distributions: Random and Annular. Demonstrates how environmental complexity affects foraging efficiency and defense effectiveness.

7   Conclusion

This study substantially advances the field of swarm robotics by addressing vulnerabilities in foraging algorithms exposed to deceptive strategies such as misleading pheromone trails. Our investigations were articulated through a series of experiments.

In Experiment 1, we observed that different pheromone trail laying rates significantly affected foraging efficiency and forager capture rates (Fig. 3a). The most notable findings were at the lay rates of 4 and 12, which represented the extremes in forager capture rates and resource collection efficiency, respectively. Experiment 2 demonstrated the effectiveness of our defense mechanism, which significantly mitigated the impact of attacks at various detractor levels (Fig. 3b). The defense strategy effectively improved foraging performance by up to 16.07% and reduced forager capture rates by up to 65.69%, even as the detractor percentage increased. This highlights the defense's adaptability and robustness in maintaining swarm functionality under attack scenarios (Fig. 4a and Fig. 4b).

Experiment 3 tested the defense's resilience against environmental challenges and showed that while obstacles affected foraging efficiency, they did not compromise the defense's effectiveness (Fig. 5a and Fig. 5b). This indicates the defense mechanism's capability to operate effectively in complex environments.

These results not only underline the efficacy of our defense mechanism in rapidly and efficiently isolating detractors but also emphasize the importance of designing swarm robotics systems that are secure, resilient, and adaptable to various operational threats. The manipulation of virtual pheromone trails poses significant challenges; however, our approach offers a potent countermeasure, establishing a foundational strategy for enhancing the robustness of foraging algorithms against pheromone-based attacks. This research contributes to a broader understanding of the integration of safety and security measures into swarm robotics at the developmental stage, ensuring that these systems are prepared to face real-world operational challenges. Future work will focus on enhancing the scalability and efficiency of our defense mechanisms, exploring applications in more dynamic environments, and possibly integrating advanced machine learning techniques to further refine adaptability and threat detection capabilities.

References

1. Akino, T.: Chemical strategies to deal with ants: a review of mimicry, camouflage, propaganda, and phytomimesis by ants (hymenoptera: Formicidae) and other arthropods. Myrmecological News **11**(8), 173–181 (2008)
2. Arnold, R.D., Yamaguchi, H., Tanaka, T.: Search and rescue with autonomous flying robots through behavior-based cooperative intelligence. J. Int. Humanitarian Action **3**(1), 1–18 (2018). https://doi.org/10.1186/s41018-018-0045-4
3. Arvin, F., Turgut, A.E., Bazyari, F., Arikan, K.B., Bellotto, N., Yue, S.: Cue-based aggregation with a mobile robot swarm: a novel fuzzy-based method. Adapt. Behav. **22**(3), 189–206 (2014)
4. Arvin, F., Turgut, A.E., Krajník, T., Yue, S.: Investigation of cue-based aggregation in static and dynamic environments with a mobile robot swarm. Adapt. Behav. **24**(2), 102–118 (2016)
5. Aswale, A., López, A., Ammartayakun, A., Pinciroli, C.: Hacking the colony: on the disruptive effect of misleading pheromone and how to defend against it. In: Proceedings of the 21st International Conference on Autonomous Agents and Multiagent Systems, pp. 27–34 (2022)
6. Bagnères, A.G., Lorenzi, M.C., et al.: Chemical deception/mimicry using cuticular hydrocarbons. Insect hydrocarbons: biology, biochemistry and chemical ecology, pp. 282–323 (2010)
7. Billen, J., Morgan, E.D.: Pheromone communication in social insects: Sources and secretions. Ants, Wasps, Bees, And Termites, Pheromone Communication In Social Insects (1998)
8. Brady, S.G.: Evolution of the army ant syndrome: the origin and long-term evolutionary stasis of a complex of behavioral and reproductive adaptations. Proc. Natl. Acad. Sci. **100**(11), 6575–6579 (2003)
9. Cheraghi, A.R., Peters, J., Graffi, K.: Prevention of ant mills in pheromone-based search algorithm for robot swarms. In: 2020 3rd International Conference on Intelligent Robotic and Control Engineering (IRCE), pp. 23–30 (2020)
10. Crist, T.O., MacMahon, J.A.: Individual foraging components of harvester ants: movement patterns and seed patch fidelity. Insectes Soc. **38**(4), 379–396 (1991)
11. Czaczkes, T.J., Grüter, C., Ratnieks, F.L.: Trail pheromones: an integrative view of their role in social insect colony organization. Annu. Rev. Entomol. **60**, 581–599 (2015)

12. Ester, M., Kriegel, H., Sander, J., Xu, X.: A density-based algorithm for discovering clusters in large spatial databases with noise. In: Simoudis, E., Han, J., Fayyad, U.M. (eds.) Proceedings of the Second International Conference on Knowledge Discovery and Data Mining (KDD-96), Portland, Oregon, USA, pp. 226–231. AAAI Press (1996)

13. Gauci, M., Chen, J., Li, W., Dodd, T.J., Groß, R.: Self-organized aggregation without computation. Int. J. Robot. Res. **33**(8), 1145–1161 (2014)

14. Hecker, J.P., Moses, M.E.: Beyond pheromones: evolving error-tolerant, flexible, and scalable ant-inspired robot swarms. Swarm Intell. **9**, 43–70 (2015)

15. Hecker, J.P., Moses, M.E.: Beyond pheromones: evolving error-tolerant, flexible, and scalable ant-inspired robot swarms. Swarm Intell. **9**(1), 43–70 (2015)

16. Higgins, F., Tomlinson, A., Martin, K.M.: Survey on security challenges for swarm robotics. In: 2009 Fifth International Conference on Autonomic and Autonomous Systems, pp. 307–312 (2009)

17. Inwood, M., Morgan, P.: Chemical sorcery for sociality: exocrine secretions of ants (hymenoptera: Formicidae). Myrmecological News Myrmecol. News **11**, 79–90 (2008)

18. Jin, B., Liang, Y., Han, Z., Ohkura, K.: Generating collective foraging behavior for robotic swarm using deep reinforcement learning. Artif. Life Robot. **25**, 588–595 (11 2020)

19. Khaldi, B., Harrou, F., Cherif, F., Sun, Y.: Self-organization in aggregating robot swarms: a DW-KNN topological approach. Biosystems **165**, 106–121 (2018)

20. Lee, M.F.R., Chien, T.W.: Artificial intelligence and internet of things for robotic disaster response. In: 2020 International Conference on Advanced Robotics and Intelligent Systems (ARIS), pp. 1–6 (2020)

21. Font Llenas, A., Talamali, M.S., Xu, X., Marshall, J.A.R., Reina, A.: Quality-sensitive foraging by a robot swarm through virtual pheromone trails. In: Dorigo, M., Birattari, M., Blum, C., Christensen, A.L., Reina, A., Trianni, V. (eds.) ANTS 2018. LNCS, vol. 11172, pp. 135–149. Springer, Cham (2018). https://doi.org/10.1007/978-3-030-00533-7_11

22. Lončar, I., et al.: A heterogeneous robotic swarm for long-term monitoring of marine environments. Appl. Sci. **9**(7) (2019)

23. Lu, Q., Hecker, J.P., Moses, M.E.: The MPFA: a multiple-place foraging algorithm for biologically-inspired robot swarms. In: 2016 IEEE/RSJ International Conference on Intelligent Robots and Systems (IROS), pp. 3815–3821 (2016)

24. Pinciroli, C., et al.: ARGoS: a modular, parallel, multi-engine simulator for multi-robot systems. Swarm Intell. **6**(4), 271–295 (2012)

25. Lu, Q., Hecker, J.P., Moses, M.E.: Multiple-place swarm foraging with dynamic depots. Auton. Robot. **42**(4), 909–926 (2018). https://doi.org/10.1007/s10514-017-9693-2

26. Robinson, E.J., Jackson, D.E., Holcombe, M., Ratnieks, F.L.: No entry signal in ant foraging. Nature **438**(7067), 442–442 (2005)

27. Sargeant, I., Tomlinson, A.: Modelling malicious entities in a robotic swarm. In: 2013 IEEE/AIAA 32nd Digital Avionics Systems Conference (DASC) (2013)

28. Sasaki, T., Hölldobler, B., Millar, J.G., Pratt, S.C.: A context-dependent alarm signal in the ant temnothorax rugatulus. J. Exp. Biol. **217**(18), 3229–3236 (2014)

29. Schneirla, T.C.: A unique case of circular milling in ants, considered in relation to trail following and the general problem of orientation. In: American Museum Novitates (1944)

30. von Thienen, W., Metzler, D., Choe, D.-H., Witte, V.: Pheromone communication in ants: a detailed analysis of concentration-dependent decisions in three species. Behav. Ecol. Sociobiol. **68**(10), 1611–1627 (2014). https://doi.org/10.1007/s00265-014-1770-3

31. Vardy, A.: Accelerated patch sorting by a robotic swarm. In: 9th Conference on Computer and Robot Vision, CRV 2012, pp. 314–321 (2012)

32. Vardy, A., Vorobyev, G., Banzhaf, W.: Cache consensus: rapid object sorting by a robotic swarm. Swarm Intell. **8**, 61–87 (2014)

33. Wenig, K., Bach, R., Czaczkes, T.J.: Hard limits to cognitive flexibility: ants can learn to ignore but not avoid pheromone trails. J. Exp. Biol. **224**(11), 242–454 (2021)

34. Wikipedia: Buffalo jump (2023). https://en.wikipedia.org/wiki/Buffalo_jump. Accessed 23 Jan 2024

35. Wyatt, T.D.: Breaking the code: illicit signalers and receivers of semiochemicals. Pheromones and animal behavior: chemical signals and signatures, pp. 244–259 (2014)

36. Xiaoning, Z.: Analysis of military application of UAV swarm technology. In: 2020 3rd International Conference on Unmanned Systems (ICUS), pp. 1200–1204 (2020)

Predator-Prey Q-Learning Based Collaborative Coverage Path Planning for Swarm Robotics

Michael Watson[1], Hanchi Ren[2], Farshad Arvin[1], and Junyan Hu[1(✉)]

[1] Department of Computer Science, Durham University, Durham, UK
{michael.watson,farshad.arvin,junyan.hu}@durham.ac.uk
[2] Department of Computer Science, Swansea University, Swansea, UK
hanchi.ren@swansea.ac.uk

Abstract. Coverage Path Planning (CPP) is an effective approach to let intelligent robots cover an area by finding feasible paths through the environment. In this paper, we focus on using reinforcement learning to learn about a given environment and find the most efficient path that explores all target points. To overcome the limitations caused by standard Q-learning based CPP that often fall into a local optimum and may be in-efficient in large-scale environments, two methods of improvement are considered, i.e., the use of a robot swarm working towards the same goal and the augmenting of the Q-learning algorithm to include a predator-prey based reward system. Existing predator-prey based reward systems provide rewards the further away an agent is from its predator, the paper adapts this concept to work within a robot swarm by simulating each agent of the swarm as both predator and prey. Simulation case studies and comparisons with the standard Q-learning show that the proposed method has a superior coverage performance in complicated environments.

Keywords: Coverage Path Planning · Reinforcement Learning · Swarm Robotics

1 Introduction

The autonomous coverage of unknown environments is a very hot and important research field in robotics. This task aims to completely cover the entirety of an environment which means the mobile agents must visit each location safely and efficiently [1]. Many coverage path planning (CPP) algorithms have been designed for mobile robots during the past decade, and they have already been applied to various real-world applications, such as precision agriculture [2,3], autonomous vehicle navigation [4], automated harvesting [5,6], search and rescue [7,8]. Furthermore, CPP also plays an important role in some emerging research fields, e.g., a multi-arm manipulator could be used to interact with bee hives and follow the trajectory of the queen honey bee [9].

M. N. Huda et al. (Eds.): TAROS 2024, LNAI 15052, pp. 320–332, 2025.
https://doi.org/10.1007/978-3-031-72062-8_28

There exist several different methods to achieve autonomous coverage, but this paper aims to improve upon one of the most known reinforcement learning algorithms, Q-learning [10,11]. Q-learning is a long-standing algorithm and is popular due to its simple implementation allowing it to be easily adapted to several situations. However, this simpleness also leads to several limitations such as easily falling into a local optimum or struggling in more complex environments. These are the main limitations the solution aims to overcome; it is important to overcome these issues as in more complex use cases such as exploring dangerous terrain and environments it is vital that the agents can find optimal paths to save on resources. A shorter path means that more of an environment can be explored, both in total and within a given period. Furthermore, the algorithm must be good in more complex environments where there may exist obstacles that the agents must avoid as it is extremely unlikely that in real-world use cases, the environment will be unobstructed. Another key technology used throughout this paper is swarm robotics [12], swarms are a collection of multiple robots all working together towards the same goal. They are highly effective in exploration tasks as they multiply how much of the environment can be explored at once. Therefore, this paper focuses on implementing the upgraded algorithm into a swarm so that they can be used together to provide a superior method of exploitation.

The paper aims to answer the question: Can you adapt Q-learning to be more effective in Coverage Path Planning problems? To attempt to answer this question the paper provides a completed algorithm and explains the method used to obtain the results displayed at the end of the paper. It will explain the Q-learning method used as well as the adaptions made to improve it for the necessary purpose. A complete program was made to test the algorithms against each other simulating custom environments, as well as easily allowing differing swarm sizes. An implementation of a standard algorithm was also implemented to be compared. The adapted solution aims to improve upon the existing method in several ways, it aims to improve performance in complex scenarios in which obstacles exist increasing the difficulty of exploration. Another vital result is to ensure that full coverage always occurs within the given movement budget, it also aims to improve the efficiency of the algorithm by reducing the average number of steps needed to fully cover. Finally, the solution aims to be consistent so that it produces reliable results.

2   Related Work

There are multiple solutions to the CPP problem which can be categorized into offline and online algorithms [13]. Offline algorithms perform coverage in a static and known environment, whereas online algorithms work in dynamic environments. The online algorithm is preferred as most use cases will involve dynamic environments in which the agent will have to adapt [14]. However, even online algorithms have their limitations one being they require agents with sensors to learn about their surroundings. Another is that they lack the same efficiency as offline algorithms which can find an optimal path and then stick to it. CPP algorithms can also take many different approaches [15] such as machine learning [16], greedy search [17], graph search [18], frontier-based exploration [19] as well as many others. Each method has its strengths and weaknesses and the best method to use often depends on the use case. For example, the greedy algorithm (Dijkstra's algorithm) is simple, easy to implement and generally fast, but it does not guarantee a global optimum. It is also an offline algorithm which means it cannot be used in dynamic environments.

Rerinforcement learning is the method which this paper focuses on. Piardi et al. [20] present a Q-learning algorithm which uses a grid-based map to optimize the CPP trajectory, this implementation has some limitations. Its simple nature means it doesn't work in environments with unknown obstacles. However, it does provide a good basis for improvement by adapting this algorithm with further methods. Zhou et al. [21] provide numerous improvements over a standard Q-learning algorithm to create an optimized algorithm, optimized Q-learning. It includes improvements such as novel Q-table initialization, a new action-selection policy, and a new reward function. The limitations of this paper however are that its reward function is still simplistic, and it also lacks integration for multiple robots. Puente-Castro et al. [22] make use of a swarm of agents to accomplish the task as a group. This method effectively demonstrates the usefulness of a swarm over an individual agent, it shows how using a swarm reduces the individual load of each agent whilst maintaining the overall result of the solution. The limitation of this method however is the reward structure, it is very simple and therefore leaves room for improvement. One method to improve the reward structure is through the use of a predator-prey based reward structure [23]. This reward structure combines three different reward functions to produce a reward for the agent. These rewards given are the predator-prey reward, a boundary reward, and a smoothness reward, in combination this provides the agent with a much more responsive representation of its environment. The limitation of this method comes from its lack of swarm implementation, it focuses on only one agent and with multiple agents, collisions can reduce the effectiveness of the rewards.

3   Methodology

This solution aims to create an efficient reinforcement learning algorithm for use in a coverage path planning scenario. Robot agents are used to complete the coverage. The multiple agents mean more of the environment can be explored concurrently; it also provides redundancies in the scenario that an agent is lost. However, there are some drawbacks in the way that the swarm interacts such as collisions as well as issues in how the swarm explores. The solution to this is to adapt a reinforcement learning algorithm around the swarm to overcome these issues. The reinforcement learning method Q-learning was selected as it is an easy-to-implement method which can be used easily in a swarm as if needed the agents can act individually.

3.1   Foundation Work

An implementation of Watkins Q-learning [10] was used for this solution. Q-learning is a simple reinforcement learning algorithm which makes use of a data structure called a Q-table and a function known as the Q-function. Q-learning works by initializing the agent's Q-table which holds a predicted reward for each state the agent can be in, and each action it can take from this state. The estimated rewards are then continually improved using a specialized Q-function (1) to produce rewards which represent how beneficial each action is.

The function takes the observed reward and the estimated future value and updates the value in the table according to the function. The function is formulated as follows:

$$Q(s,a) = Q(s,a) + \alpha[r + \gamma \max Q(s') - Q(s,a)], \tag{1}$$

where $Q(s,a)$ represents the current Q-value for the given state and action, s' refers to the next state (the state the current action will lead to), α=Learning Rate, r=reward and γ=Discount Factor. The function can be altered through the use of the parameters α, γ. These values can range from 0 to 1. The learning rate determines how much each learning increment will affect the Q-value, the greater the learning rate the faster the Q-values will converge. It is therefore important to have a learning rate small enough that it does not converge too early but also ensure that it is not so small it takes too long to converge. The discount factor determines how much importance is placed on the future reward. The greater the discount factor the more important the function finds the future reward; it is also important to balance this parameter so that the correct amount of weighting is placed on the potential future.

Another significant strategy implemented was the Epsilon-Greedy method. This is a method of introducing some randomness into the agent's selection of an action, which allows it to explore and find better paths and actions it can use. It works by using a value denoted as epsilon, the agent will then choose a random action with probability ϵ and the best action with probability $1 - \epsilon$. The value of epsilon then decays and reduces after each cycle so fewer random actions are taken. Performance can be heavily impacted by both the starting value and the decay rate of epsilon. If epsilon remains too high the algorithm will take too many random actions and never be able to converge on a suitable answer. However, if epsilon starts to low, no random actions are taken, and the agent gets stuck in a local optima never branching off to explore potentially better alternatives. A combination of these methods can be seen in Algorithm 1.

The basic reward structure used for standard Q-learning is simple, a positive reward is given the first time a location is seen, and a negative reward is given each time the agent moves. This negative reward acts as a movement penalty, it is given so that the agent will aim to cover the environment in a minimal number of steps as the more it moves the smaller the reward it gains.

$$r = \begin{cases} 1, & \textit{Undiscovered Location} \\ -0.05, & \textit{Movement Penalty.} \end{cases}$$

Algorithm 1. Q-learning

1: Initialize $Q(s,a)$ as 0 ;
2: **for** each episode **do**
3:     Initialize s;
4:     **for** each step of the episode **do**
5:         Random p;
6:         **if** $p < \epsilon$ **then**
7:             Any action a;
8:         **else**
9:             $a = \max Q(s)$;
10:        **end if**
11:        Take action a, gather r, s';
12:        $Q(s,a) = Q(s,a) + \alpha[r + \gamma \max Q(s') - Q(s,a)]$;
13:        $s = s'$;
14:    **end for**
15:    Until s is terminal.
16: **end for**

Backtracking is another method implemented into the solution to assist the agent in what to do when it sees no unvisited neighbours. If the agent reaches a position where it has no unvisited neighbours, it will begin to retrace its steps until it sees an unvisited location and it will then return to normal. The use of backtracking is to overcome situations where the agent becomes stuck in a loop moving between locations it has already seen until it terminates. Although backtracking is not efficient it is effective at ensuring the environment will always be completely covered and the agent will not get stuck. Coverage is guaranteed as anywhere the agent can reach it must have passed at some point and this method of backtracking ensures it is seen again and can be explored this time. This does create some limitations however as this form of backtracking is not very efficient because it retraces its steps. This occurs each time the agent starts backtracking. It could be improved by implementing a better method which remembers the location where the agent stopped backtracking and returned to exploration, this location could then be returned if the agent needs to backtrack again.

3.2 Swarm Structure

This solution makes use of multiple agents working together to create a collaborative swarm. The implementation of the swarm allows for the use of a changing number of agents, the swarm can contain any number of agents and still function correctly. Using more agents will allow for faster exploration, but it requires greater computational resources. The agents of the swarm all use the same algorithm however they use individual Q-tables so they can determine their best paths individually. This is necessary to take full advantage of the swarm as with a shared Q-table all agents would attempt to follow the same path which would negate the advantages of using a swarm in the first place. The use of one Q-table would also lead to agents disrupting other agents' Q-values leading to worse results than using a single agent. There is also a central controller of the swarm which holds information about the swarm. This includes information about the locations of all swarm members, which allows any member to know the position of the other members. This can be used for both avoiding collisions as well as in the reward function (2). The controller also holds how much coverage the swarm has completed so it can monitor progress and tell the members to stop once coverage is complete.

There exist some potentially problematic interactions with agents in the swarm, one of which is how the agents avoid crashing. In the simulation, a crash occurs when two agents move into the same location. This is overcome by restricting the agent's movement to not allow actions which will result in the agents overlapping. This can reduce performance sometimes as agents may have to wait for other agents to move first before they can, but it is vital to ensure all agents remain safe.

Another problem is the fact that with multiple agents exploring different locations, the agents cannot know when they have collectively covered the entirety of the environment. Without some shared information of what each agent has covered the agents will run until they have covered the entire environment individually. This is overcome using a simulated server that holds the amount of the environment covered, this allows any agent to access the collective memory to share what they have covered and know when to stop. The server also holds the specific locations that have been explored, this avoids duplicate counting as well as allowing the correct rewards to be given to the agent for unexplored and explored locations.

3.3   Predator-Prey Improvement

Predator-Prey is a method which simulates the animal relationship between predator and prey. Specifically, it mimics the behaviour of the prey, in which they aim to get as far from the predator as quickly as possible. This method was adapted into a reinforcement learning scenario by Zhang et al. [23]. The method implemented in their paper focuses on using one agent to fully explore an environment. This is done by simulating a predator at one end of the environment so that the agent will move quickly away from it, the method is effective at getting the agent to move in straight lines as it is its optimal pathing for maximum distance. This behaviour is especially useful in a coverage scenario as moving in straight lines avoids any small pockets of unvisited locations being missed and needing to be found later. Once a boundary is reached the predator can then be moved to encourage the agent to move in a direction of choice, by doing this repeatedly the agent can sweep the entirety of the environment very effectively. The solution is built upon this method; however, it is altered to work in a swarm. The key difference is that instead of simulating a predator for the agent to run from all agents in the swarm simulate both predator and prey. An individual agent aims to move as far as possible from the other agents in the swarm, simulating the prey with every other member of the swarm acting as its predator. This method is effective when the entire swarm begins in the same region of the environment as it means the swarm spreads out which allows for different regions of the environment to be explored faster. A reward is given depending on how far the agent is from its closest predator (2). The reward for moving to neighbour x is formulated as follows:

$$r_p(x) = \frac{D(x) - D_{min}(x_n)}{D_{max}(x_n) - D_{min}(X_n)}, \tag{2}$$

where $D(x)$ gives the distance from x to its nearest predator. $D_{max}(x_n)$ gives the maximum distance between one of x's neighbours and any predator, and $D_{min}(x_n)$ gives the minimum distance between one of x's neighbours and any predator.

Two further functions are also used which give rewards based on the smoothness of the path travelled and the boundary of the next state. The smoothness reward gives a reward based on the direction of the agent's travel (3), if it moves in the same direction multiple times it receives a greater reward. This is done to entice the agent into moving in straight lines, as well as discourage moving back along its path when unnecessary. The reward for moving to neighbour x is formulated as follows:

$$r_s(s) = \begin{cases} \angle x_{k-1}x_k x = 0°, & reward = 1 \\ \angle x_{k-1}x_k x = 90°, & reward = 0.5 \\ \angle x_{k-1}x_k x = 180°, & reward = 0 \end{cases} \tag{3}$$

where x_{k-1} and x_k are the positions covered at $k-1$ and k, respectively. The reward is given based on the angle between the vectors $(x_{k-1} - x_k)$ and $(x_k - x)$.

The boundary reward gives a reward based on the number of unvisited neighbours a given location has (4). The fewer the neighbours the greater the reward. This helps the agent to move into locations which may otherwise get cut off and left behind. The reward for moving to neighbour x is formulated as follows:

$$r_b(x) = \frac{N_{max} - x_N}{N_{max}}, \tag{4}$$

where N_{max} is the total number of neighbours and x_N is the number of unvisited neighbours x has.

The rewards are then combined to give a complete reward. They are combined with different weights to manipulate the importance of each individual reward. It is formulated as follows:

$$r(x) = w_p(r_p(x)) + w_s(r_s(x)) + w_b(r_b(x)), \tag{5}$$

where w_p, w_s and w_b represent the weights for the predation reward, smoothness reward and boundary reward, respectively. $r(x)$ is the final reward which is returned to the agent. The weights can be changed to differing values to change how rewards are given to the agent. A larger weight means more emphasis is placed on that reward; this allows for the algorithm to work in different ways such as focusing on spreading out as quickly as possible or trying to move in only straight lines. Using powerful values for the weights which are much higher than the others can lead to a worse performance as it overpowers the other reward functions and only one is truly considered.

The pseudocode for the complete algorithm can be seen in Algorithm 2. This combines all improvements including the use of a swarm, the use of backtracking and the use of the adapted reward function.

Algorithm 2. Improved Predator-Prey Swarm Q-learning

1: Initialize **Swarm**;
2: **for** each agent in **Swarm do**
3:      Initialize $Q(s, a)$ as 0;
4:      Initialize **Exploration Count**;
5: **end for**
6: **for** each episode **do**
7:      **for** each agent in **Swarm do**
8:           Initialize s;
9:      **end for**
10:     **for** each step of the episode **do**
11:          **for** each agent in **Swarm do**
12:               Random p;
13:               **if** $p < \epsilon$ **then**
14:                    Any action a;
15:               **else**
16:                    **if** there is an unvisited location **then**
17:                         $a = \max Q(s)$
18:                    **else**
19:                         $a = $ Previous action
20:                    **end if**
21:               **end if**
22:               Take action a, gather s', gathering r using (5);
23:               $Q(s, a) = Q(s, a) + \alpha[r + \gamma \max Q(s') - Q(s, a)]$;
24:               **if** s' is unvisited **then**
25:                    **Exploration Count** += 1;
26:               **end if**
27:               $s = s'$;
28:          **end for**
29:          Until s is terminal.
30:     **end for**
31: **end for**

---

4   Results and Evaluation

The results are simulated in a custom-made environment, which is a square grid that can vary in size. Throughout the testing, the weights used in (5) are $w_p = 2$,

$w_s = 1$, $w_b = 1.5$. The learning rate was 0.001, the discount factor was 0.9, the starting epsilon value was 0.2 and there were 1000 learning episodes carried out. There was a movement budget set based on the environment size, this was the maximum number of steps allowed for each coverage attempt.

4.1   Coverage Performance in Cluttered Environments

The following plots illustrate an example of coverage in an environment of size 12 with a swarm size of 5 agents. Some obstacles are randomly placed in the environment. These results will show how the agents navigate in more complex environments.

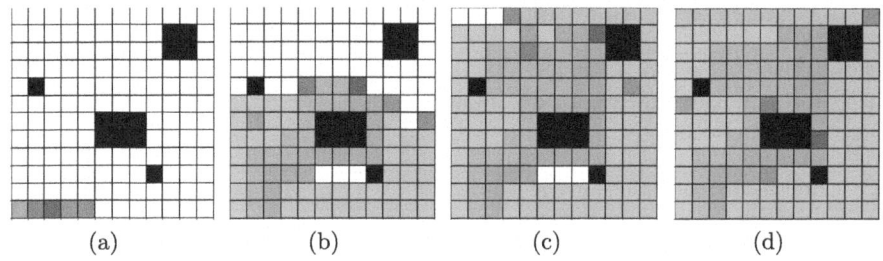

Fig. 1. The trajectories of 5 agents using the proposed method on an environment size 12 with random placed obstacles at (a) 0 s, (b) 14 s, (c) 28 s, and (d) 42 s.

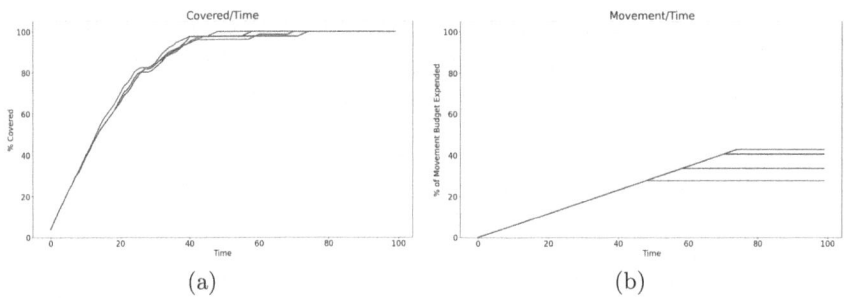

Fig. 2. The (a) Coverage/Time and (b) Movement/Time of the proposed method with swarm size 5 on an environment size 12 with random placed obstacles.

In Fig. 1, the trajectories of the agents can be seen. It demonstrates how each agent explores their section of the environment and also demonstrates the predator-prey methodology with the agents moving apart to explore different edges of the environment.

Figures 2(a) and 2(b) show that even when obstacles are added the algorithm maintains its consistency with tight groupings of the simulations. It also shows that even with obstacles the algorithm always ensures that the environment is covered.

4.2   Coverage with Different Swarm Sizes

To test the performance of our strategy when using different number of robots, the next plots show the coverage in an environment of size 12 with differing swarm sizes, i.e., 3, 5, and 10 agents all using the same predator-prey learning.

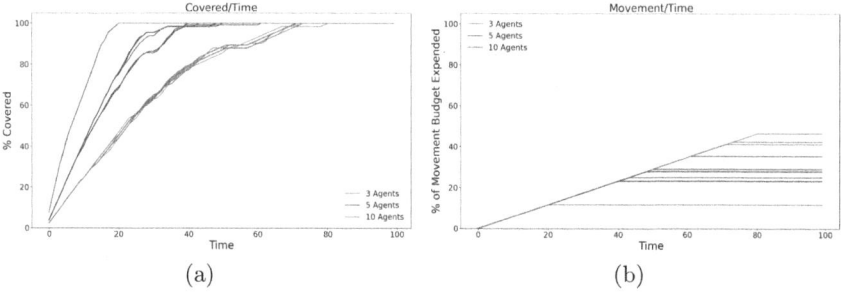

Fig. 3. The (a) Coverage/Time and (b) Movement/Time of the proposed method with different swarm sizes.

Figures 3(a) and 3(b) show the comparisons of different swarm sizes. This provides evidence of the effectiveness of the swarm as by adding additional agents the speed of the solution is drastically improved. These plots show that reliable results can be provided by any size swarm however the larger the swarm the better the results. This is useful to know as depending on the use case there may be different restrictions on the number of agents a swarm can have.

4.3   Comparison with Standard Q-Learning

The following graphs demonstrate the comparison between the standard Q-learning [10] and the version created in this paper. Each algorithm was run for 50 loops and then the average results of the simulations were plotted so they could be compared. A swarm of 5 agents in the same environment was used for both algorithms to ensure a fair comparison.

It can be seen in Fig. 4(a) and Fig. 4(b) that the predator-prey algorithm performs very well ensuring that the environment is always covered. Whereas with the standard Q-learning, there are some cases in which coverage fails and the entire movement budget is expended. This can be seen in Fig. 4(a), not every line (simulation) reaches 100% before the end of the time. This was an important criterion to meet for the solution ensuring that the environment can always be completely covered in a small number of steps, ensuring that the algorithm was efficient. This could potentially be further improved by improving the backtracking part of the algorithm.

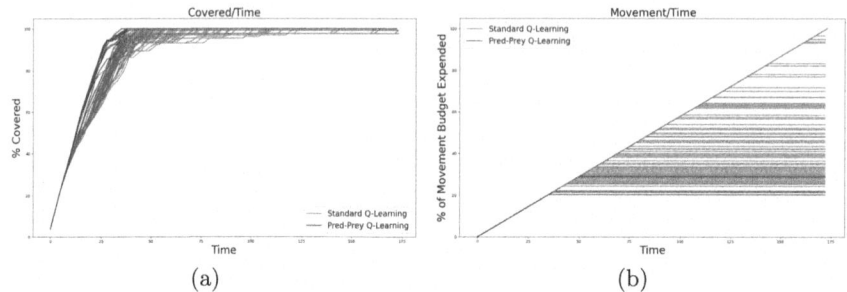

Fig. 4. Comparison between Predator-Prey and Standard algorithms with (a) Coverage/Time and (b) Movement/Time.

Figure 4(a) and Fig. 4(b) also show the improved algorithm is much more consistent providing similar results for all simulation loops, whereas the standard algorithm results vary drastically in its results. This was also important to ensure that the swarm would act predictably so that the user could know how the swarm would perform before using it, this is a very important factor in some potential use cases.

5   Conclusion

This paper utilizes a robot swarm using an adapted Q-learning algorithm to learn about their environment and plan an efficient path to explore it. The solution managed to successfully extend a standard predator-prey Q-learning to allow agents of the swarm to simulate both prey and predator, this led to a more effective coverage as agents aimed to stay apart and therefore covered different sections of the environment. The solution was much more efficient than a standard Q-learning swarm, ensuring that the environment was fully covered in fewer steps. The efficiency and effectiveness were then validated by various simulation case studies. For future work, we will implement the proposed method in real mobile robots.

Acknowledgement. This work was supported by EU H2020-FET-OPEN Robo-Royale project [grant number 964492].

References

1. Cao, Z.L., Huang, Y., Hall, E.L.: Region filling operations with random obstacle avoidance for mobile robots. J. Robot. Syst. **5**(2), 87–102 (1988)
2. Höffmann, M., et al.: Coverage path planning and precise localization for autonomous lawn mowers. In: 2022 Sixth IEEE International Conference on Robotic Computing (IRC), pp. 238–242. IEEE (2022)
3. Huang, K.C., Lian, F.L., Chen, C.T., Wu, C.H., Chen, C.C.: A novel solution with rapid voronoi-based coverage path planning in irregular environment for robotic mowing systems. Int. J. Intell. Robot. Appl. **5**(4), 558–575 (2021)
4. Xie, S., Hu, J., Ding, Z., Arvin, F.: Cooperative adaptive cruise control for connected autonomous vehicles using spring damping energy model. IEEE Trans. Veh. Technol. **72**(3), 2974–2987 (2023)
5. Hameed, I.A.: Intelligent coverage path planning for agricultural robots and autonomous machines on three-dimensional terrain. J. Intell. Robot. Syst. **74**(3), 965–983 (2014)
6. Wang, L., Wang, Z., Liu, M., Ying, Z., Xu, N., Meng, Q.: Full coverage path planning methods of harvesting robot with multi-objective constraints. J. Intell. Robot. Syst. **106**(1), 17 (2022)
7. Rekabi-Bana, F., Hu, J., Krajník, T., Arvin, F.: Unified robust path planning and optimal trajectory generation for efficient 3D area coverage of quadrotor UAVs. IEEE Trans. Intell. Transp. Syst. **25**(3), 2492–2507 (2024)
8. Wu, K., Hu, J., Li, Z., Ding, Z., Arvin, F.: Distributed collision-free bearing coordination of multi-UAV systems with actuator faults and time delays. IEEE Trans. Intell. Transp. Syst. (2024)
9. Rekabi-Bana, F., Stefanec, M., Ulrich, J., et al.: Mechatronic design for multi robots-insect swarms interactions. In: 2023 IEEE International Conference on Mechatronics (ICM), pp. 1–6. IEEE (2023)
10. Watkins, C.J.C.H.: Learning from delayed rewards (1989)
11. Clifton, J., Laber, E.: Q-learning: theory and applications. Ann. Rev. Stat. Appl. **7**, 279–301 (2020)
12. Navarro, I., Matía, F.: An introduction to swarm robotics. ISRN Robot. **2013**, 1–10 (2013)
13. Galceran, E., Carreras, M.: A survey on coverage path planning for robotics. Robot. Auton. Syst. **61**(12), 1258–1276 (2013)
14. Champagnie, K., Arvin, F., Hu, J.: Decentralized multi-agent coverage path planning with greedy entropy maximization. In: 2024 IEEE International Conference on Industrial Technology, pp. 1–6 (2024)
15. Tan, C.S., Mohd-Mokhtar, R., Arshad, M.R.: A comprehensive review of coverage path planning in robotics using classical and heuristic algorithms. IEEE Access **9**, 119:310–119:342 (2021)
16. Xing, B., Wang, X., Yang, L., Liu, Z., Wu, Q.: An algorithm of complete coverage path planning for unmanned surface vehicle based on reinforcement learning. J. Mar. Sci. Eng. **11**(3), 645 (2023)
17. Jia, Y., et al.: The UAV path coverage algorithm based on the greedy strategy and ant colony optimization. Electronics **11**(17), 2667 (2022)
18. Nasirian, B., Mehrandezh, M., Janabi-Sharifi, F.: Efficient coverage path planning for mobile disinfecting robots using graph-based representation of environment. Front. Robot. AI **8**, 624333 (2021)

19. Zhou, B., Zhang, Y., Chen, X., Shen, S.: FUEL: fast UAV exploration using incremental frontier structure and hierarchical planning. IEEE Robot. Autom. Lett. 6(2), 779–786 (2021)
20. Piardi, L., Lima, J., Pereira, A.I., Costa, P.: Coverage path planning optimization based on q-learning algorithm. In: AIP Conference Proceedings, vol. 2116. AIP Publishing (2019)
21. Zhou, Q., Lian, Y., Wu, J., Zhu, M., Wang, H., Cao, J.: An optimized Q-learning algorithm for mobile robot local path planning. Knowl. Based Syst. 286, 111400 (2024)
22. Puente-Castro, A., Rivero, D., Pazos, A., Fernandez-Blanco, E.: UAV swarm path planning with reinforcement learning for field prospecting. Appl. Intell. 52(12), 14101–14118 (2022)
23. Zhang, M., Cai, W., Pang, L.: Predator-prey reward based q-learning coverage path planning for mobile robot. IEEE Access 11, 29673–29683 (2023)

Path Planning for Multi-agent Systems Using Deep Q-Networks Reinforcement Learning

Ibrahim Alispahić$^{(\boxtimes)}$ and Adnan Tahirović

Faculty of Electrical Engineering, Department of Automatic Control and Electronics, Sarajevo, Bosnia and Herzegovina
{ialispahic1,atahirovic}@etf.unsa.ba

Abstract. This work presents a comprehensive study of the application of multi-agent reinforcement learning ($MARL$) based on deep Q-networks (DQN), aiming to enhance the cooperation and coordination of multiple agents in complex environments. The core problem addressed is the multi-agent traveling salesman problem, i.e. the effective collaboration and target-reaching by multiple agents. This challenge is inherently present in a variety of applications, including inspection and maintenance tasks based on autonomous aerial robotic systems. The research delves into the intricacies of multi-agent systems, emphasizing the dynamic interaction between agents and the environment. Although the proposed algorithm is developed and trained in a planar space, its principles and methodologies are easily extendable to any operational environment. The proposed approach is capable of training on a number of different scenarios, using various numbers of agents and targets, regardless of their initial positions, making it highly adaptable to scenarios like aerial inspections, where the number of agents, the number of inspection sites and their locations are unknown a priori. Furthermore, the work outlines how the enhanced DQN algorithm can be tailored for scenarios with numerous autonomous agents and targets, ensuring efficient path planning and target acquisition in constrained and unstructured spaces. This work also provides a solid foundation for future research in integrating advanced reinforcement learning techniques into the design and operational strategies for solving tasks, which can be described by the multi-agent traveling salesman problem, including inspection and maintenance with an autonomous multi-agent system.

Keywords: Multi-agent reinforcement learning ($MARL$) · Deep Q-networks (DQN) · Path Planning · inspection and maintanance

1 Introduction

The domain of aerial robotics faces the challenge of efficiently coordinating multiple autonomous agents within dynamic environments. This paper intro-

M. N. Huda et al. (Eds.): TAROS 2024, LNAI 15052, pp. 333–344, 2025.
https://doi.org/10.1007/978-3-031-72062-8_29

duces a novel path planning strategy using Multi-Agent Reinforcement Learning (MARL) combined with Deep Q-Networks (DQN) to enhance path optimization and coordination among agents. Our research focuses on the multi-agent traveling salesman problem (mTSP), where traditional methods often fall short in adaptability and real-time decision-making. By implementing MARL and DQN, this work facilitates dynamic solutions crucial for environments like aerial inspections and maintenance. This model extends the foundational framework by Busoniu et al. [1] and integrates advancements in deep reinforcement learning for routing problems as discussed by Kool et al. [4], shifting from static strategies to adaptive, agent-centric coordination. However, challenges remain, such as the computational complexity of training MARL systems and the need for rapid decision-making in dynamic environments. Ensuring robust communication among agents amid environmental uncertainties is also critical. The subsequent sections detail our contributions to MARL and mTSP, describe the methodologies employed, and present experimental validations of our proposed solutions.

2 Contribution

This section highlights the significant contributions of our paper within the domains of *MARL* and *TSP*, showing how our research extends and differentiates from existing works. By comparing our findings with recent advancements, we underscore the novelty and applicability of our approaches in enhancing multi-agent systems and complex path optimization.

2.1 MARL

Foundational insights into multi-agent systems in aerial robotics have been provided by Busoniu et al. [1]. Our research builds on this by enhancing *MARL* with a *DQN* model for real-time adaptability and decision-making in dynamic environments, transitioning from theoretical frameworks to practical applications in multi-agent path planning. Our *MARL*-based *DQN* model, designed for the *mTSP*, manages up to four agents and 20 targets, extending beyond the two-agent scenarios discussed by Tampuu et al. [2]. This model innovates through calibrated hyperparameters and a nuanced reward function, enhancing strategic coordination in complex settings. Extensive training underscores our commitment to refining the model's robustness in unpredictable scenarios. This refined *DQN* implementation advances the field by addressing the complexities of *mTSP* in multi-agent systems, demonstrating the practical impact of evolved *MARL* strategies.

2.2 TSP

The Traveling Salesman Problem (*TSP*) is fundamental in optimization, focusing on finding the most efficient path to visit each city once. The seminal work by

Dantzig et al. [3] established critical strategies for *TSP*, influencing subsequent innovations.

The Multi-Agent Traveling Salesman Problem (*mTSP*) extends *TSP* by incorporating multiple agents, increasing complexity due to the need for inter-agent coordination. This variant highlights the challenge of multi-agent systems, especially in real-time, adaptive path planning. Our research shifts focus to the application of *TSP* in aerial robotics, addressing challenges posed by dynamic environments and multiple agents. Recent advancements, such as Neural Networks by Kool et al. [4] for combinatorial optimization and policy gradient methods by Deudon et al. [5], inform our approach and offer new perspectives that enhance traditional *TSP* solutions. Distinctively, our work integrates these advanced computational strategies into multi-agent path planning, emphasizing real-time adaptability and strategic coordination, significantly extending the scope and applicability of *TSP* solutions.

3   Methodology

3.1   Evolution of Reinforcement Learning in Multi-agent Systems

Reinforcement Learning (*RL*) has transitioned from a single-agent framework, where the goal is to optimize behavior in a Markov Decision Process (*MDP*), to a more complex Multi-Agent Reinforcement Learning (*MARL*) paradigm. In single-agent *RL*, agents learn to map states to actions to maximize rewards, a concept explored in detail by Sutton and Barto [8]. *MARL* introduces additional complexities as multiple agents interact within the same environment, requiring not only optimal action learning but also adaptation to other agents' behaviors. This shift adds layers of complexity including coordination and strategic planning among agents, extensively discussed by Busoniu, Babuska, and De Schutter [1]. In *MARL*, the policy function of each agent (π_i) accounts for both individual and collective behaviors, enhancing the capability to navigate dynamic scenarios. Innovative *DQN* strategies by Mnih et al. [17] and advancements in policy optimization by Fujimoto et al. [16] demonstrate how *MARL* can effectively synchronize individual goals with group dynamics.

The following subsections will detail the methodologies of two *MARL* algorithms used in this study: Policy Optimization (*PO*) and Deep Q-Networks (*DQN*), showcasing the current state of *MARL* strategies and their applications in complex environments.

3.2   Policy Optimization in Multi-agent Systems

PO is a strategic approach in multi-agent systems, particularly effective for complex problems like the multi-agent variant of the *TSP*. At its core, *PO* utilizes a Policy Network, a deep neural network that interprets the environmental state and outputs action probabilities. This network is instrumental in shaping agent behaviors, ensuring that actions are not just executed, but are also based on a thoughtful process involving continuous learning and adaptation from accumulated experiences.

The *PO* algorithm operates through distinct phases within each episode: initialization of the Policy Network, action prediction and execution using an epsilon-greedy strategy for balance between exploration and exploitation, and meticulous analysis of each action's outcomes. Rewards from these actions serve as a direct performance indicator, guiding the crucial phase of Policy Network updates. In this phase, the network weights are adjusted through gradient descent, using calculated losses based on the rewards and action probabilities. Optimizers like the Adam optimizer are crucial in this stage, fine-tuning the network to ensure efficiency and precision in learning, and consequently, in the decision-making process of agents in dynamic, multi-agent environments.

3.3   Deep Q-Network Algorithm for Multi-agent Systems

After exploring the *PO* algorithm, the *DQN* approach is also analysed. *DQN* integrates deep learning with Q-learning, offering better generalization and efficiency in multi-agent systems, particularly for complex scenarios involving numerous agents and targets.

Algorithm Structure. *DQN* combines Q-networks and a replay buffer method, featuring:

- **Q-Networks and target networks:** Estimating Q-values for each action-state pair.
- **Epsilon-Greedy approach:** Balancing exploration and exploitation for action selection.
- **Replay Buffer:** Storing experiences for stable and efficient learning.
- **Network Updating:** Regularly updating Q-networks based on collected experiences and target Q-values.

The pseudo-code (1) illustrates the learning process, highlighting the interaction of these components.

Algorithm Analysis. The initialization involves setting up Q-networks and target networks for each agent, establishing the learning environment, and setting key hyperparameters (Table 1). The chosen values for the hyperparameters were based on standard practices in the literature and tuned to ensure stable learning and effective exploration-exploitation balance.

Training with *DQN*. During each episode, agents select actions using the epsilon-greedy strategy, similar to the *PO* algorithm. Executed actions lead to new states and rewards, which are then stored in the replay buffer. The Q-network is periodically updated based on samples from the replay buffer, following the update rule based on the Bellman equation:

Algorithm 1. Deep Q-Network (*DQN*)

1: Initialize Q-networks and target networks for each agent
2: Set algorithm parameters (e.g., learning rate, epsilon)
3: Initialize replay buffer
4: **for** each episode **do**
5:     Reset environment and receive initial state
6:     **while** episode not finished **do**
7:         **for** each agent **do**
8:             Select action using epsilon-greedy approach
9:             Execute action and receive reward and new state
10:            Store experience in replay buffer
11:        **end for**
12:        Update Q-network using experiences from replay buffer
13:    **end while**
14:    Adjust parameters (e.g., decrease epsilon)
15: **end for**

Table 1. Hyperparameter definitions.

Hyperparameter	Definition
replay_buffer	Size of the replay buffer (1,000,000)
epsilon_start	Initial value for epsilon (1.0)
epsilon_end	Final value for epsilon (0.05)
epsilon_decay	Decay rate for epsilon (0.9975)
gamma	Discount factor (0.8)
tau	Parameter for soft update of the target network (0.01)
num_episodes	Number of training episodes (5,000)
batch_size	Batch size for training (64)

[1] Number of episodes may vary based on the number of agents and targets.

$$Q_{\text{new}}(s,a) = Q(s,a) + \alpha \left(r + \gamma \max_{a'} Q(s',a') - Q(s,a) \right) \qquad (1)$$

where α denotes the learning rate, r is the reward, γ is the discount factor, and $Q(s',a')$ are the Q-values for the new state s' and actions a'. The update process involves:

1. **Sampling from the Replay Buffer:** A batch of experiences (states, actions, rewards, new states, and episode termination info) is sampled from the replay buffer.
2. **Calculating Target Q-Values:** Target Q-values are computed as:

$$Q_{\text{target}}(s',a') = r + \gamma \max_{a'} Q_{\text{target}}(s',a') \qquad (2)$$

3. **Calculating Loss:** The loss is calculated using the Mean Squared Error between the target Q-values and the current Q-values:

$$L = \frac{1}{N} \sum (Q_{\text{target}}(s', a') - Q(s, a))^2 \tag{3}$$

4. **Backpropagation and Optimization:** Backpropagation is used to update the weights in the Q-network, optimizing the network parameters:

$$\theta \leftarrow \theta - \alpha \nabla_\theta L \tag{4}$$

This iterative process ensures the continuous improvement of the Q-network, learning more effective decision-making strategies across various *TSP* scenarios. The implementation of soft target updates and the replay buffer are key to enhancing the performance of the *DQN* algorithm. These techniques play a pivotal role in stabilizing and improving the learning process.

Soft Update of Target Network. To stabilize the learning process, a soft update is applied to the target network after updating the Q-network:

$$\theta_{\text{target}} \leftarrow \tau \theta_Q + (1 - \tau) \theta_{\text{target}} \tag{5}$$

where τ represents the update rate (typically a small fraction, e.g., 0.001), θ_Q are the weights of the main Q-network, and θ_{target} are the weights of the target network. This technique ensures gradual updates to the target network, contributing to the stability of the learning targets over time.

Replay Buffer. The replay buffer is a crucial learning enhancement mechanism in *DQN*. It stores and later randomly samples experiences to form training mini-batches, described by:

$$P(sample) = \frac{1}{number\ of\ stored\ experiences} \tag{6}$$

where $P(sample)$ indicates the selection probability of each experience. This random sampling reduces correlations between sequential experiences, promoting more robust and generalized learning outcomes. Advanced replay buffers also prioritize experiences with higher temporal difference (TD) errors, enhancing the efficiency of learning from significant experiences.

The adoption of *DQN* over *PO* notably enhanced stability, quality, and speed of training, even in simple one agent-one target scenarios. In more complex setups involving multiple agents and targets, adjustments such as hyperparameter tuning, replay buffer enhancements, and soft target updates have been crucial for achieving effective training convergence.

3.4   Environment Architecture

This subsection outlines the network and environment architecture of our simulations.

The *QNetwork*, crucial for *DQN*, features an input layer matching the environment's state dimensions, two hidden layers with 128 neurons each using ReLU activation, and an output layer generating Q-values. This design balances efficient learning and strategy optimization while preventing overfitting, enhanced by double precision weight conversion for accuracy in complex scenarios.

The *MaTsEnvironment* simulates the *TSP* in a multi-agent context, with a straightforward graphical interface that displays agents as blue dots and targets in red. This layout is designed to maximize movement space and strategic flexibility while avoiding edge constraints.

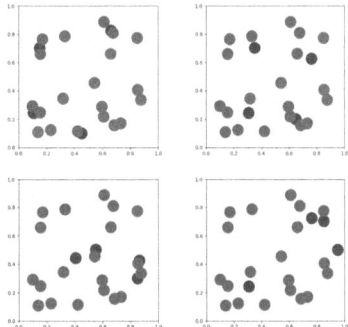

Fig. 1. The image depicts the progression of four agents (in blue) as they locate 20 targets. Unvisited targets are marked in red, while visited targets are highlighted in green. (Color figure online)

Visual monitoring within the *MaTsEnvironment* is vital for evaluating agent interactions and strategy effectiveness, as shown in Fig. 1. It supports dynamic, adaptable simulations by initializing agents and targets, defining parameters like agent count and target locations, and randomizing positions to ensure variability.

The environment employs a detailed observation vector that provides agents with extensive insights into the state, including positions, velocities, and distances, reducing computational demands on agents. The `step` function translates actions into environmental changes and manages reward calculations, focusing on target proximity and efficient navigation. This streamlined reward system, refined from earlier complex versions, directly links actions to outcomes, improving performance and learning efficiency.

3.5   Algorithm Validation

Validating the algorithm is crucial to verify its efficacy in multi-agent environments. A Python script, utilizing pre-trained Q networks, rigorously evaluates the algorithm's performance. The script employs *softmax* action selection, which, by considering the Q values, significantly reduces the likelihood of agents getting stuck in local minima, especially in complex scenarios with numerous agents and targets.

The validation involves initializing the *MaTsEnvironment*, loading the Q networks, and executing multiple episodes. During these episodes, the agents' actions and the environmental changes are closely monitored. The *softmax* function selects actions based on a probabilistic model, enhancing the decision-making process. Validation provides insights into the algorithm's performance and pinpointing areas for potential enhancement. This comprehensive approach ensures a deep understanding of the algorithm's operational efficiency and its capability in effectively navigating and solving the multi-agent variant of complex problems.

4   Experimental Results

This section presents an analysis of the experimental outcomes from testing the *PO* and *DQN* algorithms on the *mTSP*. Special emphasis is placed on the *DQN* algorithm due to its notable performance, analyzing how different hyperparameters and modifications affect its efficiency and training quality. The comparative analysis benchmarks our *MARL* approach against other methods, highlighting its relative efficiency, convergence speed, and adaptability in dynamic environments. The results are substantiated through performance monitoring with `TensorBoard` from the `PyTorch` library, centering on metrics such as *Total Reward*. Notably, the metric *Episode Length* is consistently employed across all figures presented in the paper, offering a uniform standard for evaluating and comparing the performance outcomes.

4.1   Comments on the *PO* Algorithm

The experimental outcomes using the *PO* algorithm revealed its limitations in solving the *mTSP*. Despite adjustments in hyperparameters, reward system modifications, and state vector experimentation, improvements in training quality and algorithm efficiency were marginal. Changes in learning rate and reward adjustments showed slight improvements in agent behavior and episode length reduction. However, the complexity of scenarios, especially with more targets, posed challenges that the *PO* algorithm struggled to overcome efficiently. These findings prompted the exploration of the *DQN* algorithm to better address the environment's intricacies and dynamics.

4.2   Comparison of the Results

Transitioning from *PO* to the *DQN* algorithm significantly improved performance. Initial *DQN* trials showed remarkable outcomes, with substantial decreases in average episode length and execution time compared to *PO*, as shown in Fig. 2. However, training quality decreased with an increase in the number of agents or targets.

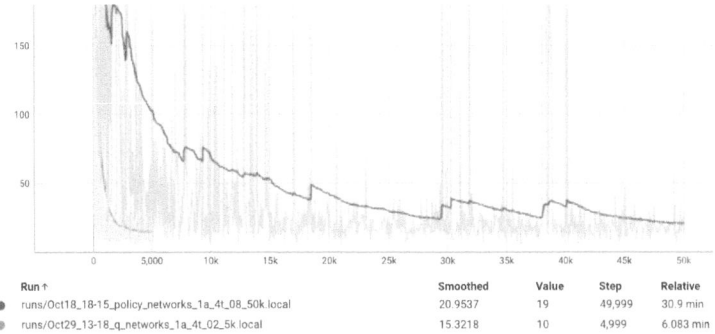

Fig. 2. Comparison of *PO* (gray) and *DQN* (orange) algorithm for the case of 1 agent 4 targets (Color figure online)

Enhancements to the *DQN* algorithm for larger agent or target numbers included network target adjustments and addressing local minima issues. Expanding the environment and adjusting the target network update frequency showed mixed results. Introducing a more complex reward system and soft target updates (Fig. 3) improved training stability and efficiency, yet challenges with local minima persisted.

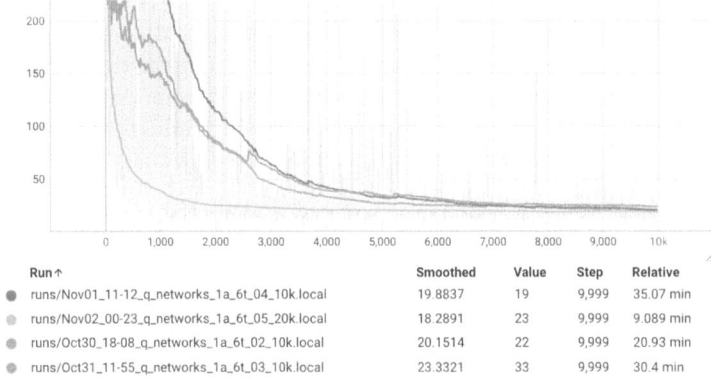

Fig. 3. Comparison of different experiments with the *DQN* algorithm for 1 agent 6 targets, including the effect of soft target network updates (orange) (Color figure online)

Applying softmax action selection during validation significantly reduced local minima occurrences, resulting in more stable and improved validation results (Fig. 4). Additionally, using weight decay in the Adam optimizer led to better training outcomes, especially in complex scenarios with more agents and targets (Fig. 5).

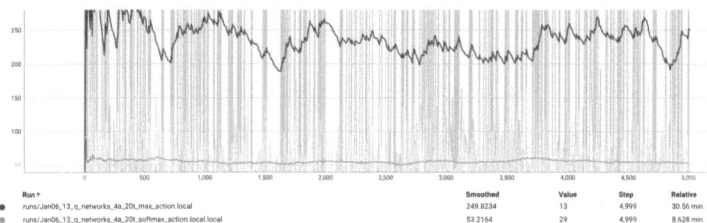

Fig. 4. Comparison of validation average episode length of the *DQN* algorithm for 4 agents and 20 targets, with (blue) and without (gray) *softmax* action selection (Color figure online)

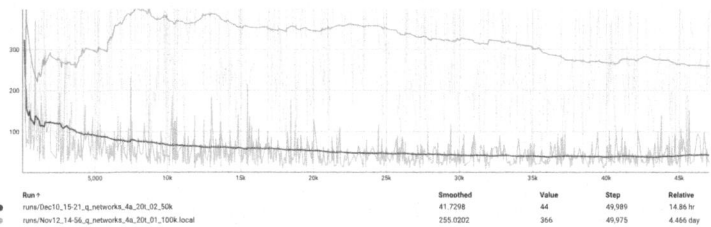

Fig. 5. Comparison of training the *DQN* algorithm for 4 agents and 20 targets, with (gray) and without (yellow) *weight decay* factor (Color figure online)

Further enhancements included targeted training for specific configurations, which proved to be significantly faster and more focused compared to general model training. This approach, illustrated in Fig. 6, demonstrates the potential for achieving optimal paths for fixed positions after the model is partly trained on a general scenario. The progress of four agents in finding 20 targets is documented, showcasing the practical application of the *DQN* algorithm in complex multi-agent scenarios.

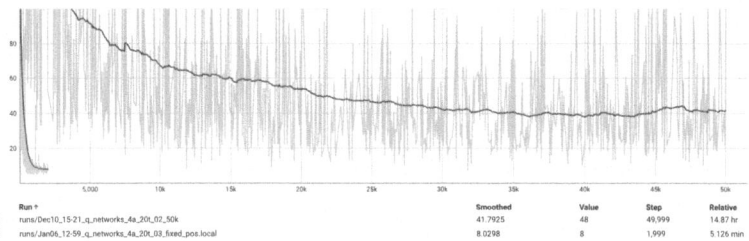

Fig. 6. Comparison of training the *DQN* algorithm for 4 agents and 20 targets for general scenario (gray) and fixed positions of agents and targets (purple) (Color figure online)

Summary of Results. The results showed that the *DQN* algorithm outperformed the *PO* algorithm, particularly in terms of episode length and training efficiency. Modifications such as soft target updates, softmax action selection, and weight decay in the optimizer significantly improved performance. The *DQN* algorithm demonstrated robust adaptability to different scenarios, highlighting its potential for complex multi-agent systems.

5   Conclusion and Future Work

This research has significantly advanced multi-agent path planning by addressing a specifically designed multi-agent *TSP* (*mTSP*) challenge, focusing on enhancing the *DQN* algorithm. It tackled the dynamic complexities of multi-agent, multi-target environments, providing a detailed analysis and experimental evaluation of *DQN* aspects like training speed, scalability, and adaptability. Our modifications, such as soft target updates, softmax action selection, and weight decay in the optimizer, led to marked improvements in performance. The *DQN* algorithm demonstrated superior efficiency, convergence speed, and adaptability over the *PO* algorithm, particularly in scenarios involving many agents and targets, demonstrating clear benefits over traditional *TSP* solutions. The findings propose several directions for future research to enhance path optimization and reinforcement learning applications:

1. Implement the Twin Delayed DDPG (*TD3*) algorithm to improve learning and solution quality in complex *mTSP* scenarios, inspired by advancements in continuous control by Lillicrap et al. [14].
2. Apply Curriculum Learning for more efficient training in complex settings with multiple agents and targets.
3. Innovate neural network structures by exploring varied layers, activation functions, and regularization techniques, guided by the deep learning foundations set by Ioffe and Szegedy [15].
4. Extend these methodologies to other optimization problems or similar dynamic domains, following the innovative approaches by Kool et al. [4] for solving routing challenges.
5. Continuously integrate the latest advancements in path optimization and reinforcement learning to improve the solutions developed in this research.

These strategic directions are expected to propel forward the theoretical and practical capabilities in addressing complex *mTSP* and related challenges, thereby contributing to the ongoing evolution in the field.

Acknowledgement. This work has been supported in part by the scientific project "Strengthening Research and Innovation Excellence in Autonomous Aerial Systems - AeroSTREAM," supported by the European Commission HORIZON WIDERA-2021-ACCESS-05 Programme through the project under G.A. number 101071270.

References

1. Busoniu, L., Babuska, R., De Schutter, B.: A comprehensive survey of multi-agent reinforcement learning. IEEE Trans. Syst. Man Cybern. Part C (Applications and Reviews), **38**(2), 156–172 (2008)
2. Tampuu, A., Matiisen, T., Kodelja, D., Kuzovkin, I., Korjus, K., Aru, J., et al.: Multiagent cooperation and competition with deep reinforcement learning. PLoS ONE **12**(4), e0172395 (2017)
3. Dantzig, G.B., Fulkerson, D.R., Johnson, S.M.: Solution of a large-scale traveling-salesman problem. J. Oper. Res. Soc. Am. **2**(4), 393–410 (1954)
4. Kool, W., van Hoof, H., Welling, M.: Attention, learn to solve routing problems! In: International Conference on Learning Representations (2018). https://api.semanticscholar.org/CorpusID:59608816
5. Deudon, M., Cournut, P., Lacoste, A., Adulyasak, Y., Rousseau, L.-M.: Learning heuristics for the TSP by policy gradient. In: International Conference on the Integration of Constraint Programming, Artificial Intelligence, and Operations Research (CPAIOR 2018), pp. 170–181 (2018)
6. Sutton, R.S., Barto, A.G.: Reinforcement Learning: An Introduction. MIT Press (2018)
7. Zhang, Y., Yang, M.: Multi-Agent Reinforcement Learning: A Selective Overview of Theories and Algorithms. Springer, Handbook of Reinforcement Learning and Control (2021)
8. Sutton, R.S., Barto, A.G.: Reinforcement Learning: An Introduction. MIT Press, Cambridge, MA (2018)
9. Russell, S., Norvig, P.: Artificial Intelligence: A Modern Approach. Pearson, Upper Saddle River, NJ (2010)
10. Khatib, O.: Real-time obstacle avoidance for manipulators and mobile robots. In: Autonomous Robot Vehicles, Springer, pp. 396–404 (1986)
11. Puterman, M.L.: Markov Decision Processes: Discrete Stochastic Dynamic Programming. John Wiley & Sons (2014)
12. Kaelbling, L.P., Littman, M.L., Cassandra, A.R.: Planning and acting in partially observable stochastic domains. Artif. Intell. **101**(1–2), 99–134 (1998)
13. Ottoni, A.L.C., Nepomuceno, E.G., Oliveira, M.S.d., et al.: Reinforcement learning for the traveling salesman problem with refueling. Complex Intelligent Syst. **8**, 2001–2015 (2022). https://doi.org/10.1007/s40747-021-00444-4
14. Lillicrap, T.P., et al.: Continuous Control with Deep Reinforcement Learning. Google DeepMind, London, UK (2016)
15. Ioffe, S., Szegedy, C.: Batch Normalization: Accelerating Deep Network Training by Reducing Internal Covariate Shift. Google, 1600 Amphitheatre Pkwy, Mountain View, CA 94043 (2015)
16. Fujimoto, S., van Hoof, H., Meger, D.: Addressing Function Approximation Error in Actor-Critic Methods (2018)
17. Mnih, V., et al.: Human-level control through deep reinforcement learning. Nature **518**(7540), 529–533 (2015)
18. Lin, L.-J.: Self-improving reactive agents based on reinforcement learning, planning and teaching. Mach. Learn. **8**(3-4), 293–321 (1992). Springer

Author Index

M. N. Huda et al. (Eds.): TAROS 2024, LNAI 15052, pp. 345–347, 2025.
https://doi.org/10.1007/978-3-031-72062-8

The manufacturer's authorised representative in the EU is Springer
Nature Customer Service Centre GmbH, Europaplatz 3, 69115 Heidelberg,
Germany. If you have any concerns regarding our products, please
contact ProductSafety@springernature.com

Printed and bound by CPI Group (UK) Ltd, Croydon, CR0 4YY
29/04/2026
02099537-0006